Marine Hydrocarbon Spill Assessments

From Baseline Information through to Decision Support Tools

Marine Hydrocarbon Spill Assessments

From Baseline Information through to Decision Support Tools

Edited by

Oleg Makarynskyy
MetOcean Dynamic Solutions Pty Ltd,
Darwin, NT, Australia

ELSEVIER

Elsevier
Radarweg 29, PO Box 211, 1000 AE Amsterdam, Netherlands
The Boulevard, Langford Lane, Kidlington, Oxford OX5 1GB, United Kingdom
50 Hampshire Street, 5th Floor, Cambridge, MA 02139, United States

Notices

Knowledge and best practice in this field are constantly changing. As new research and experience broaden our understanding, changes in research methods, professional practices, or medical treatment may become necessary.

Practitioners and researchers must always rely on their own experience and knowledge in evaluating and using any information, methods, compounds, or experiments described herein. In using such information or methods they should be mindful of their own safety and the safety of others, including parties for whom they have a professional responsibility.

To the fullest extent of the law, neither the Publisher nor the authors, contributors, or editors, assume any liability for any injury and/or damage to persons or property as a matter of products liability, negligence or otherwise, or from any use or operation of any methods, products, instructions, or ideas contained in the material herein.

British Library Cataloguing-in-Publication Data

A catalogue record for this book is available from the British Library

Library of Congress Cataloging-in-Publication Data

A catalog record for this book is available from the Library of Congress

ISBN: 978-0-12-819354-9

Publisher: Candice Janco
Acquisitions Editor: Louisa Munro
Editorial Project Manager: Tracy Tufaga
Production Project Manager: Sruthi Satheesh
Cover Designer: Christian Bilbow

Typeset by MPS Limited, Chennai, India

Contents

José Ramón Bergueiro López Department of Chemical Sciences, University of Santiago de Compostela, Santiago, Spain; Department of Chemical Engineering, University of the Balearic Islands, Palma, Spain; Research group: CONSEMAR - Pollution and Marine Safety, University of La Laguna, San Cristóbal de La Laguna, Spain

Jake Nelson Oak Ridge Institute for Science and Education, National Energy Technology Laboratory, Oak Ridge, TN, United States; Department of Geosciences, Auburn University, Auburn, AL, United States

Tor Nordam SINTEF Ocean, Trondheim, Norway; Department of Physics, Norwegian University of Science and Technology, Trondheim, Norway

Trond Nordtug SINTEF Ocean AS, Trondheim, Norway

Hari Radhakrishnan MEDSLIK (Mediterranean Oil Spill and Trajectory Prediction Model), Nicosia, Cyprus

Jagoš R. Radović Department of Geoscience, University of Calgary, Calgary, AB, Canada

Lucy Romeo National Energy Technology Laboratory, Morgantown, WV, United States

Michael Sabbatino NETL, Albany, OR, United States

Mattia Serra School of Engineering and Applied Sciences, Harvard University, Cambridge, MA, United States; Department of Physics, University of California San Diego, La Jolla, CA, United States

Jørgen Skancke SINTEF Ocean, Trondheim, Norway

Sarantis Sofianos Ocean Physics and Modelling Group, University of Athens, Athens, Greece

Katerina Spanoudaki Coastal & Marine Research Lab, Institute of Applied and Computational Mathematics, Foundation for Research and Technology-Hellas, Heraklion, Greece

Joaquín Tintoré SOCIB and IMEDEA (CSIC-UIB), Palma, Spain

Patrick Wingo National Energy Technology Laboratory, Albany, OR, United States

George Zodiatis Coastal & Marine Research Lab, Institute of Applied and Computational Mathematics, Foundation for Research and Technology—Hellas, Heraklion, Greece; MEDSLIK (Mediterranean Oil Spill and Trajectory Prediction Model), Nicosia, Cyprus

Introduction

Oceans and seas may be receiving petroleum substances, including oil, from two sources: natural leaks (or seeps) through the seabed sediments or just through geological fractures, and from anthropogenic activities, such as hydrocarbon exploration, production, and transportation. Appreciating that those two may not be sometimes clearly identifiable/differentiable and are hard to be assessed separately, this book focuses on the hydrocarbons received by the marine environment from the latter source. Specific attention in this book is paid to assessing the fate of those hydrocarbons and their impacts on the marine and coastal environments. Environments here are understood in the widest possible sense and include ecosystems and their components, social environment, and economic environment.

The written history of the anthropogenic marine petroleum spills spans more than a century, with multiple spill locations present in each of the oceans, as well as in marginal and internal seas. Some of the first spill records are related to petroleum transport from Borneo to Australia (SS Petriana, around 1300 tons spilled in Port Phillip Bay (Petriana, 1903), and from the United States of America to England (Thomas W. Lawson, around 7400 tons spilled near Isles of Scilly (Larn, 1992). Much more recently, in 2018, all 136,000 tons of natural-gas condensate cargo being transported from Iran to South Korea were spilled in the East China Sea (Madrigal, 2018).

Since the late forties of the 20th century, the advancements of exploration methods, of constructing and maintaining shelf and deepwater drilling platform technologies, and, since the late seventieth, floating production storage and offloading units brought petroleum pollution threats to the offshore areas, which previously would be located far away from shipping routes. Pertaining to the difficulties of capping pressurized underwater wells in a case of blowout, the largest (in terms of the amount of spilled hydrocarbon) marine spills are contributed to offshore exploration and production. The latest incidents of this kind, which had international impact and consequences, were Montara in the Timor Sea in 2009 (up to 30,000 tons spilled; https://www.environment.gov.au/marine/marine-pollution/montara-oil-spill) and Deepwater Horizon in the Gulf of Mexico in 2010 (up to 627,000 tons spilled; https://www.epa.gov/enforcement/deepwater-horizon-bp-gulf-mexico-oil-spill).

Despite building the hazard awareness and investing into the respective training of personnel, into the latest technology and the most advanced petroleum exploration and transportation equipment, as well as into the development of spill prevention regulations and procedures, spills continue occurring due to either or both human error/lapse of judgment and equipment failure. Thus we should do our best to be prepared to respond quickly and effectively to such events, which—if unprepared—may very quickly become catastrophically damaging to humans and more broadly to the marine and coastal environments.

This book focuses on the state-of-the-art desktop and laboratory assessments that as a rule can and should be undertaken at the stage of preparation for a spill event, which may be taking place in the marine environment. The authors contributing for this book look at the modern data collection and computational techniques and methodologies, numerical algorithms, and laboratory procedures, which can help us with understanding the present-day environmental conditions of the area of interest or concern, hydrocarbon properties and behavior when spilled, how an oil slick may move in the marine environment horizontally and—if spilled from seabed—vertically, how hydrocarbon toxicity may be estimated and which oil components may pose the greatest risk in terms of toxicity, how we may prepare and mount spill response, and, finally, which decision support tools and hardware are available and may be used in a spill event.

The authors believe that the book fills the gap in the current market because it provides a comprehensive and up-to-date source of information on a wide range of spill assessment topics, each of which is of critical importance for painting a consistent and coherent picture of a spill event and its probable consequences. As such, the book is intended to be read by subject matter specialists—researchers and students in the respective fields—as well as oil and gas industry and maritime safety professionals, regulators, decision makers, first responders, contingency, and response planning software developers. Though intended to mainly be read by people with a professional interest in marine spill assessments, everybody else with general interested in the subject (e.g., environmental activists and members of environmentally oriented nongovernmental organizations) will find it easy to read, learn, and understand, thanks to the thorough explanations of the used terms and terminology and presented practical examples.

In accordance with the above major topics, the book is structured as follows:

The first chapter focuses on the baseline data, which are a crucial component at the stages of spill prevention, preparedness, and response. The chapter discusses types of baseline information, data availability, and limitations as well as guides the reader through the process and requirements of building baseline datasets relevant to marine spills. Improving data availability and quality through better networking between research and industry groups, ease of data acquisition and validation, and timely, automated where possible, updates are noted as necessary future steps in forming comprehensive baseline datasets.

Spill weathering processes are central to the second chapter, which states that assessing spill behavior is not rocket science but, in fact, is much more complex. This statement is reinforced by describing factual difficulties facing everybody who is dealing with characterizing hydrocarbon properties, and itemizing and parameterizing weathering phenomena. The chapter provides an advice on how those difficulties can be overcome when delving into the realm of numerical algorithm and model development for assessing how different hydrocarbons would weather in the marine environments.

Understanding how spilled oil moves horizontally on and with the ocean surface and within a water layer, looking at the mathematics of the motion and then at the solutions of how such a motion can be represented numerically are the topics of the third chapter. Importantly, both the Lagrangian and Eulerian frameworks (the former used for simulating trajectories of individually moving particles, which can be considered as oil parcels, and the latter used for simulating changes of oceanographic fields in time and space) are introduced in this chapter. The chapter discusses different levels of diffusion in the ocean, what model diffusion coefficients are for and how they can be determined. Remote current monitoring approaches, such as high-frequency radars and satellite measurements, impacts of windage and Stokes drift are presented. Examples of the model tool implementations are given, and aspects of marine spill hindcast and forecast which require further improvements are itemized in this chapter.

The subject of the fourth chapter is vertical movement of the spilled/released from the seabed oil. The chapter discusses the impacts of rising through and dispersing and dissolving in the water column oil. It focuses on the turbulent diffusion and mixing processes in the ocean, and on how those can be modeled. Particular attention is paid to how oil vertical mixing/entrainment may be initiated by breaking waves, on determining the entrainment rate and droplet size distribution, droplet rise processes and parameterizations, and on how those can be affected by chemical dispersants. The Eulerian and Lagrangian approaches to solving oil transport in the vertical direction are presented. The chapter identifies a number of useful pitfall examples of what should be avoided in oil spill model development and presents case studies of subsurface oil transport from known maritime incidents. A few advanced topics and further reading are offered to inquisitive and willing-to-learn readers.

The fifth chapter talks about the topics considered in the previous chapters from the point of view of operational model development and practical implementation for the purpose of spill response at the levels of both national and international government response agencies. Some of the marine spill events, which led to the introduction and adoption of international protocols and conventions on sea protection from pollution, and numerical spill models based on the Lagrangian approach are itemized here. Of particular interest in this chapter are descriptions of the hydrocarbon biodegradation processes and the pseudo-component approach used to model it, the governing equations for modeling of wellhead blowouts, the considerations of spill assessments when sea ice is present, and the sources of marine data including remote sensing from which information can be drawn by operational response agencies.

The challenges of oil toxicity assessments are discussed in the sixth chapter. The range of properties and variability of crude and processed petroleum compositions as well as impacts of different degrees of weathering all play roles in what fraction of toxic oil components may become bioavailable. For consistent toxicity assessments, standardization of the laboratory procedures is therefore desirable but difficult to achieve. For better results the chapter proposes to pay specific attention to the following: preparation of exposure solutions, characterization of exposure, proper assessment of oil component bioavailability, and establishment of consistent end points. A method to control both the concentration and size of droplets is then described and some application results are presented.

The seventh chapter introduces definitions of petroleum substances from the point of view of geo-chemistry and showcases research centered on oil detection in marine environments. The implementation of the gas chromatography-based laboratory approaches is intercompared with such novel techniques as multidimensional chromatography and (ultra)high-resolution mass spectrometry. Several recent case studies demonstrate that the newer techniques offer a range of improvements including identification of spill residues over multiyear and up to multidecade periods. The implementation of the novel methods also leads to revisiting our understanding of what happens to the spilled petroleum in the marine environment. The role of photooxidation in the weathering process is reassessed in this chapter; it is stated that the rate of surface slick photooxidation is comparable with the rate of evaporation. This has a direct impact on how the petroleum substance further degrades in general and biodegrades in particular. The marine-oil-snow sedimentation and flocculent accumulation phenomenon caused by aggregation of oil droplets with marine particulate matter (e.g., marine snow) are described here. Importantly, both the photooxidation and marine-oil-snow phenomena showed implications for related toxicity studies and thus for consideration and prediction of spill impacts.

An overview of the foundations, both theoretical and practical, of spill risk assessments, as well as impact and response analyses are presented in the eighth chapter. These are critical components for enhancing spill preparedness and for an adequate mounting of spill response. The chapter demonstrates how such assessments and analyses can be performed when preparing for or in an actual emergency situation. Impacts are proposed to be assessed as a function of site vulnerability and degree of oiling, and risks—as a function of impact and oiling probability. Response analysis is then based on the required level of response (e.g., strategic or tactical) and spatial optimization solution. The chapter presents real-world examples of how the proposed methodology can be applied using purpose-developed software tools. It identifies knowledge gaps and how those can be bridged in the future.

The topic of decision support tools for contingency and response planning in island environments is considered in detail in the final chapter of the book. The chapter highlights a number of numerical software tools used for the purposes of planning and response as well as briefly describes applied field techniques and equipment used in spill combat operations. It gives examples of how those software tools are used for better placement and implementation of the mentioned applied techniques and equipment. An intercomparison is presented of how a few different hydrocarbon spill assessment and trajectory models work and perform in the case of a hypothetic spill near an island. The chapter proposes to use an integrated management of coastal area framework for responding to a real-world spill, if or when such occurs.

In conclusion, it must be noted that the book has been prepared for publication over a few-year period and endured changes in authors' personal circumstances and countries' governments, natural disasters of various scales, and lately a pandemic. The persistence of the authors, many of whom worked on the book outside their ordinary "office hours" and actual offices, will be definitely appreciated by the grateful readers.

Thanks go to MetOcean Dynamic Solutions Pty Ltd, and personally to Managing Director Dr. Dina Makarynska, who was very understanding and generously supported the book preparation efforts during the entire journey.

References

Petriana (28 November 1903). Australian Maritime Safety Authority.

Larn, R. (1992). Shipwrecks of the Isles of Scilly. Nairn: Thomas & Lochar.

Madrigal, A. (19 January 2018). The world has never seen an oil spill like this. The Atlantic.

Chapter 1

Baseline data for spill assessments: ambient conditions, socioeconomic data, sensitivity maps

Lucy Romeo[1], Patrick Wingo[2], Michael Sabbatino[3] and Jennifer Bauer[2]

[1]*National Energy Technology Laboratory, Morgantown, WV, United State,* [2]*National Energy Technology Laboratory, Albany, OR, United States,* [3]*NETL, Albany, OR, United States*

1.1 Why and for what baseline data are needed?

Efforts to effectively prevent, prepare for, and respond to natural or anthropogenic disasters often rely heavily on the availability and accuracy of baseline data. Baseline data are measurements and information collected prior to an event or specific activity, such as a disaster, restoration project, management change, or industry-driven development. Baseline data are typically spatial or temporal in nature, representing a state of existence in the real world. In marine and coastal ecosystems, baseline data can encompass a variety of ambient, environmental, social, economic, and human health variables. This chapter will highlight the value of baselines and identify means to collect and build representative databases for marine and coastal ecosystems to aid in oil spill preperation, assessment, and response.

Major marine oil spill events, such as the Gulf War oil spill, the Deepwater Horizon blowout, and the Sanchi collision have stressed the value of baseline data to evaluate the magnitude of environmental, economic, and human health effects (Carswell, 2018; Goldstein, Osofsky, & Lichtveld, 2011; Joyner & Kirkhope, 1992; Nelson, Bauer, & Rose, 2014; Zhang et al., 2020). During a spill, baseline data are commonly used to guide cleanup and response efforts to protect vulnerable and sensitive communities and environments through comparison with information on the spill itself, such as the amount of oil, the trajectory and movement of oil, duration oil expected to be present, and its physical properties (Nelson, Grubesic, Sim, & Rose,

Marine Hydrocarbon Spill Assessments. DOI: https://doi.org/10.1016/B978-0-12-819354-9.00007-7

spatially structured or unstructured along the $x-y$ or longitude–latitude plane (NOM Group, 2003). Curvilinear structured grids divide the model region into quadrilateral units, which can be linear, or curved (Thompson et al., 1999). The grid cells in a curvilinear grid are distorted along one or more parallel curves (Thompson et al., 1999), which is useful for fitting data in extreme polar regions. Unstructured grids typically take the form of a tri-mesh; that is, each region is shaped as a triangle, although there are other geometric meshes occasionally used for building unstructured grids, such as orthogonal, hexagonal, or mixed shape meshes (Thompson et al., 1999). Regions in an unstructured grid can vary wildly in size, with smaller cells used in regions of high geographic complexity, such as coastlines, inland seas, and dense archipelagos (Thompson et al., 1999). Examples of these gridding methods are shown in Fig. 1.2.

With ocean models that also capture depth, there are several approaches to take as the depth regions may follow a regular interval or may become shorter near the surface and the ocean floor. Shorter intervals typically represent depth regions of higher complexity (NOM Group, 2003). Additionally, the contours between regions may follow specific depths or follow the approximate contours of the underlying bathymetry (NOM Group, 2003). There are a number of ways to represent ambient attributes across space; it is important for a spill modeler to understand these representations as appropriate for the ocean models used to derive baseline simulated ambient data.

Data originating from an ocean model can be used as surrogates for static or dynamic conditions in other physical models, such as those that simulate oil transport throughout the water column (Duran, Beron-Vera, & Olascoaga, 2018). Many of the traits mentioned above are used to provide data to drive the ambient conditions, which are critical for simulations (Sim et al., 2015). Oil spill simulation models, such as the US National Energy Technology Laboratory's (NETL) Blowout and Spill Occurrence Model™ (https://edx.netl.doe.gov/blosom/) or the National Oceanic and Atmospheric Administration's (NOAA) General NOAA Operational Modeling Environment (https://response.restoration.noaa.gov/oil-and-chemical-spills/oil-spills/response-tools/gnome.html), frequently take advantage of the simulated ambient characteristics produced by ocean models. This can increase the complexity of processes being simulated without the associated computational costs of running a full-scale ocean model simulation in addition to simulated oil spill fate and transport processes.

As previously discussed, ambient data are the driving force behind the transport and destination of each representative spill parcel within many oil spill simulation models. Without baseline data, there will be little variation in simulation runs, and there would be no defensible connection to observable real-world phenomena. While large-scale ocean models may produce data in a custom format (Wallcraft, Carroll, Kelly, & Rushing, 2003), data are often shared using file formats which are universally available to the research population. One of the most widely used formats for distributing

FIGURE 1.2 Examples of the various gridding methods applied. (A) Cartesian grids, (B) curvilinear grids, and (C) unstructured trimesh.

model-derived ambient data is Network Common Data Form, also referred to as netCDF (Unidata, https://www.unidata.ucar.edu/software/netcdf/). The netCDF file format was designed with scientific data in mind, with data stored as series of platform-independent arrays annotated with metadata describing their representations (Unidata). Due to the large nature of datasets generated by ambient models, several specialized data transfer protocols exist for pulling down slices of available data; one of the most common protocols is the Open Network Data Access Protocol (OPeNDAP, https://www.opendap.org/).

While not all ocean modeling teams make their data public, many do, particularly those hosted by public institution, such as NOAA's National Centers for Environmental Information (NOAA; https://www.ncei.noaa.gov/). Model output data can usually be downloaded directly, packaged in files (often netCDF), or accessed using a special connection protocol, such as the previously mentioned OpenDAP. Often times, the model data are available through a variety of means via a Thematic Real-Time Environmental Distributed Data Services (THREDDS) data server (Unidata; https://www.unidata.ucar.edu/software/thredds/current/tds/). Data produced by both the NCOM and HYCOM models are provided on their own THREDDS data server; at the time of this writing, NCOM-derived data can be retrieved from https://www.ncdc.noaa.gov/data-access/model-data/model-datasets/navoceano-ncom-glb by clicking on one of the TDS links in the table found on this page. Similarly, HYCOM results can be found on https://hycom.org/dataserver.

1.2.2 Socioeconomic

Socioeconomic data can represent an array of civil, industrial, and commercial factions and are valuable to quantify the potential costs of oil spill events. Socioeconomic data can also represent vulnerability measurements, monetary values, including cost of the oil lost or cost of resources or activities impacted, or indices representing the demographics of an area. These data can originate from surveys, industry reports, previously published datasets, papers, news, and models.

Understanding the cultural and economic importance of marine sector industries and activities, such as fishing, tourism and recreation, and the energy industry, is key for building a representative baseline. This information can be used to assess an area's spatial vulnerability, predict potential economic costs, and build better response plans. The use of socioeconomic data in a baseline varies regionally and is dependent on present industries and activities, data availability and accuracy, and spatial scale. For example, when evaluating future oil spill response preparedness at a spatially explicit port complex in Brazil, a study mapped area vulnerability using digital maps and satellite images, and gathered survey information on neighborhood income, education, and dependence on fishing (de Andrade, Szlafsztein,

Souza-Filho, dos Reis Araújo, & Gomes, 2010). At a global scale, the International Oil Pollution Compensation Fund analyzed oil spill cleanup cost and total cost to build a baseline for the International Maritime Organization (IMO) to better understand how to evaluate environmental risk (Kontovas, Psaraftis, & Ventikos, 2010). This large-scale study applied regression analysis on past spill statistics, including the total amount of oil spilled and number of spills, to quantify historical cost and estimate costs moving forward.

The formats and availability of socioeconomic data vary. Formats include data already spatially structured, with geometric or coordinate information tied to areas or locations on the earth, such as comma-separated values (CSV) files and shapefiles. These structured formats can be more easily reformatted, transformed, and analyzed. Less accessible data formats include Portable Document Format (PDF) and hardcopy papers, both of which require additional processing to convert the data into a format suitable for analysis. These formats are especially common for archiving historical socioeconomic data records.

Depending on the region of interest and spatial scale, socioeconomic data can be accessed through online databases, government websites, industry reports, or literature on past spills or spill cleanup reports. These data are typically already processed into products such as publications and static maps, which take additional sleuthing to find individual, spatial, and temporal values. Socioeconomic data can also be obtained as outputs from models. For example, to better understand response costs and socioeconomic damages of offshore oil spills in the US, the Environmental Protection Agency (EPA) developed the Basic Oil Spill Cost Estimation Model (BOSCEM) (Etkin, 2004). BOSCEM incorporates data on the amount spilled, type of oil, location-specific socioeconomic values, environmental vulnerability, and response effectiveness (Etkin, 2004). Building an understanding of models like BOSCEM, users can leverage the framework for their own areas of interest and purposes.

For areas where credible publicly available data are hard to come by, researchers would either need to collect and assemble data, which is dependent on funding and local interest, or rely on what resources are available and derive a socioeconomic proxy from what information they have. In a 2015 study assessing the potential socioeconomic and environmental impacts of oil spills in the US Gulf of Mexico, researchers were unable to acquire credible, spatial, publicly available socioeconomic data relating to tourism, which is a common and valuable indicator of coastal economies (Bauer et al., 2015). Coastal tourism in the Gulf of Mexico contributed 11% to the overall Gross Domestic Product of the US economy in 2010, with the industry employing 50.6% of people working in ocean sectors (National Oceanic

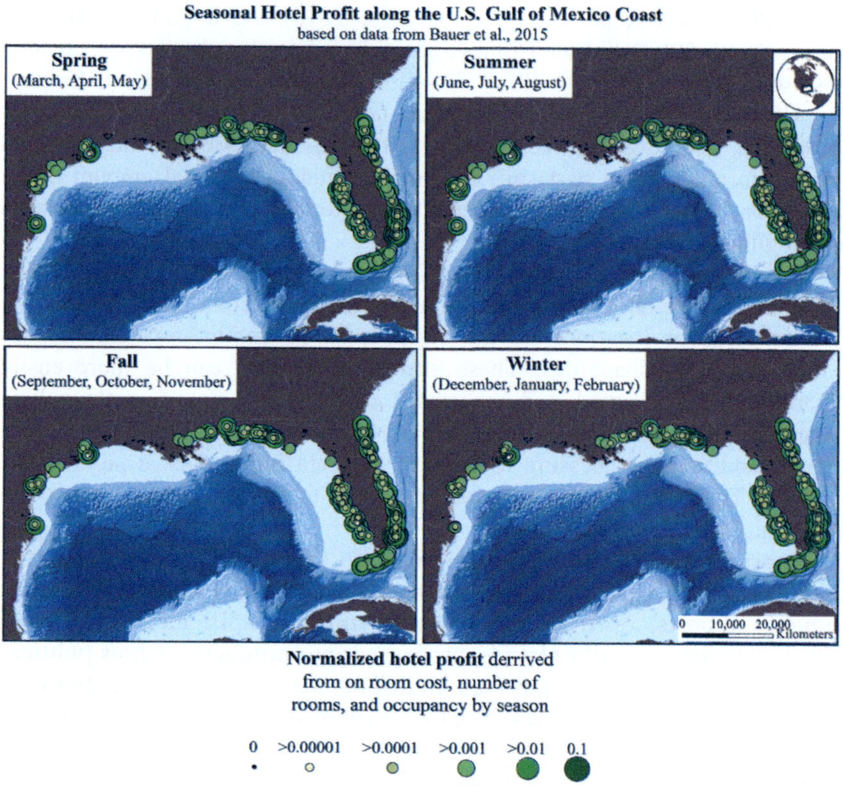

FIGURE 1.3 Maps illustrating the seasonal fluctuations of tourism along the Gulf of Mexico's US coastline. The maps shown above applied normalized hotel profit differences season-by-season as a proxy for coastal tourism, the data of which were created as a socioeconomic representation of the baseline for the US Gulf of Mexico (Bauer et al., 2015). Data were derived from locational information, state-wide seasonality obtained from literature and hotel interviews, room rates, and number of rooms per hotel.

Atmospheric Administration, 2013). To overcome this hurdle, researchers leveraged available hotel data and seasonal information by conducted interviews with hotel staff to calculate profit associated with room occupancy as a proxy for tourism (Bauer et al., 2015). Results of this data collection process are shown in Fig. 1.3, which displays the seasonal normalized profit per hotel in the US along the Gulf of Mexico's coastline.

1.2.3 Sensitivity maps

Sensitivity maps contain data representing areas that are environmentally vulnerable to oil spills, including shorelines, critical habitats, and natural resources. Information derived from these maps is critical for preparation and response mitigation to reduce the potential environmental consequences of oil spills. Sensitivity maps vary regionally, and in some cases, need to be created using environmental and natural resource data.

The most widely used approach to sensitivity mapping throughout the world is Environmental Sensitivity Index (ESI) (Jensen, Halls, & Michel, 1998). ESI are spatial information that are comprised of three factors: shoreline types, oil-sensitive biological resources, and commercial, recreational, and human-use resources (Fig. 1.4). ESI maps were first applied in 1979, when ESI maps of the Texas coast were prepared to evaluate impacts from the Ixtoc I well blowout in the Gulf of Mexico (Jensen et al., 1998). Since then, sensitivity maps, including ESI, have been built, updated, and applied in many regions. The ESI approach has been implemented throughout the world including the US, Brazil, Canada, United Arab Emirates, Greenland, India, Israel, Jordan, El Salvador, Germany, South Africa, Mauritius, Nigeria, and New Zealand.

Along Brazil's coast sits one of the world's most continuous mangrove habitats, which are considered by NOAA to be a sensitive cover-type flora to oil spills (NOAA, 1997). ESI maps were produced along this region for oil spill contingency planning, applying remotely sensed elevation (i.e., topography and bathymetry), geomorphological, meterological and oceanographic, biological, and socioeconomic data (Souza Filho, Goncalves, de Miranda, Beisl, & de Faria Almeida, 2004).

Nigeria is the largest oil producer in Africa, with the maximum crude oil production capacity of 2.5 million barrels a day according to the Nigerian National Petroleum Corporation (2018). The Nigerian government estimates that approximately 7,000 oil spills have occurred between 1970 and 2000 (Baird, 2010). In an effort to reduce the impacts of oil spills along Nigeria's coast, the Federal government, nongovernment agencies, and oil firms developed oil spill management policies, which included ESI mapping developed by the Environmental Systems Research Institute (Nwilo & Badejo, 2006).

A risk assessment with sensitivity maps was conducted along India's Chennai coast to better prioritize resources more likely to be impacted by oil spills (Kankara, Arockiaraj, & Prabhu, 2016). This assessment combined coastal resource information to build ESI maps and outputs from oil spill models to better understand the potential movement of oil spills in proximity to coastal resources.

Where sensitivity maps are available, they can take multiple forms. Static and web maps are two products of sensitivity mapping. Static maps, which can be printed or available digitally as PDF files or images, comprise standalone visual information that can be used to understand proximities but does not link features, attributes, and boundaries for deeper analyses. Because of their static nature, it is important for users to gather metadata on these maps to make sure they are not out of date. Time permitting, users can take static maps and digitize them into spatial data, which can then be updated and

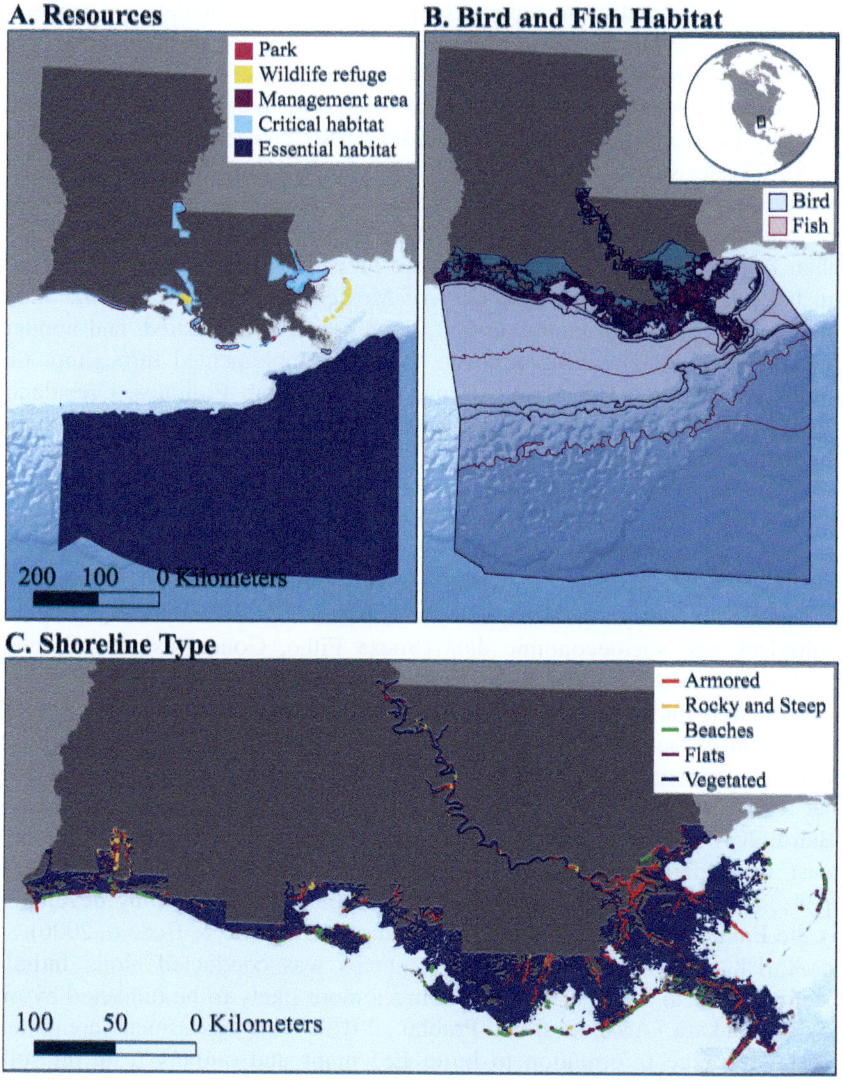

FIGURE 1.4 Example of Environmental Sensitivity Index data for the state of Louisiana (the US), including (A) a subset of resources, (B) a subset of habitats, and (C) shoreline types. Source: *Data were procured from Navy Coastal Ocean Model's Office of Response and Restorations online Environmental Sensitivity Index data page: https://response.restoration.noaa.gov (National Oceanic, 2014).*

analyzed, leveraging additional data sources. As the name implies, web maps are hosted online and enable users to visualize the data in multiple ways. Web maps are more easily updated as the data can be automatically fed to the platform through background processes or have linked sources.

Furthermore, data displayed in web maps can be made available for download in spatial formats.

Like socioeconomic data, the availability of sensitivity maps and the data that go into them are regionally dependent. Sensitivity maps can be found within literature and through government agency websites and records (i.e., NOAA). In many cases, if sensitivity maps are available, updates are required to ensure accuracy. Where sensitivity maps are not readily available, users can build their own using credible online data sources. Shoreline type and vulnerability can be derived from remotely sensed imagery. Species distributions and critical habitats are sometimes available online as spatial datasets or through static maps that require digitization. Species distribution modeling and habitat modeling (e.g., generalized additive or multiplicative models) can also be applied but require knowledge of species–environment, spatial–temporal relationships. Human-use coastal and offshore resources, including commercial and recreational fishing areas, and marine protected area are also sometimes available online as spatial datasets or static maps.

1.2.4 Data availability, limitations, and expectations

Developing robust, comprehensive baselines can be difficult due to the sheer volume of sources and variety of types of data that often need to be integrated. Attempts to build holistic and comprehensive baselines can require the integration of hundreds or thousands of unique datasets collected and maintained by numerous government agencies, universities, nonprofits, and commercial entities. This can become even more complex if the baseline's spatial footprint overlays multiple political boundaries or regulatory jurisdictions.

Baselines in marine and coastal ecosystems often combine static and dynamic data that span a range of spatial and temporal scales to characterize and understand the potential short- and long-term effects or impacts. Depending on the region of interest and the underlying dynamics and relationships, baselines may be in constant flux and require more frequent updates. In some situations, "shifting" baselines may also need to be considered, due to cyclical patterns exhibited in the ecosystem or within the relationships between variables, such as seasonal influences for species and habitats, as well as the "boom and bust" cycles exhibiting in offshore oil and gas and offshore fisheries (Berkes et al., 2006; Leslie & Kinzing, 2009).

Integrating data from numerous sources can require significant effort to process and clean the data. Data preparation efforts can account for up to 80% of the time spent on a project (Crowdflower 2016, 2017). These efforts comprise collecting, cleaning, and integrating diverse resources into a standardized and structured data format. This process ensures consistent units of measure, rectifies different date/time formats, and determines if and how datasets can be integrated when they were collected using different sampling procedures or technologies. The use of numerous integrated datasets from multiple sources can also introduce a great deal of uncertainty surrounding the quality of baseline data.

Efforts to create baselines and supporting infrastructure available before a disaster can provide significant value and ensure enough time exists to overcome the challenges mentioned above (Amin and Goldstein, 2008). During a spill event, having a baseline in place prevents the need to assign critical resources to collect and produce data, whereas those resources could instead remain focused on response needs. Prior development that takes place outside of the stress and urgency during a disastrous event helps afford time to identify, collect, process, and integrate data from disparate sources (Amin and Goldstein, 2008). Frequently, more baseline data are being made available through a variety of credible sources, including different federal, state, local, and tribal governments, nonprofits, and organizations. Baseline development should include detailed metadata and the use of common, standard, open-source formats. Outside of the creation of baselines, prior effort to develop a high-availability and accessible system to access baselines is extremely valuable during a spill event. Well-designed, online systems and applications enable users to search, discover, access, and share the most current, authoritative baseline data.

There are an increasing number of web portals, online repositories, data catalogs, and web-mapping applications that serve up baseline marine and coastal data. Some examples include NOAA's Integrated Ocean Observing System (https://ioos.noaa.gov/) and the National Research Infrastructure of Australia's Integrated Marine Observing System (https://imos.org.au/). These systems define resources, catalog data, and provide additional content pertaining to physical oceanography information. Data catalogs and repositories include Data.gov and GeoPlatform (https://www.geoplatform.gov/), NOAA's Data Catalog (https://data.noaa.gov/datasetsearch/), SEA ScieNtific Open Data Edition (https://www.seanoe.org/), the National Science Foundation's Biological and Chemical Oceanography Data Management Office (https://www.bco-dmo.org/), the Pan-European Infrastructure for Ocean and Marine Data management's platform SeaDataNet (https://www.seadatanet.org/), and the Australian Ocean Data Network portal (https://portal.aodn.org.au/). Web-mapping tools for visualization and analysis include NOAA's Environmental Response Management Application (https://response.restoration.noaa.gov/resources/maps-and-spatial-data/environmental-response-management-application-erma) and NETL's GeoCube (https://edx.netl.doe.gov/geocube/). Whether developing a new baseline or using existing baselines from existing platforms, it is important to understand the quality of the data and limitations of their use. Key questions should be asked around the authority or credibility of the data sources, the availability of detailed metadata, and metrics that evaluate or help evaluate, data quality.

Metadata, or data describing data, should allow for the evaluation and determination of data quality and any limitations or caveats regarding their use. Metadata documentation pertaining to the data source, in relation to who collected, processed, generated the data, as well as where the data are hosted, can provide insight into the credibility and scientific reliability of a source. Credible sources often meet or exceed current data standards and ensure accessibility to their data, as well as documentation on the methods or technologies used during data collection and processing. Ensuring robust integration and development of baselines can be challenging if specific details about the technologies or methods used are not provided with the metadata. Barriers to receiving comprehensive metadata include unavailability of documentation, intellectual property concerns, and staff turnover. Additional red flags include sources that rely on submission from unknown users, have little to no data structure, fail to follow a quality assurance/quality control (QA/QC) process, or offer limited information on how the data were acquired, processed, and generated.

In addition, metadata offers additional context for a data source, such as author, citation, date created, date published, data updated, and content summary. For spatial data, metadata includes pertinent information such as spatial reference system, extent, spatial accuracy or scale, and processing history (Table 1.1). The use of metadata standards offers value in understanding data integrity and usability. The US Federal Geographic Data Committee Geospatial Metadata Standards (https://www.fgdc.gov/metadata/) recommends using a consensus or industry standard over a unique government standard, as well as established international standards for metadata. Metadata should be formatted to meet common international standards, as it increases the ability to search and identify the data. One standard format includes International Standards Organization (ISO) 19139 Metadata Standard, which is very commonly used for spatial data.

Metadata can be external to the file or database containing the data and is often stored in a common text document format such as plain text files, extensible markup language (XML), hypertext markup language (HTML), and javascript object notation (JSON). Metadata can also be stored internally and is often included in the data structure or in the header of the file containing the data. Nearly all image file formats, like Tagged Image File Format (TIFF) files include this type of metadata within their file headers. Spatial data also include metadata that is both external and internal to the files storing the data.

Along with capturing the who, what, where, when, and how in metadata, there is also an increased demand for the inclusion of information related to the uncertainty of the provided data. Uncertainty can stem from a variety of sources, including collection methods, errors introduced through measurement, equipment, and human error. Uncertainty can also be introduced during processing, analysis, and data integration, and upload to electronic sources. As a result, there is no standard way to evaluate and communicate uncertainty. However, approaches and metrics to assess data quality are

TABLE 1.1 Example attributes, their formats, and example inputs of a data catalog, the data within which represents a baseline.

Attribute	Description	Example
Unique ID	Numeric value that could act as primary key among tables	0, 1, 2, 3
Dataset	Name of dataset	Oil and gas wells
Description	Description of dataset	This is a dataset representing offshore oil and gas wells in the Gulf of Mexico.
Category	Sector of baseline of which dataset represents	Infrastructure
Source website	Website where dataset was accessed	http://www.DataSource.org
Citation	Full citation	Jon Snow. (2019). Offshore Wells. *Data Source Organization*. Acquired on December 30, 2019 from http://www.DataSource.org
Licensing	Open-source or licensing requirements based on source	Requires written permission from Data Source Organization
File location	Local or online directory path	C://Temp//Infrastructure
File name	Exact name of file as it appears	OG_Wells.shp
Date updated	Date last updated by source	12/1/2019
File format	Comma-separated values, shapefiles, Postgres database, Portable document format (PDF), Tag image file format (TIFF)	Shapefile
Spatial reference system	Projected or geographic coordinate systems (i.e. datum) the data are in	World Geodetic System 1984
Temporal extent	Extent of time represented by data	2000–2019, 19 years
Spatial extent	Region or spatial area represented by data	Gulf of Mexico, 598,458 miles^2
Quality metrics	Ranking or weighted values representing spatial, temporal, or source quality	Quality scale is 1–5, 5 being highest quality. Spatial quality = 5, temporal quality = 3, source quality = 5
Comments	Text describing how data were processed	Data were queried to represent only oil and gas wells

being evaluated across the scientific community as data volume and demand increase. Currently, many online systems that host baseline data are relying a range of QA/QC processes to ensure data quality. Processes range from security permissions that only allow approved users to create or contribute data to reporting of specific metrics related to source quality and data completeness (i.e., percentage of incomplete or missing metadata entries or attributes in a data record).

The compilation of data into portals, repositories, catalogs, and applications helps to facilitate the development of data-driven models and tools that are trustworthy, reproducible, and reusable. These systems help structure the data and metadata, as programmed user interfaces present a standardized representation of information. In addition, these online applications are often able to quickly adapt to meet community and user demands, such as the inclusion of additional information like collection methods or data quality metrics. Many modern online systems and applications serve baseline data from high-availability server architecture that can handle high-demand and rapid access. Furthermore, modern systems increasingly offer access to data though the use of application programming interfaces, or APIs, wherein data can be accessed through protocol-specific access points and uploaded directly into applications, tools, and models. Direct access and real-time updates help baselines remain current. With the inclusion of user accounts and permissions, systems can restrict access to specified data that are sensitive or proprietary and can dictate how data are shared between users under a variety of use cases, such as data development. Knowledge of the credibility, benefits, limitations, and expectations of data being collected for a baseline, and understanding the applications currently offering data, is crucial to ensure the design and development of comprehensive, representative baselines.

1.3 Building a baseline

A proper baseline represents as accurate and as complete of a picture as possible of the socioeconomic and environmental factors that exist within the region of interest and during the time of concern. This section provides a framework for building a complete baseline. Since the technological and data landscape are constantly changing, this will be a high-level framework that can be applied for future efforts (Fig. 1.5).

1.3.1 Identifying knowledge needs

When building a baseline, a researcher must have a solid understanding of what data are needed and how data will be used. Assessing the potential impacts of oil spills requires information on the current state of knowledge, identified data, and technology gaps. A thorough literature review of

A machine learning approach to data searching is implemented by the Smart Search Tool©, an automated web crawler designed to find data resources (Rose et al., 2018). Developed by NETL and Mid-Atlantic Research, Innovation and Technology Center, the Smart Search Tool scans the user-defined "seed" resources and returns a list of prevalent key words and phrases, which are then offered up for manual QA/QC process. The tool then takes the list and completes an extensive web crawl utilizing an Internet search engine. This process returns a typically large and thorough list of data resources that users can query for useful or relevant data.

With both manual and automated data collection methods, users can employ scripting languages, such as Python (http://www.python.org), to scrape a web page for relevant information, potentially reducing the time, costs, and efforts of capturing the data in a usable format. Similar processes can be applied to scanned documents, where high-quality data need to be reformatted prior to use in modeling and analysis.

Regardless of the approach used, a data catalog should be built as data are collected, with each record representing metadata for a dataset that has been identified or downloaded. The documentation of the data catalog is critical for recording identifying metadata, including name and source. An example catalog would include the attributes as shown in Table 1.1.

1.3.3 Integrating, analyzing, and publishing data

Socioeconomic and environmental data are often disparate, differing in type, source, and format. Acquired data should be reformatted and processed into a database, applying the data toward building a more consistent baseline. To prepare the data catalog to be utilized and searched, several steps should be taken. Nonspatial data should be transformed into spatial datasets, if spatial information is available (e.g. spatial coordinates, addresses, etc.). All spatial datasets should be projected into a common spatial reference system, which includes a projection, datum, and unit, to maintain spatial accuracy during analyses. Attributes of each dataset should be listed and compared to establish how to best integrate them with the catalog. Data should then be scrubbed to remove any redundancies and errors. Each step throughout this process should be recorded to ensure repeatability.

Data can be integrated into a useful product once all data have been processed into a common, or at least compatible, format. The resulting product should provide a spatial understanding of the current baseline. Further analysis will potentially uncover patterns, trends, and hotspots within the baseline (i.e., where socioeconomic risk is highest). The resulting product should be standardized based on the data it contains, potentially as a database or a set of datasets. For example, researchers at NETL combined over six million records of global oil and gas infrastructure into one geodatabase (Rose et al., 2018) for the Environmental Defense Fund, making the data easily accessible and suitable for subsequent analyses. Another example data product is a collection of vector grids containing multivariate information on the potential impacts incurred across multiple US regions, which when compared with actual or simulated oil spill events can help predict where response resources are likely to be needed (Romeo et al., 2015; Nelson et al., 2018). Fig. 1.6 demonstrates an output of these efforts in the US Gulf of Mexico, including coastlines, where data from over 40 datasets representing socioeconomic and environmental variables were spatial summarized by dataset presence (e.g., 1 = 1 dataset). The data represented in Fig. 1.6 were virtually released on the web-mapping platform GeoCube (Romeo et al., 2020).

Frequently, baseline data and analyses thereof are published as part of a manuscript or report through maps, tables, and results. The rigors of publishing are valuable for evaluating the validity and scientific contribution of the work accomplished. In addition to producing papers, baseline datasets should be published as databases or through some other standard of accessible data packaging, with the associated metadata and processing information. Publishing data ensures future usability and reproducibility of any scientific assertions, a key part of the scientific method.

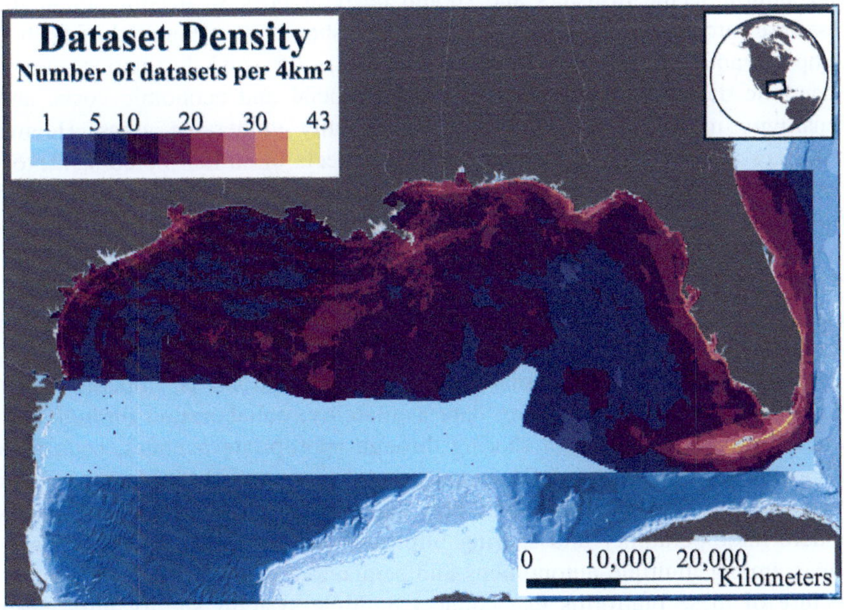

FIGURE 1.6 Mapped baselines showing the density of 43 datasets representing socioeconomic and environmental vulnerability in the US Gulf of Mexico and adjacent coastal communities in the datasets were summarized by presence to understand data coverage and included specific species distributions, critical habitats, commercial fishing areas, transportation tracklines, tourism and recreation data, public infrastructure, and oil and gas infrastructure. This figure was made using the Cumultative Spatial Impact Layer™ tool (Romeo et al., 2019).

As previously mentioned, baselines can be published online as part of a data catalog or repository. Like the data in Fig. 1.6, baseline databases can also be uploaded to web-mapping applications to aid in data visualization, access, and analysis. If a new portal is being developed to host baseline data, it is important to design a system that addresses the limitations and expectations previously mentioned. If baselines are to be hosted through an existing source, the same information can help identify appropriate platforms given the targeted use and audience for the baseline. For example, GeoCube is NETL's web-mapping interface that contains a series of data collections varying in topics pertaining to fossil energy research. GeoCube is hosted on the data and science curation platform Energy Data eXchange® (EDX) (Rose et al., 2020). EDX has been tailored for its research-driven fossil energy audience by offering public- and private-facing options, enabling data sharing among research groups internally or released with a digital object identifier (DOI) number externally.

1.4 What can we do to improve baselines

Marine and coastal baselines are critical to adequately prepare for natural and anthropogenic disasters and should be built prior to needing them for response purposes. At a minimum, baselines should be comprised of spatiotemporal ambient data for modeling spills trajectory and destination, socioeconomic data for evaluating the potential social and economic costs, and sensitivity mapping to identify environmentally vulnerable areas. Having baselines at the ready can expedite other projects or shifts, including development and restoration projects and management changes. Lessons learned from past spill events have signified the value of baselines for decision makers. In addition, these lessons have helped identify enhancements to the handling of data availability, acquisition, and processing that could ultimately build better baselines, strengthening our ability to prepare for and respond to oil spills.

Collaboration among researchers and organizations is imperative for improving baselines. Increasing data availability, collaboration through networking, projects, and data sharing through appropriate avenues, such as a Cooperative Research and Development Agreement (CRADA), supersedes many of the issues faced when building a baseline. Federated platforms enable interoperability and sharing information technology, including baselines, among multiple autonomous and semiautonomous sources. The architecture of these platforms that connect multiple systems significantly eases the process of finding, acquiring, and updating data, as they can act as one-stop shops. Because baselines are regionally specific and situational, not all data shared are applicable; however, the methods, tools, and models applied might be leverageable across geographies and time (Romeo et al., 2015). Examining what has already been accomplished, what has worked, and understanding why certain baselines are more successful than others is a valuable step in driving future improvements.

In addition to data sharing and an awareness of what's available, there are advanced data acquisition approaches and methods that could be applied to improve baselines. The application of automated approaches through machine learning can accumulate data at a rate greater than any one researcher. These practices, when used correctly, can save time and money but require an understanding of the process and a thorough QA/QC procedure to ensure what is discovered and collected is within the scope of the baseline.

Thanks to increased data availability and the advanced practices of data acquisition, it is becoming easier to collect a large quantity of quality data. But with more data comes more effort, energy, and computational ability required to navigate it. Furthermore, the higher the quality of the data, the subtler the errors are. Simply put, the ability to collect quality data has outpaced user ability to parse, manage, and store it. For data to be useful, there needs to be a flexible strategy for data processing, management, and maintenance. Like all data-driven efforts, there are issues and errors to be aware of, addressed, and improved upon. Issues and errors include data redundancy, human error, lacking or nonexistent metadata, and absence of protocol. How can the data acquired be better processed to improve the baseline product?

The use of validated, and in some cases novel, approaches, tools, and models to transform data are imperative for handling big data and overcoming these issues. One example of an approach to quickly sort big data is the machine learning method of Natural Language Processing (NLP). NLP can sort data by applying linguistic analyses to the extracted data from textual sources (Ray, Johnny, Trovati, Sotiriadis, & Bessis, 2018). An example of a tool that can help with QA/QC is the Variable Grid Method© (VGM). This novel tool enables users to quantify and visualize spatial data trends and underlying uncertainty (Bauer & Rose, 2015). Communicating the relationship between uncertainty and data, VGM helps to effectively guide and inform research and decision makers, while supporting advanced computational analyses. This tool is just one example of a valuable resource for effectively understanding large volumes of baseline data. Once data are appropriately parsed, they can be stored using a object relational database platform, such as PostGIS. Object relationship database storage is useful, as databases are specifically designed for processing queries and serving up data. Such a configuration simplifies the tasks of processing, managing, and maintaining data.

Lastly, baseline maintenance is crucial for a baseline to remain as true as possible to representing real-world ambient, socioeconomic, and environmental phenomena. Maintenance might include utilizing automated processes to complete real-time data updates, wherein the database periodically pulls and processes data directly from the data source and makes the updated data directly available for query. If these processes are unavailable, alternative solutions, such as a manually updating datasets, will be needed. These elements are all dependent on computational power, time, and money but are critical steps to building and improving upon baselines.

Acknowledgments

This work was performed in support of the US Department of Energy's Fossil Energy, Oil and Natural Gas Research Program. It was executed by NETL's Research and Innovation Center, including work performed by Leidos Research Support Team staff under the RSS contract 89243318CFE000003.

Disclaimer

This work was funded by the Department of Energy, National Energy Technology Laboratory, an agency of the United States government, through a support contract with Leidos Research Support Team (LRST). Neither the United States government nor any agency thereof, nor any of their employees, nor LRST, nor any of their employees, makes any warranty, expressed or implied, or assumes any legal liability or responsibility for the accuracy, completeness, or usefulness of any information, apparatus, product, or process disclosed, or represents that its use would not infringe privately owned rights. Reference herein to any specific commercial product, process, or service by trade name, trademark, manufacturer, or otherwise, does not necessarily constitute or imply its endorsement, recommendation, or favoring by the US government or any agency thereof. The views and opinions of authors expressed herein do not necessarily state or reflect those of the United States government or any agency thereof.

References

Alonso, G., Casati, F., Kuno, H., & Machiraju, V. (2004). *Web services. Web services* (pp. 123–149). Berlin, Heidelberg: Springer.

Amin, S., & Goldstein, M. (Eds.), (2008). *Data against disasters: Establishing effective systems for relief, recovery, and reconstruction*. The World Bank.

Baird, J. (2010). Oil's shame in Africa. https://en.wikipedia.org/wiki/Newsweek.

Bauer, J. R., Nelson, J., Romeo, L., Eynard, J., Sim, L., Halama, J., et al. (2015). A spatio-temporal approach to analyze broad risks and potential impacts associated with uncontrolled hydrocarbon release events in the offshore Gulf of Mexico. NETL-TRS-2-2015; EPAct Technical Report Series (U.S. Department of Energy, National Energy Technology Laboratory), 60, http://www.netl.doe.gov/File Library/Research/onsite research/NETL-TRS-2-2015_CSIL_BroadImpacts.20150219.pdf.

Bauer, J. R., & Rose, K. (2015). Variable grid method: An intuitive approach for simultaneously quantifying and visualizing spatial data and uncertainty. *Transactions in GIS, 19*(3), 377–397.

Berkes, F., Hughes, T., Steneck, R., Wilson, J., Bellwood, D., Crona, B., et al. (2006). Globalization, roving bandits, and marine resources. *Science*, *311*(5767), 1557–1558. Available from https://doi.org/10.1126/science.112280.

Carswell, C. (2018). Unique oil spill in East China Sea frustrates scientists. *Nature*, *554*(7690). Available from https://doi.org/10.1093/10.1038/d41586-018-00976-9.

Cockburn, I. M., Henderson, R., & Stern, S. (2018). *The impact of artificial intelligence on innovation*. National Bureau of Economic Research, No. w24449.

Crowdflower. (2016). Data science report 2016. San Francisco, CA: Figure Eight Inc. Available at https://visit.figure-eight.com/Data-Scientist-Report_T.

Crowdflower. (2017). Data scientist report 2017. San Francisco, CA: Figure Eight Inc. Available at https://www.figure-eight.com/download-2017-data-scientist-report/.

de Andrade, M. M. N., Szlafsztein, C. F., Souza-Filho, P. W. M., dos Reis Araújo, A., & Gomes, M. K. T. (2010). A socioeconomic and natural vulnerability index for oil spills in an Amazonian harbor: A case study using GIS and remote sensing. *Journal of Environmental Management*, *91*(10), 1972–1980.

Duran, R., Beron-Vera, F. J., & Olascoaga, M. J. (2018). Extracting quasi-steady Lagrangian transport patterns from the ocean circulation: An application to the Gulf of Mexico. *Scientific Reports*, *8*(1), 5218.

Duran, R., Romeo, L., Whiting, J., Vielma, J., Rose, K., Bunn, A., & Bauer, J. (2018). Simulation of the 2003 Foss Barge-Point Wells Oil Spill: A comparison between BLOSOM and GNOME oil spill models. *Journal of Marine Science and Engineering*, *6*(3), 104.

Etkin, D. S. (2004). Modeling oil spill response and damage costs. In *Proceedings of the Fifth Biennial Freshwater Spills Symposium* (Vol. 15, p. 15).

Goldstein, B. D., Osofsky, H. J., & Lichtveld, M. Y. (2011). The Gulf oil spill. *New England Journal of Medicine*, *364*(14), 1334–1348.

Jensen, J. R., Halls, J. N., & Michel, J. (1998). A systems approach to Environmental Sensitivity Index (ESI) mapping for oil spill contingency planning and response. *Photogrammetric Engineering and Remote Sensing*, *64*, 1003–1014.

Joyner, C. C., & Kirkhope, J. T. (1992). The Persian Gulf War Oil Spill: Reassessing the Law of Environmental Protection and the Law of Armed Conflict. *Case Western Reserve Journal of International Law*, *24*(29). Available from https://scholarlycommons.law.case.edu/jil/vol24/iss1/2.

Kankara, R. S., Arockiaraj, S., & Prabhu, K. (2016). Environmental sensitivity mapping and risk assessment for oil spill along the Chennai Coast in India. *Marine Pollution Bulletin*, *106*(1–2), 95–103.

Kontovas, C. A., Psaraftis, H. N., & Ventikos, N. P. (2010). An empirical analysis of IOPCF oil spill cost data. *Marine Pollution Bulletin*, *60*(9), 1455–1466.

Leslie, H. M., & Kinzing, A. P. (2009). Resilience science. In K. McLeod, & H. Leslie (Eds.), *Ecosystem-based management for the oceans* (1st ed., pp. 55–73). Washington, DC: Island Press.

Maximenko, N., Niiler, P., Rio, M., Melnichenko, O., Centurioni, L., Chambers, D., ... Galperin, B. (2009). Mean dynamic topography of the ocean derived from satellite and drifting buoy data using three different techniques. *Journal of Atmospheric and Oceanic Technology*, *26*, 1910–1919.

McLeod, K., & Leslie, H. (2009). *Ecosystem-based management for the oceans* (Island, Washington, DC).

National Oceanic and Atmospheric Administration. (1997). *Environmental sensitivity index guidelines*. Version 3.0. NOAA: Seattle, Tech. Memorandum NOS ORandR, 11, 1997.

National Oceanic and Atmospheric Administration. (2013). *Economics: National Ocean Watch (ENOW)*. Available from http://www.csc.noaa.gov/digitalcoast/data/enow.

National Oceanic and Atmospheric Administration, Office of Response and Restoration. (2014). Louisiana and Lower Mississippi River. Environmental Sensitivity Index (ESI) Maps and Data. https://response.restoration.noaa.gov/esi. Accessed 01.01.20.

Nelson, J. R., Bauer, J. R., & Rose, K. (2014). Assessment of geographic setting on oil spill impact severity in the United States—Insights from two key spill events in support of risk assessment for science-based decision making. *Journal of Sustainable Energy Engineering*, *2*(2), 152–165.

Nelson, J. R., Grubesic, T. H., Sim, L., & Rose, K. (2018). A geospatial evaluation of oil spill impact potential on coastal tourism in the Gulf of Mexico. *Computers, Environment and Urban Systems*.

Nigerian National Petroleum Corporation. (2018, November 29). Oil production. http://www.nnpcgroup.com/nnpcbusiness/upstreamventures.oilproduction.aspx.

Nwilo, P. C., & Badejo, O. T. (2006). Impacts and management of oil spill pollution along the Nigerian coastal areas. *Administering Marine Spaces*, *119*(International Issues), 1–15.

Ray, J., Johnny, O., Trovati, M., Sotiriadis, S., & Bessis, N. (2018). The rise of big data science: A survey of techniques, methods and approaches in the field of natural language processing and network theory. *Big Data and Cognitive Computing*, *2*(3), 22.

Romeo, L., Bauer, J., Mark-Moser, M., Nelson, J., Dyer, A., & K. Rose. (2020). Offshore Gulf of Mexico data collection. https://edx.netl.doe.gov/dataset/offshore-gulf-of-mexico-data-collection, https://doi.org/10.18141/1644085.

Romeo, L., Rose, K., Bauer, J. R., Disenhof, C., Sim, L., Nelson, J., ... Barkhurst, A. (2015). *Adapting the National Energy Technology Laboratory's offshore integrated assessment modeling approach for the offshore Arctic; NETL-TRS-3–2015. EPAct Technical Report Series* (p. 40) Morgantown, WV: United States Department of Energy, National Energy Technology Laboratory.

Romeo, L., Nelson, J., Wingo, P., Bauer, J., Justman, D., Rose, K., et al. (2019). Cumulative spatial impact layers: A novel multivariate spatio-temporal analytical summarization tool. *Transactions in GIS*, *23*(5), 908–936. Available from https://doi.org/10.1111/tgis.12558.

Rose, K., Bauer, J., Baker, V., Bean, A., DiGiulio, J., Jones, K., ... Tong, A. (2018). *Development of an open global oil and gas infrastructure inventory and geodatabase; NETL-TRS-6–2018. NETL Technical Report Series* (p. 594) Albany, OR: United States Department of Energy, National Energy Technology Laboratory. Available from https://doi.org/10.18141/1427573.

Rose, K., Rowan, C., Baker, D. V., Jones, T. J., Barkhurst, A., McFarland, D., & EDX Team. (2020). Energy Data eXchange® (EDX). National Energy Technology Laboratory (NETL). https://edx.netl.doe.gov.

Sim, L., Graham, J., Rose, K., Duran, R., Nelson, J., Umhoefer, J., & Vielma, J. (2015). *Developing a comprehensive deepwater blowout and spill model. NETL-TRS-9–2015. EPAct Technical Report Series* (p. 44) Albany, OR: United States Department of Energy, National Energy Technology Laboratory.

Souza Filho, P. W. M., Goncalves, F. D., de Miranda, F. P., Beisl, C. H., & de Faria Almeida, E. (2004). Environmental sensitivity mapping for oil spill in the Amazon coast using remote sensing and GIS technology. In *Proceedings of the 2004 IEEE international geoscience and remote sensing symposium, IGARSS'04* (Vol. 3, pp. 1565–1568). IEEE.

Stelzenmüller, V., Lee, J., South, A., Foden, J., & Rogers, S. I. (2013). Practical tools to support marine spatial planning: A review and some prototype tools. *Marine Policy*, *38*, 214–227.

The NOM Group. (2003). Navy operational ocean circulation and tide models. http://www.oc.nps.edu/nom/modeling/. Accessed 30.11.18.

Thompson, J., et al. (Eds.), (1999). *Handbook of grid generation*. Boca Raton: CRC Press.

Wallcraft, A., Carroll, S. N., Kelly, K. A., & Rushing, K. V. (2003). Hybrid Coordinate Ocean Model (HYCOM) Version 2.1 User's Guide. https://hycom.org/attachments/063_hycom_users_guide.pdf. Accessed 13.11.18.

White, C., Halpern, B. S., & Kappel, C. V. (2012). Ecosystem service tradeoff analysis reveals the value of marine spatial planning for multiple ocean uses. In *Proceedings of the National Academy of Sciences*, 201114215.

Unidata: netCDF. https://www.unidata.ucar.edu/software/netcdf/ (Accessed 7 Jun 2021).

Zhang, R., Wingo, P., Duran, R., Rose, K., Bauer, J., Ghanem, R., et al. (2020). Environmental Economics and Uncertainty: Review and a Machine Learning Outlook. *Oxford Research Encyclopedia of Environmental Science*. Available from https://doi.org/10.1093/acrefore/9780199389414.013.572.

2.2 Characterizing oil—enter the Tower of Babel

In Christian literature, God punishes men for building a tower to Heaven by causing them to speak different and mutually incomprehensible languages. Supernatural powers were not required for the oil industry to put the outside observer in a similar situation. Similar names may refer to different properties and the same property may have multiple names. Generally speaking, oil can be separated into two large groups: crude oil extracted from reservoirs and refined products constructed through distillation from crude oils for specific purposes. A large online database of crude oils and refined products is maintained by the Emergencies Science and Technology Section of Environment and Climate Change, Canada (ESTD, 2019) that lists hundreds of oils and refined products.

Nearly all crude oils have a fairly narrow range of elemental composition. Speight (2005) reports the breakdown as carbon(83%–87%), hydrogen (10%–14%), nitrogen (0.1%–2%), oxygen (0.05%–1.5%), and sulfur (0.05%–6%). For reference, wood elemental composition is approximately 50% carbon, 42% oxygen, 6% hydrogen, 1% nitrogen, and 1% other elements. Trace metals may also be present in crude oil and are occasionally used for "fingerprinting" mystery oil samples (Yang, Gao, & Casey, 2008). Crude oils are often given names that refer to the geographical source, for example, Louisiana sweet crude, Arabian heavy. Sweet or sour does not refer to the oil taste but the amount of sulfur in the oil. Crude oil with sulfur content of more than a half percent is considered to be sour. Oils extracted from the reservoir may contain dissolved gases and are referred to as "live" oils while those without this characteristic are called "dead" oils.

Crude oils contain hydrocarbons that have boiling points that range from −160°C to greater than 600°C. Refineries use this variation to process crude into useful products. Riazi (2005) divides petroleum fractions are commonly their boiling points (Table 2.1).

Lubricating oil typically has a boiling point >400°C. Oil products used for fuel have traditionally been divided into five grades (Schmidt, 1986). Mixtures of these grades form intermediate fuel oils (IFO) that are labeled by their resulting viscosity. For example, IFO 180 is a mix of a 12% distillate (fuel oil #1 or #2) and 88% residual oil (fuel oil #6) with a viscosity of 180 cSt at 50°C. Definition and units of viscosity are discussed later in this chapter.

2.3 Bulk oil properties—considering the forest rather than individual tree

Industry distributes oil by volume or mass. The traditional volume of oil is the barrel (bbl), which is equal to 42 US gallons or 159 L. Crude oils have between 6.5 and 8 bbls per metric ton, depending on the oil specific gravity.

TABLE 2.1 Boiling point ranges for common petroleum products (adapted from Riazi, 2005).

Fraction	Boiling range (°C)
Gasoline	−1 to 200
Naptha	−1 to 200
Jet fuel	150 to 255
Kerosene	205 to 255
Diesel fuel	205 to 290
Light gas oil	255 to 315
Heavy gas oil	315 to 425
Vacuum gas oil	425 to 600
Residium	>600

Certain bulk oil properties are not dependent on the individual hydrocarbons in the oil but still have a major impact on the fate and behavior of the spill. The two most important are density and viscosity. The former is often characterized by the equivalent specific gravity. Industry reports specific gravity in API degrees

$$\mathrm{API} = \left(\frac{141.5}{sg}\right) - 131.5 \tag{2.1}$$

where sg is the specific gravity at stock tank conditions (100 kPa, 16°C). The number constants are selected so that pure water has an API of 10. Oils with API less than 10 would be nonbuoyant in freshwater but might be slightly buoyant in seawater.

Both density and viscosity will change due to interaction with the environment and weathering of the oil. For example, the density of oil increases as the temperature decreases. The increase parameter is a nonlinear function (ASTM, 2007) but can be approximated as linear over typical ocean surface temperature ranges. Density will also increase as lighter hydrocarbons are preferentially removed due to evaporation or dissolution. For buoyant oils, emulsification will also increase net surface slick density.

A traditional formulation for oil density change would be (Lehr, 2001)

$$\rho_{oil} = \rho_{ref}\left[1 - c_{\rho 1}\left(T - T_{ref}\right)\left(1 + c_{\rho 2}(f_{evap} + f_{solv})\right]\right. \tag{2.2}$$

for nonemulsified oil and

$$\rho_{emul} = f_w \rho_w + (1 - f_w)\rho_{oil} \tag{2.3}$$

for emulsified oils. Reasonable values, with temperature expressed in Kelvin, for $(c_{\rho 1}, c_{\rho 2})$ might be 0.008 1/°K and 0.18.

Viscosity is an ambiguous term when applied to oil since there are actually two different quantities that are called viscosity. Kinematic viscosity has dimensions of area divided by time with its SI unit being the stoke, named after the Irish mathematician Sir George Gabriel Stokes who helped develop the Navier-Stokes equation for fluid flow. Dynamic viscosity, sometimes called absolute viscosity, is kinematic viscosity multiplied by the oil density, giving it dimensions of mass divided by the product of time and distance. Its SI unit is the poise. The oil industry traditionally measured kinematic viscosity by the time it took for an oil sample to flow through a certain type of measuring viscometer (e.g., Saybolt universal second). Fortunately, this is less common today and most large oil property libraries store kinematic viscosity/ dynamic viscosity in one hundredth of stoke/poise, that is, centistoke/ centipoise. Most oils are more viscous than water, which has a kinematic viscosity of about 1 cSt. Kerosene has a kinematic viscosity of 10 cSt while many crude oils have viscosities of several hundred centistokes. Oils that have emulsified may no longer follow Newton's law of viscosity, which states that shear stress between adjacent fluid layers is linearly proportional to the negative of the velocity gradient between them. Such non-Newtonian fluids do not have a single fixed viscosity since the answer depends on the velocity gradient and may be no longer describable by a simple scalar. Nevertheless, most spill models treat even highly viscous oils as if they were Newtonian.

Viscosity measurement accuracy is less sensitive than density measurement. While doubling an oil density would likely make a buoyant oil become nonbuoyant (2.1), would be difficult in the field to detect the difference between 100 and 200 cSt oil. Canceling this observation, however, is the significantly larger numerical change in viscosity value that occurs as the oil interacts with the environment. For example, viscosity, unlike density, varies widely as oil temperature changes. While the most widely used ASTM reference temperature for viscosity measurement is 15°C, a traditional industry reference temperature in laboratory measurements can be 100°F (311°K). This is not a common oil slick temperature. The earliest empirically generated viscosity temperature adjustment formula was produced by Reynolds (1886) but most spill models today use a version of MacKay et al. (1982), loosely based on the Arrhenius equation,

$$\ln\left(\frac{\upsilon_{oil}}{\upsilon_{ref}}\right) = c_{\upsilon 1}\left(\frac{1}{T_{oil}} - \frac{1}{T_{ref}}\right) \tag{2.4}$$

Widely different estimates exist for $c_{\upsilon 1}$ (Lehr, 2001) and the fit with field data is generally poor, especially as oil viscosity approaches the Newtonian limit. More complex models exist (e.g., Masuko & Magill, 1988) but have not been widely adopted for spill behavior forecasting, probably because most industry research has concentrated on viscosity measurements at the higher temperature and pressure of oil reservoirs (Rahuma, 2016), not those at typical spill environmental conditions.

Oil evaporation will also greatly increase viscosity as the lighter components preferentially leave the slick. Mackay et al. (1982) suggested an empirical fit

$$\ln\left(\frac{\upsilon_{oil}}{\upsilon_{ref}}\right) = c_{\upsilon 2} f_{evap} \tag{2.5}$$

Field measurement indicates that $c_{\upsilon 2}$ cannot be a simple constant for all oils. Most models either fit $c_{\upsilon 2}$ to the initial oil API or viscosity. More recently, Onyelucheya, Osaka, Onyelucheya, and Kamen (2014) proposed that the entire right side of the equation should be replaced by a 2-parameter, empirically determined, exponential function of f_{evap}.

As noted previously, emulsification will often increase slick viscosity beyond the Newtonian limit. Widely used weathering models adopt some form of the Mooney equation (e.g., Schramm, 1992) to estimate emulsion viscosity.

$$\ln\left(\frac{\upsilon_{emul}}{\upsilon_{oil}}\right) = \frac{c_{\upsilon 3} f_w}{(1 + c_{\upsilon 4} f_w)} \tag{2.6}$$

The most common values for the empirical constants $c_{\upsilon 3}, c_{\nu 4}$ are those of MacKay et al. (1982) but alternative values exist (Pajouhandeh, Kavousi, Schaffie, & Ranjbar, 2016). Pal and Rhodes (1989) claim that their correlation [*P*–*R*] gives better results than the straight Mooney equation when the viscosity approaches non-Newtonian conditions

$$[P-R] \qquad \ln\left(\frac{\upsilon_{emul}}{\upsilon_{oil}}\right) = \left(1 + \frac{c_{pr1} f_w}{\left(c_{pr2} - c_{pr1} f_w\right)}\right)^{c_{pr3}} \tag{2.7}$$

The reader is referred to the original paper for the determination of the empirical constants.

Another important bulk oil property is oil–water interfacial surface tension, σ_{o-w}. Surface tension is the tendency of fluid surfaces to shrink into the minimum surface area possible. It has units of force per unit length and is usually expressed in spill response as dyn/cm. The surface tension at the air–water interface at normal temperatures is around 72 dyn/cm. Almost all untreated (no dispersant added) oils have an oil–water surface tension between 10 and 35 dyn/cm. Dispersants lower this value significantly—as low as 0.1 dyn/cm with a gradual increase over time. Along with gravity, oil–water surface tension causes spilled oil to spread on calm waters (Fay, 1971). Air–oil interfacial tension reduces wind-induced frictional stress on oil-covered waves, slashing the number of breaking waves. The absence of small wind waves (capillary waves) in oil-covered waters is used in radar detection of oil slicks. Surface tension is a weak inverse function of temperature and more strongly dependent on the state of slick weathering. MacKay et al. (1982) proposed that surface tension increases linearly as a fraction of the net oil evaporated although the author is not aware of the inclusion of this algorithm in commonly used weathering models.

A handful of other bulk properties have been included in existing oil databases or show potential for future weathering models. Environment Canada (ESTD, 2019) has measured for certain selected oils an "adhesion" parameter that calculates the ability of oil to stick to a test needle inserted into an oil sample. As mentioned earlier, reservoir oil often contains dissolved gas. The gas–oil ratio is usually reported in the industry by the nonmetric units of cubic feet of gas at standard pressure and conditions (scf) to stock tank barrels (stb). For a handful of oils, laboratory measurements have been performed on the evaporative fraction necessary to be lost to cause emulsification to begin and the resulting maximum water mass fraction of the stable emulsion. The latter can be as high as 0.9 for many oils.

Crude oil and most refined products are mixtures of a large number of different hydrocarbons with different chemical structures. It is not practical to provide a comprehensive list of each unique hydrocarbon so industry has traditionally separated them into SARA—four broad structural groups standing for saturates, aromatics, resins, and asphaltenes. Saturates are nonpolar oil molecules without double bonds that include linear, branched, and cyclic saturated hydrocarbons. The simplest saturate is methane, with one carbon atom attached to four hydrogen atom. Aromatics contain one or more benzene rings, chicken wire-shaped structures with a changing mixture of single and double bonds among the six carbon atoms in the ring. Consensus definition of the last two categories is less established. According to the Society of Petroleum Engineers (SPE, 1990), resins are large hydrocarbon molecules with one to three sulfur, oxygen, or nitrogen atoms per molecule. The basic structure is composed of three or more rings, primarily aromatic. Resins readily dissolve in petroleum. Asphaltenes are very large hydrocarbon molecules of one to three sulfur, oxygen, or nitrogen atoms per molecule that do not readily dissolve in oil.

Because oil is a mixture, certain properties of pure substances do not apply. For example, crude oil does not have a definite melting point. Instead, industry uses a counterpart, "pour point" (PP), as a substitute. Pour point is commonly defined as the lowest temperature at which oil will flow at a given rate under specified conditions. Unfortunately, these specified conditions are not applied uniformly and pour point measurements of the same oil at different laboratories can vary by more than 10°C (ESTD, 2019). Unlike melting point, pour point changes when the oil is subject to weathering. Since evaporation is usually the chief factor in altering hydrocarbon mixture ratios for most spills, a first-order correction to fresh oil pour point (PP_0), based on mass fraction evaporated (MacKay, Stiver, & Tebeau, 1983), has been adopted by many weathering models.

$$PP = PP_0(1 + c_{PP} f_{evap}) \tag{2.8}$$

The empirically determined constant, c_{PP}, varies according to the type of oil with a typical value being about 1/3. Emulsification also affects pour point, so many spill models incorporate a similar linear adjustment for pour point increase based on water fraction of the emulsion.

One bulk property important for the safety of first responders is the oil flash point; the temperature to which the oil must be heated to produce a vapor ignitable when exposed to an open flame under specified test conditions (Jokuty et al., 1999). Flash point increases due to weathering by the same mechanisms as pour point, and, as in the former case, is dependent on the method used to measure it.

Another bulk property related to flash point for flammable hydrocarbon mixtures is the bulk vapor pressure, the pressure the evaporating oil exerts on its surroundings. More voluble hydrocarbons have higher vapor pressure and typically larger flammability risk. Early evaporation models (MacKay & McAuliffe, 1989) were based on the oil bulk vapor pressure value although most current models consider vapor pressure of individual distillation cuts when computing evaporative mass loss.

2.4 Oil weathering estimation—the two (or maybe three) philosophical schools

Those fortunate enough to have financial capital to invest will quickly discover that there are competing philosophies on how to select stocks and bonds. On the one hand, there are the traditionalists who value a stock price by using classical economic and accounting techniques to assess the underlying corporate assets. However, other advisers have concluded that such approaches are naïve since the complexities of the actual market are too great to be captured by simple balance sheet approaches. Instead, they argue that one should choose what to invest based upon past stock prices and related indices. Surprisingly, oil weathering behavior forecasting has likewise bifurcated into expert camps with the majority researchers favoring classical analysis where the specific weathering process is reduced to a set of equations derived from fundamental chemistry and physics. A different approach is to recognize that the necessary simplifications and approximations required are likely to not capture all the important subprocesses. Alternative "statistical" models have used specific curve fits (e.g., Fingas, 2014) to separate weathering behavior components and plot laboratory-measured behavior for many oils. It has been found through many years of study that each behavior process has its own unique characteristics that can often be estimated by standard statistical software.

A spinoff of the statistical approach is the adoption of neural networks from artificial intelligence research (Yetilmezsoy, Fingas, & Fieldhouse, 2012). Classical analysis and laboratory measurement are used to assign initial values to the neural networks. A set of internal nodes then processes the data and assigns salient weights to various combinations of the input parameters, based upon resulting actual field results. The system "learns" from experience which weights provide the closest result to actual observations. While the network results are constrained by the underlying science (e.g., emulsion water fraction must be nonnegative), the final set of rules may show little resemblance to traditional algorithms. It seems possible that trends in information technology will lead to the use of such artificial intelligence to resolve difficult weathering processes such as emulsification.

2.5 Weathering processes—those that do not alter the spill's chemical mixture

The major weathering processes can be divided into two groups. Certain processes such as spreading, dispersion, oil–particle aggregation, tar ball formation, and emulsification do not change the underlying hydrocarbon mixture although the processes themselves are greatly impacted by any chemical structural changes. Evaporation, dissolution, and photooxidation, on the other hand, do alter the chemical nature of the slick. For these latter processes, it is not sufficient to estimate mass balance and physical characteristics of the spilled oil but rather to also predict changes to the chemical balance.

2.5.1 Oil spreading

With regard to the first group, spreading of the oil is the most immediate phenomenon after the initial release. The classic work by Fay (1971) provided the formulas used by most early models. According to Professor Fay, oil slick spreading on calm water could be separated into three phases, each phase defined by its advancing and retarding force. The first phase balances gravitationally caused spreading with the inertia of the oil and lasts for only few minutes or less in all instances except the largest spills. The second phase replaces inertia with viscous drag of the surrounding water as the retarding force. The time till transition to this second phase, t_{i-v}, is calculated to be

$$t_{i-v} = c_{Fay1}\left(\frac{V_{oil}}{g\Delta_{\rho}\upsilon_{w}}\right)^{1/3} \tag{2.9}$$

where a common model choice of $c_{Fay1} = 2.6$ yields a t_{i-v} for a 1000 cubic meters of API 26 of approximately a half hour. Therefore it is common for spill models to ignore the first phase and take the area at transition from phase II to Phase II as the initial spill area,

$$A_0 = c_{Fay2}\left(\frac{\Delta_{\rho} g V_{oil}{}^{5}}{\upsilon_{w}^{2}}\right) \tag{2.10}$$

with $c_{Fay2} \simeq 10$. Spill area increases under gravity-viscous spreading as time to the 3/2 power, stopping after the slick becomes so thin that surface tension replaces gravity as the driving force in the third and final phase. However, real spills are unlikely to be contiguous by the third phase so few models include this phase.

Since actual spills seldom happen on calm water surfaces, the Fay formulas have performed poorly in the field, usually underestimating observed coverage (Jeffery, 1973; Korinenko & Malinovsky, 2014; Lehr, Fraga, Belen, & Cekirge, 1984). Fig. 2.1 illustrates the discrepancy between real spreading and the Fay formulas for an experimental spill of 50 bbls of Arabian crude oil (Lehr et al., 1984).

Oil does not spread with uniform thickness as hypothesized in the Fay formulas. Part of the oil (often about 10% of the total slick area) will be much thicker than most of the slick. Responders have developed a so-called 10–10 rule for this difference in coverage. The thick oil represents 10% of the slick area but contains 10 times the amount of oil than surrounding sheen, which derives from this thick portion. Hence, cleanup concentrates on this thicker portion. Mackay, Buist, Mascarenhas, and Patterson (1980) attempted to correct for the discrepancy between actual observation and Fay by using two different rules for the thick oil spreading and the sheen spreading. The thick portion spreads by an application of Fay-type spreading while the sheen derives it extent from the thick part. Garcia-Martinez, Mata, and Flores-Tovar (1986) subsequently suggested modification to Mackay to correct what they perceived as an error in the original Mackay rules. Lehr et al. (1984) suggested that a final minimum thickness of 0.1 mm is a good rule of thumb for the thick portion of oil slicks when initial oil viscosity is greater than 100 cSt, while less viscous oils might have a terminal thickness an order of magnitude smaller. Venkatesh (1990) observed, however, that oil spills in Arctic waters sometimes exceed this suggested final thickness. As not only the final thickness but also spread rate can depend on the viscosity of the oil (a factor absent in Fay), Venkatesh (1988) has suggested an alteration to standard Fay spreading that adjusts for oil viscosity, slowing the spread rate for more viscous products.

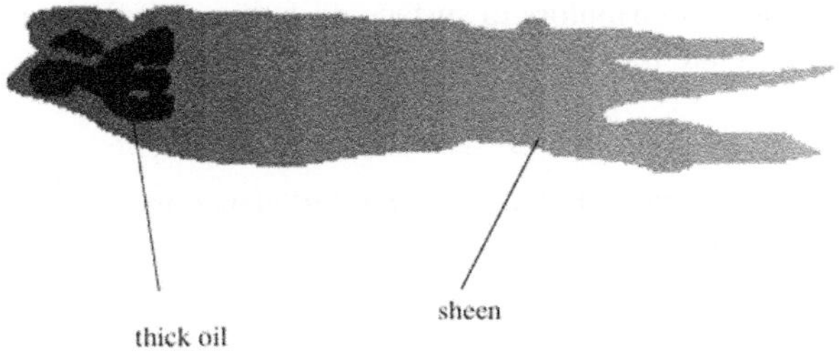

FIGURE 2.1 Image of 50-barrel experimental oil spill showing the special features of oil spreading.

Environmental factors can play a significant role in slick spreading. Obviously nearby shoreline may restrict the surface area. Simecek-Beatty and Lehr (2017) note that early spreading is better described by an ellipse in the wind direction rather than a circle, as assumed in Fay. Gjosteen and Loset (2004) found that broken and brash ice restricts surface spreading, provided that the ice concentration exceeds 30%. Yapa and Weerasuriya (1997) have proposed alternative rules to calculate oil spreading under floating broken ice. While these modifications have strong arguments in their favor, few are included in most major spill models.

As seen in Fig. 2.1, thick oil initially collects chiefly in the downwind direction, giving the slick an overall "comet" shape. Surface wind (usually measured at 10 m elevation),U_{10}, induces a surface current u_c given by (Shemden, 1972)

$$u_c \simeq \sqrt{\rho_{air}/\rho_w} \cdot U_{10} \simeq 0.03 U_{10}. \tag{2.11}$$

Elliott (1986) assumed a logarithmic reduction based on water depth in the strength of this current and its operational disappearance at depth of approximately 1.5 significant wave heights for fully developed seas. Significant wave heights are discussed later in this chapter along with the accompanying wave action breaking surface oil into different size droplets that are driven into different depths and experience a varying reduced horizontal current while submerged. According to Stokes law, the droplets will rise to the surface as a function of their size with the bigger droplets having a buoyancy velocity ratio to smaller droplets that is equivalent to their surface area ratio. The end result is the formation of the comet appearance as the bigger droplets rejoin the surface slick first, with smaller drops, containing less oil, following later. There is permanent submersion of droplets smaller than 50–70 μm (Elliot & Hurford, 1989; Geng et al., 2016). The resurfaced oil can be modeled as a non-Fickian horizontal turbulence process (Elliot, Hurford, & Penn, 1986) with diffusion parameter

$$D_{eddy}(\mathrm{m}^2/\mathrm{s}) \sim 0.033 t^{0.16} \tag{2.12}$$

This unsurprising result can be understood by observing that, as the slick increases in area, small gyres in current flow cease to be transport mechanisms and instead become contributors to surface oil horizontal dispersion. Eq. (2.11) can be shown to relate the diffusion parameter to surface oil coverage as

$$D_{eddy} \propto A_{slick}^{0.86} \tag{2.13}$$

Ahlstrom (1975) suggested combining turbulence-induced spreading with Fay spreading by re-writing Fay gravity-viscous spreading as a pseudo-diffusion process

$$D_{Fay}(\mathrm{m}^2/\mathrm{s}) = 0.13 \left(\frac{\Delta_\rho g}{\sqrt{\upsilon_w}} \right)^{1/3} \frac{A_{slick}}{\sqrt{t}} \tag{2.14}$$

Surface horizontal diffusion would dominate when $D_{eddy} > D_{Fay}$.

However, the surface oil may no longer remain a single cohesive slick by this time but will rather be described by separated distinct patches over a larger area. This will be more common for viscous crude spills rather than light refined products.

Also, another important surface phenomenon will impact surface coverage by concentrating oil into linear collection areas. Langmuir circulation (Fig. 2.2) is produced through a nonlinear interaction of surface currents and Stokes drift due to waves. This results in vortices of alternating helicity and spatially oscillating convergence and divergence zones on the water surface.

Langmuir cells form counter rotating vortices aligned with the wind when surface wind exceeds 4–6 knots (Phillips, 2002). Similar to breaking waves, Langmuir cells tend to mix the upper ocean, sometimes even deep enough to entrain water across the mixed layer base. Langmuir forces importantly tend to collect the oil into windrows. As such, they counter other forces such as water turbulence and gravitational spreading that cause the oil to form a thin sheen. Experiments have shown that the cell spacing is linear to surface wind speed. Zedel and Farmer (1991) used sonar observations of bubble cloud patterns to estimate row spacing based on surface wind speed

$$L_{lang} = c_{lang1} U_{10} + c_{lang2} \tag{2.15}$$

If length is measured in meters, $c_{lang1} = 0.079$ s and $c_{lang2} = 1.76$ m. Oil collection bands are separated by twice this distance, as the oil prefers spacing with a convergence valley over those with a divergent surface flow. Oil will hence collect in the convergence rows and move away from the divergent rows, countering the gravitational spreading described by the first two Fay spread regimes. A reasonable estimate of the maximum horizontal velocity during this collection process is $\sim 0.005 U_{10}$ (Lehr & Simecek-Beatty, 2000). Oil that arrives in the collection zones will then move

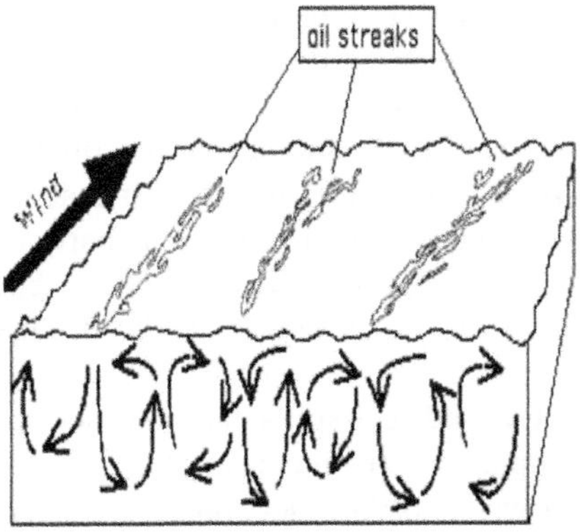

FIGURE 2.2 Cross-section of idealized slick showing Langmuir effects on surface oil (US National Oceanic and Atmospheric Administration).

downwind at speeds faster than the typical surface transport factor of 3% wind speed, further stretching the slick. Any particular windrow has a limited lifetime, ranging from 2 to 120 minutes according to Thorpe (2004). As old Langmuir cells coalesce, new windrows arise between existing ones. The net effect is a reduction of the number of oiled windrow bands (Simecek-Beatty & Lehr, 2017).

Some fraction of the oil will be transported below the surface by Langmuir processes. Li and Garrett (1993) estimated a maximum downward velocity of approximately 1% of the wind speed. Oil droplets, along with other floating debris, can theoretically be trapped subsurface in Stommel retention zones where buoyancy is balanced by the Langmuir downward velocity component.

Galt and Overstreet (2009) describe a spreading model that explicitly includes Langmuir influences. Similar Langmuir model variants have been included in several spill behavior and transport forecast programs. However, given the difficulty of consistently incorporating all of the above factors into classical oil-spreading formulas, a few researchers have proposed substituting the statistical approach, discussed earlier, instead of constructing a complex analytical model (Giwa & Jimoh, 2010). Others have resorted to the use of cellular automata (Karafyllidis, 1997) or fractional Brownian motion (Guo, Wang, Xie, & Cui, 2009).

None of these alternative approaches have gained widespread adoption in the spill modeling community. Instead, most major spill forecast models use Lagrangian elements (LEs), representing either distinct volume or mass amounts of oil, to model simultaneously both slick spreading and transport. The locations of the LEs are then translated into actual surface coverage by various mathematical techniques (e.g., Durgut & Reed, 2017). The challenge with this approach is correctly separating those factors that represent only uncertainty in surface transport from those that directly cause surface oil dispersion (or coalescence in the case of Langmuir cells). For example, Elliot style dispersion of the surface slick will, at least at the beginning of the spill, be much smaller than LE transport diffusion components for most trajectory models. An early approach was to apply Fay spreading to each LE and then use the sum of the estimated individual areas to estimate the whole slick surface. This method suffers a critical weakness in that it neglects the impact neighboring oil has on the spread of oil at a particular location. Zheng, Kobayashi, and Yapa (1995) adjust for this latter term by maintaining a linked-list of nearby LEs. Spaulding (2017) discusses other potential solutions.

Not all hydrocarbon-based fuel spills can be described by the above analysis. LNG, for example, is an increasingly shipped hydrocarbon fuel. However, if spilled on water, it does not spread the same as normal fuel oils. Fay (2007) noted that LNG spreading would, because of a resulting vapor cushion between the two liquids, more closely resemble a liquid spreading on a smooth hard surface than a noncryogenic, buoyant liquid spreading on water.

2.5.2 Natural surface dispersion

A second major weathering process that does not directly change the underlying slick chemistry is dispersion of the oil into the water column, usually due to the action of breaking waves at the air–water interface. An important cleanup tool for large spills can be the application of surfactants to decrease surface tension and thereby increase the production of small oil droplets that are more easily dispersed. The earliest models of dispersion simply assumed a constant dispersion rate as a percentage of the surface slick, depending on sea state (Blaikely, Dietzel, Glass, & van Kleef, 1977). Current models use an advanced approach that incorporates more of the underlying physical processes. They generate not only net volume dispersed but also the rate of droplet formation and the size distribution and location of the resulting oil droplets.

In order to understand all the issues involved, it is necessary to detour briefly into the science of air–water interaction. Both kinetic and thermal energy are exchanged between the air and water, but not to an equal degree. Ocean bodies have a definite thermal effect on the atmosphere but less so the reverse. Surface winds, however, do transfer momentum to the water surface by creating gravity waves and inducing surface currents. These waves and currents can greatly change the fate of floating oil. For understanding this fate, it is important to characterize the atmospheric boundary layer, the lowest part of the atmosphere that is directly affected by (and affects) the surface of the earth. Unless one is near the coast, the atmosphere will usually exhibit neutral stability over large water bodies, an atmospheric condition that occurs when the environmental lapse rate is equal to the dry adiabatic rate. The boundary layer wind speed vertical dependence can in such a case be approximated by either a simple logarithmic or power law function.

Four major factors influence the development of wind waves on the ocean: wind speed, fetch, duration, and water depth. Waves will form at the air–sea interface when the wind speed is approximately 1 m/s or greater. Fetch refers to the length of water over which a given wind has blown. The longer the fetch, the larger the expected wave height. Duration refers to the time extent of the given wind strength and direction. If the wind is consistent in strength and direction for a prolonged time period, the wave structure will reach an equilibrium situation called "fully developed sea." Water depth becomes important in shallow areas where bottom friction will have an effect. The smallest waves have both gravity and surface tension as a restoring force while larger waves are affected mostly by gravity alone. As any sea state will have a mixture of wave heights, mariners usually characterize the surface wave strength not by the most common wave height but by the average height of the largest one-third of the waves, called the significant wave height, H_s. For fully developed seas and unlimited fetch, the Shore Protection Manual (1984) predicts the significant wave height (m) and wave period (s) based on an adjusted wind stress U_A

$$U_A = 0.71 U_{10}^{1.23} \qquad H_s = 0.0248 U_A^2 \qquad T_s = 0.83 U_A \tag{2.16}$$

Wave field description has become a major area of study in meteorology and oceanography. One very simple distribution, applicable to fully developed seas, is the Rayleigh distribution, a one-parameter distribution based on significant wave height (Fig. 2.3).

While more complex distributions are now the common practice, the Rayleigh distribution reasonably matches the observation provided that individual wave velocity components are independent and normally distributed.

Surface waves represent energy that is needed to disperse oil from the water surface into the water column. A perfectly calm water body would have no dispersion. Oceanographers assign a degree of water surface roughness that can then be used to characterize potential dispersion energy. Surface roughness is often defined by two related properties: friction velocity u_* and roughness length z_0. They are related by

$$\frac{z_0 g}{u_*^2} = \alpha_{ch} \tag{2.17}$$

Originally, the Chernock parameter, α_{ch}, was thought to be a constant but has since been seen to be a widely fluctuating function of the steepness of the dominant waves (Drennan, Taylor, & Yelland, 2005). Some researchers (Terry et al., 1996) simplify the above equation by approximating $z_0 \simeq H_s$. This uncertainty in mapping water surface roughness to surface wind may reflect an intrinsic limitation to wind-wave forecasting based on the concept of a wave energy spectrum (Liu & Schwab, 2002). In fact, field observation indicates that little oil dispersion will occur until breaking waves began and most spill behavior models therefore require breaking wave conditions for

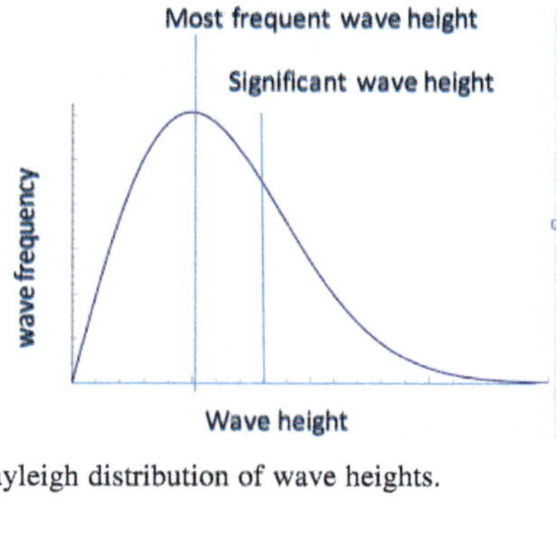

FIGURE 2.3 Sample Rayleigh distribution of wave heights.

dispersion to happen. Waves that have small height compared to their wavelength, λ, become liable to breaking when $H_w/\lambda \approx 0.08$. Waves of any wavelength will break in shallow waters when the wave height is greater than 80% of the water depth. Lehr and Simecek-Beatty (2000) relate breaking wave fraction in the open ocean (unlimited depth) to wind speed (expressed in m/s)

$$f_{bw} = \frac{0.025(U_{10} - 3)}{\tau_M} \quad 3 \leq U_{10} \leq 4$$
$$f_{bw} = \frac{0.01U_{10} + 0.01}{\tau_M} \quad U_{10} > 4 \tag{2.18}$$

where τ_M refers to the Monahan (1971) time constant (2.54 s for freshwater and 3.85 s for saltwater). Zatsepa, Ivchenko, Korotenko, Solbakov, and Stanovoy (2018) suggest a similar two-range formula but use a higher wind speed for the patch point between the two formulas. Zhao and Toba (2001) alternatively calculate f_{bw} by the use of a Reynolds-like dimensionless number

$$R_{bw} = \frac{C_D U_{10}^2}{v_{air}\varpi_{pm}} \qquad f_{bw} = 3.88 \cdot 10^{-7} R_{bw}^{1.09} \tag{2.19}$$

where the drag coefficient C_D is a linear function of surface wind speed above 5 knots, and $\varpi_{pm} = 0.86/(gU_{10})$. Inherent in these formulas is the assumption that a spatially fixed but time-varying series (breaking waves at a fixed water surface point) is statistically the same as spatially varying whitecap coverage at a fixed point in time. There is some controversy among spill modelers that this is a valid assumption. Also, estimating whitecap coverage from wind speed is even less precise than estimating surface roughness. See, for example, fig. 7 in Holthuijsen and Herbers (1986).

Breaking waves impart energy vertically into the surface water that is then dissipated with depth and time. Measured energy dissipation (per unit water mass) values range from 0.01 to 10.0 m^2/s^3. Hwang and Slatten (2008) conclude that average wave energy dissipation at the surface is proportional to the cubic of the surface wind speed. Field studies are mixed on determining energy dissipation as a function of depth. Sutherland and Melville (2015) report that, near the surface, energy dissipation decayed inversely with depth, while below approximately one significant wave height, it decayed more quickly, decreasing inversely to the square of the depth. Complicating matters, energy dissipation also varies horizontally between the crest and trough of any wave (Gemmrich & Farmer, 2004).

Breaking waves have two obvious effects on the surface oil: (1) the oil is broken into droplets of varying size and (2) the resulting droplets are displaced at varying depth in a wave affected surface layer. Larger drops will quickly rejoin the surface slick while smaller droplets will remain for an extended time within this mixed layer beneath the slick, with buoyancy countered by turbulence. A very rough conjecture of the maximum droplet size of any buoyant oil particle that remains effectively dispersed can be done by equating the terminal upward velocity from Stokes law

$$u_{Stoke} = \frac{g\delta_{oil}^2}{18v_w} \cdot \Delta\rho \tag{2.20}$$

with a turbulent velocity estimate from surface roughness. For a typical oil and common sea states, this yields a subsurface maximum permanently dispersed oil droplet size on the order of 50–100 μm. For a detailed analysis, see Li, Spaulding, and French-McCay (2017).

The classical model on oil dispersion is the work of Delvigne and Sweeney (1988) (DS) and their method is still widely used in existing spill models. However, DS has several known difficulties. For example, while dispersant chemicals create increased and smaller oil droplets by reducing oil–water interfacial surface tension, DS contains no explicit dependence on surface tension. Spaulding (2017) points out that DS model formulation is dimensional, creating a challenge in going from laboratory to field. He also notes that there is no explicit procedure to address droplet resurfacing. According to DS, the droplets will be mixed uniformly beneath the slick to a depth of 1.5 the significant wave height. Field observations often report oil droplets uniformly to a depth $\sim H_s$, so many models simply use this depth as a shortcut for their maximum oil drop intrusion. While recognizing that droplet intrusion and breakage are strong functions of the local energy dissipation rate, most current dispersion models still follow the practice of DS of employing globally an average value for energy dissipation as a function of significant wave height.

The choice for operational droplet size distribution from breaking waves has not reached consensus among spill modelers. DS, based on laboratory results, concluded that the number density, N_{oil}, of droplets of size δ_{oil} could be computed as

$$N_{oil}(\delta_{oil}) = N_0\delta_{oil}^{-2/3} \tag{2.21}$$

with N_0 itself a function of energy dissipation rate, breaking wave fraction, and oil viscosity. Noticeably, N_0 does not consider surface tension as a parameter, perhaps because all the sample oils tested in their laboratory had approximately the same surface tension numbers. As mentioned earlier, reducing surface tension is the mechanism used for chemical dispersant operations and plays a critical role in determining the resulting droplet sizes. Therefore modern revisions of DS explicitly include surface tension. Johansen, Reed, and Bodsberg (2015) and Lehr, Simecek-Beatty, Aliseda, and Boufadel (2014) assume that droplet size distribution will evolve eventually to a log-normal or log-normal type distribution and only this final, stable, distribution is critical to modeling dispersion. A length scale is required to determine such a distribution, and common scale choice is the Sauter mean droplet diameter, $\overline{\delta}_s$. The Sauter mean diameter of a collection of spherical objects of different sizes is equal to the diameter of equisized spherical objects forming the collection (Kowalczuk & Drzymala, 2016). Newer models estimate the Sauter mean often by a modified Weber number scaling that includes both surface tension and oil viscosity to overcome the limitations of DS. Dimensional analysis suggests one possible example might be

$$\overline{\delta}_s = \alpha_\delta \left(\frac{\upsilon_{oil}}{\upsilon_w} \right)^{1/3} \cdot \left(\frac{\sigma_{o-w}}{\rho_w} \right)^{3/5} \cdot \varepsilon^{-2/5} \tag{2.22}$$

The experiments of Johansen et al. (2015) and Li and Katz (2016) indicate a plausible value of the empirical, dimensionless, constant, $\alpha_\delta \sim 0.008$. Zeinstra-Helfrich, Koops, and Murk (2016) found that mean droplet size decreased with larger waves, but increased with surface oil thickness. However, the former dominates the latter in influencing droplet size. Their data indicate that, for waves generated by wind speeds greater than about 7–8 knots (beginning of breaking waves), slick thickness is a minor correction to mean droplet size.

While Babinsky and Sojka (2002) argue that an actual log-normal distribution is a stable choice for a final droplet size distribution, the computationally simpler Rosin-Rambler distribution, a form of the cumulative Weibull distribution, is often a more popular selection. Expressed as cumulative volume fraction

$$Fvol(\delta_{oil}) = 1 - \exp\left(-0.69 \left[\frac{\delta_{oil}}{\lambda} \right]^{\kappa} \right) \quad \text{with} \quad \kappa = 1.8 \quad \lambda = \overline{\delta}_s \tag{2.23}$$

An alternative school (e.g., Nissanka, & Yapa, 2017, 2018; Zhao et al., 2014) argues that constructing only an estimated final static droplet size distribution is inadequate to describe the actual dispersion process and that a true population balance model is required. Such models take account of various mechanisms causing breakup along with those resulting in coalescence of droplets or simply resisting breakup. A major advantage of numerical population balance models over straight correlation models is that the former provide a transient droplet size distribution that may be important to understanding weathering of the oil. For example, these models often show an early bimodal distribution that then evolves over time to a log-normal distribution. Disadvantages of this approach are increased computational demands and extra specifications about the environmental turbulence causing the dispersion. A standard assumption is that the turbulence important for droplet breakage is isotropic and centered around eddy sizes chiefly in the lower end of inertial sub-range, stopping at the Kolmogorov microscale.

2.5.3 Oil–particle aggregation

Dispersed oil, and to a lesser degree, surface oil, are susceptible to adhesion to solid particles in the surrounding water. These can be organic matter, mineral particles, or even brash ice. Depending on the nature of the adhering particle, the buoyancy of the particle-oil combination may be increased or decreased. In the open ocean, such aggregation is a minor factor in oil fate. However, in muddy rivers where the sediment load can easily reach 0.5 kg/m^3, it is usually a significant process. Brakstad, Lewis, and Beegle-Krause (2018) claims that significant deposition from oil–particle aggregation occurs whenever suspended sediment exceeds 10 mg/L.

Aggregation is classified into two types. For Type1 aggregation, the adhering particles are smaller than the oil droplets and coat them. Type 2 aggregation reverses the relative sizes of particle and droplet. Most aggregation models in spill behavior forecasting are designed to represent Type 1 aggregation as it produces a more stable combination with the particles, providing a barrier to future oil droplet coalescence and often creating negative buoyancy. Moreover, in most cases where oil–particle interaction occurs, there is an excess of solid particles compared to the number of available oil droplets. This allows for a very simple formula to estimate the time dependent percentage of nonattached oil droplets. The solid particle number density, N_s, is assumed constant while the nonaggregated oil droplet density, N_{oil}, of some given size decreases (Payne et al., 1987) according to

$$\frac{dN_{oil}}{dt} = -1.3\alpha_{sed}\sqrt{\varepsilon/\upsilon_w}\cdot N_s N_{oil} \tag{2.24}$$

The "sticking" coefficient, α_{sed}, varies between 0 and 1, depending on the type and size of the adhering particles. Hill, Khelifa, and Lee (2003), reviewing earlier experiments by Delvigne, gave a range α_{sed} for kaolinite of 0.006–0.041. Much of current research (e.g., Zhao et al., 2016) is directed at algorithm development for estimating α_{sed} from the size distribution and nature of both the particles and oil droplets, and also on the properties of the surrounding water.

According to Khelifa and Gamble (2006), aggregation can play a key role in tarball formation, the long-term fate of much of the larger hydrocarbons. While noting that oil viscosity and mixing energy are the dominant factors that control the formation of tarballs, these researchers say that aggregation of oil with suspended sediment can cause significant increases or decreases in tarball density, depending on the type and concentration of the solid particles.

2.5.4 Oil–water emulsion formation

Emulsification has a significant impact on response options for major surface oil spills. Emulsified oil typically has both increased volume and increased viscosity. However, not all oils emulsify, nor is the stability of the formed emulsion the same in all cases. Stability of an emulsion is defined as its ability to retain its water fraction. In order to form a stable emulsion, an interfacial film must exist between the water droplet and surrounding oil with most experts concluding that this role is most effectively accomplished by the dissolved asphaltene oil component. Some oils may not initially emulsify but may do so after a period of weathering. In most, but not all, cases, the more stable the emulsion, the larger the water fraction, which can reach 90% for some oils. Predicting when emulsification will occur, and the properties of the formed emulsion, has been a continuing challenge for the modeling community. Emulsification is a complicated physical phenomenon, further convoluted by the intricate mixture of hydrocarbons that describe the typical oil or oil product. MacKay et al. (1980) developed the first widely adopted emulsification model; a model that is still in use today in spite of its limited usefulness in many actual spills. MacKay assumed that the water fraction, f_w, of the emulsion would monotonically increase to some limiting value $f_{\max}$, based on the surrounding water turbulence (usually the result of wind waves).

$$\frac{df_w}{dt} \propto \mathbf{U}_{10}^2(f_{\max} - f_w) \tag{2.25}$$

Such an approach does not allow for the common phenomena of dewatering, found in unstable and mesostable emulsions. Xie, Yapa, and Nakata (2007) corrected this oversight by adding a term to account for emulsion water loss. The MacKay model also did not predict the constant of proportionality, which is currently found by direct empirical measurement of different oils under various sea conditions, or by analogy to those measured oils. Usually, emulsification, if it occurs, happens rapidly so that the complete solution to the Mackay equation is not required, only the end state. For an improved response, its sufficient to know if and when oil will emulsify, will the emulsion be stable, and what will be the water fraction.

Field observations indicate that minimum and maximum viscosities exist for a stable emulsion to occur. Oils with kinematic viscosities less than 10 cSt cannot retain water droplets for sufficient time to form an emulsion and oils with viscosities greater than 10,000 cSt limit most water droplets from entering the oil in the first place. Daling, Molestad, Johansen, Lewis, and Rodal (2003) report field results that show that a minimum oil thickness of 50–500 μm, depending upon oil type, is required for emulsification to commence. This probably correlates with limits on the underlying size of the water droplets in the emulsion. Hence, only the thick oil, not the sheen, can emulsify. Shaw (2003) found that there is a minimum oil–water interfacial surface tension required for a stable emulsion, usually in the range of 10–20 dynes/cm. Most oils exceed this limit unless treated by surfactants.

It is generally, but not universally, hypothesized that emulsification begins when resin-dissolved asphaltenes begin to precipitate out (McLean & Kilpatrick, 1997). The Yen-Mullins model (Mullins et al., 2012) considers asphaltene precipitation by examining clustering of asphaltenes based upon their concentration in the oil and other factors. The model requires oil information not widely available to the spill response community and is highly dependent on accurate estimation of resin and asphaltene fraction.

Fingas and Fieldhouse (2012) eschew a standard physical model in favor of a statistical approach based on agglomeration clusters in multidimensional parameter space defined by key state variables, or clusters based on one-to-one and onto functional mappings of these key state variables. Depending upon which cluster any oil maps onto, it is classified as forming a stable, mesostable, entrained, or unstable emulsion. Aske, Kallevik, and Sjoblom (2002) accomplish a similar division of oils by using principal component analysis. A review by Spaulding (2017) recommends this clustering approach as superior to the traditional MacKay style models. Yetilmezsoy et al. (2012) translate the Fingas and Fieldhouse functional cluster rules into a series of if–then statements and then apply the method of fuzzy logic to determine propensity to emulsify. Lehr (2017) parallels the Fingas and Fieldhouse clustering approach by defining modified rules that, to the extent possible, are consistent with existing emulsion science and measured values. For example, as noted previously, asphaltenes play a key role as indigenous surfactants in stabilizing emulsions (McLean et al., 1998; Xia, Lu, & Cao, 2014) while resins act as solvating agents, allowing the asphaltenes to migrate to the oil–water interface. However, too much resin content will allow the asphaltenes to remain in solution. Fingas and Fieldhouse combine resin and asphaltene fraction at 3% as a minimum lower limit for either stable or mesostable emulsions while Lehr (2017), based on past measurements, uses the asphaltene–resin ratio to estimate $f_{\max}$. If r_{a-r} is the asphaltene–resin ratio then

$$f_{\max} = 0.61 + 0.5 r_{a-r} - 0.28 r_{a-r}^2 \quad 0 \leq f_{\max} \leq 0.9 \tag{2.26}$$

The loose fit (see Fig. 2.4) of the above equation to field data shows the limit to this approach although the results are somewhat consistent with the Fingas and Fieldhouse (2012) observation that an r_{a-r} value of around 0.6 is best for a stable emulsion.

2.6 Weathering processes—those that do alter the chemical mixture

Evaporation, dissolution, and photooxidation alter the chemical nature of the slick. For these processes, it is not sufficient to estimate mass balance and physical characteristics of the spilled oil. The weathering model must also predict changes to the chemical balance.

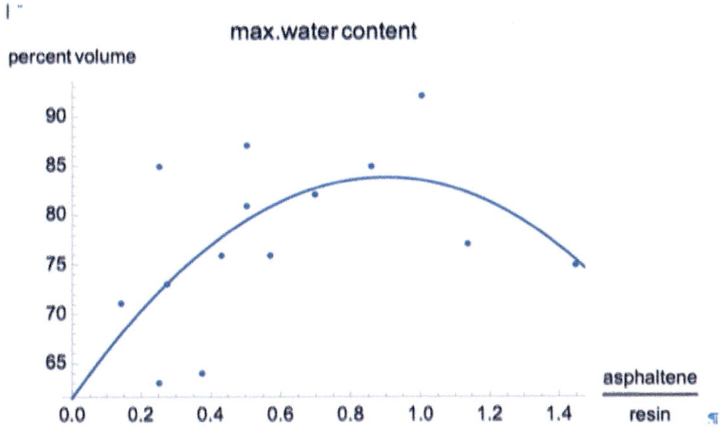

FIGURE 2.4 Maximum emulsion water content as a function of asphaltene/resin ratio.

2.6.1 Evaporation

For most surface spills, evaporation is the major removal mechanism for oil from the slick, exceeding not only other natural processes but also standard mechanical removal applications. Small spills of light refined products will typically disappear in less than a day due to evaporation unless high seas drive the product into the water column. Net evaporative loss is also one of the easiest predictions to make for most spill events if distillation information on the spilled oil is available. The tendency to evaporate can be closely linked to the boiling point of any hydrocarbon with little regard to its specific structure. Instead, the key factor is molecular weight that in turn can be related to carbon number. Fig. 2.5 shows the plot of boiling point versus carbon number for select n-alkanes. Smith and MacIntyre (1971) found that little of any distillation cut with a boiling point exceeding 270°C will evaporate during the lifetime of most surface spills. This corresponds to hydrocarbons with 15 or more carbon atoms, and in most actual events, hydrocarbons containing 12 or more carbon atoms can be considered nonevaporative for mass balance purposes.

Predicting short-term evaporative mass loss rate is a much more challenging assignment. Two classical approaches that are widely used differ on modeling the oil structure properties but agree on the general gross nature of the slick. Both assume that the slick can be treated as a vertically homogeneous mixture of uniform thickness with contiguous and compact surface coverage. The driving factor will be the effective vapor pressure above the slick and the limiting factor will be the ability of the surface wind to remove the saturated boundary layer above the slick. In the Stiver-Mackay [SM] approach, the oil is treated as a single component fluid whose chemical

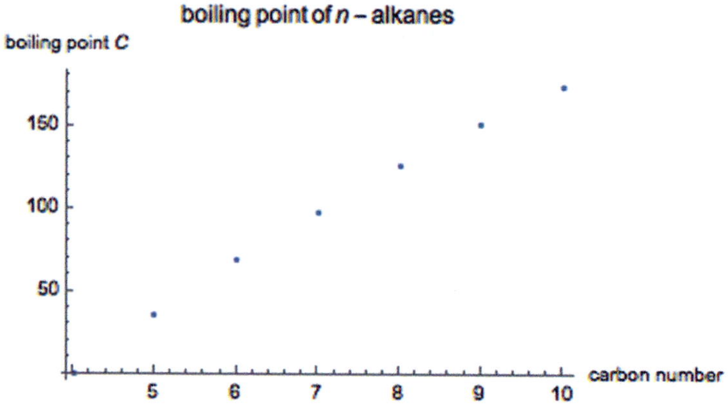

FIGURE 2.5 Boiling point of n-alkanes based on carbon number.

nature changes over time. The actual distillation history of the oil is linearized with an estimated initial boiling point T_b and distillation curve slope T_{sl}. SM then defines a dimensionless "evaporative exposure" variable that depends on time, wind, and average oil thickness

$$\Theta = K_{SM} U_{10}^{7/9} \delta_{th} t \tag{2.27}$$

The time dependence of the fraction of evaporated oil f_{evap} can be calculated, given a water temperature of T_w, according to

$$\frac{\mathrm{d}f_{evap}}{\mathrm{d}\Theta} = \exp\left(a - \frac{b}{T_w}\left[T_b - T_{sl} f_{evap}\right]\right) \tag{2.28}$$

SM empirically sets $K_{sm} = 0.002 \quad a = 6.3 \quad b = 10.3$. Others have determined different (a, b) values for specific oils (Belore and Buist, 1994). The model is highly sensitive to T_b (Overstreet et al., 1995) and may overestimate evaporation rate when compared to alternative approaches. Xie et al. (2007) suggest a correction to reduce evaporation for emulsified oils.

The second classical approach for modeling evaporation treats the oil as a limited mixture of pseudo-components, differing by their boiling points. The advantage of this approach is that it coincides with the way industry stores data on oils for refining purposes. The evaporative loss rate for each pseudo-component is a function of its vapor pressure and a mass transfer coefficient. Various suggestions have been made for constructing these coefficients (Mackay and Matsugu, 1973; Williams, Hann, & James, 1975). Employing Raoult's Law of an ideal mixture, the net evaporative loss is then the sum of the individual losses. While the chemical nature of any specific pseudo-component does not change, the relative component ratios do change as those components with higher vapor pressure evaporate more quickly.

The problem with these two classical approaches is that many of their underlying assumptions are false. Real oil slicks do not have uniform thickness or form a contiguous, compact surface on the water. In fact, the water surface is not flat but subject to breaking waves in even moderate wind conditions. The spray from the waves increases the area for the oil–air interface while the plunging waves take some of the oil into the water column and away from any evaporative loss. Places where the oil is thick are not necessarily well mixed. For these locations, diffusion through the slick may be the controlling factor for lighter hydrocarbons, not saturation of the boundary layer (Hanna and Drivas, 1993). Moreover, certain processes, such as photo-oxidation, may cause a crusting affect at the oil–air interface, reducing evaporation (Lehr, 1996). Finally, oil is not an ideal mixture as there is a heat of solution connected to the component mixing so that it may be necessary to include an adjustment to Raoult' Law.

Fingas (1996, 2013) argues that the many questionable assumptions in the classical approach make any of its predictions highly suspect. He especially rejects the concept that hydrocarbon evaporation is, like water evaporation, restricted by the saturation concentration of molecules above the slick, noting that volatile oil components such as pentane have saturation concentrations two orders of magnitude larger than water. Instead, he claims that evaporation is mostly diffusion (through the slick) limited, and depends not on wind, but on the chemical mixture of the oil and the water temperature. If diffusion through the slick is important, evaporative loss rate will be reduced (Kotzakoulakis and George, 2018). Rather than trying to model this from first principles, Fingas recommends resorting instead to a straight statistical method where an assumed functional shape of the evaporative loss is adjusted to laboratory data for each particular oil. For example, he calculates the time dependent evaporative fraction loss of Prudhoe Bay crude by (t is elapsed time in seconds from spillage)

$$f_{evap} = [0.0169 + 0.0045(T_w - 273)] \cdot \ln\left(\frac{t}{60}\right) \tag{2.29}$$

2.6.2 Dissolution

The water counterpart to oil evaporation (air the receiving medium) is oil dissolution (water the receiving medium). The old saying that oil and water do not mix is scientifically accurate so, for surface spills, dissolution mass loss is a small fraction of the mass loss due to evaporation or dispersion. Riazi and Edalat (1996) found that the rate of dissolution under normal sea surface conditions is approximately 0.1% of the rate of evaporation. However, for subsurface releases such as the Deepwater Horizon spill, dissolution can replace evaporation as a major weathering process. Unlike evaporation, where the size of the hydrocarbon molecule is more important than its chemical structure, dissolution rates are highly sensitive to the type of hydrocarbon. Despite having very similar molecular weights, benzene, an aromatic, is more than an order of magnitude more soluble in water than n-hexane, an alkane. For mass balance purposes then, only the aromatic component in SARA of the oil is viably subject to dissolution, as the other two groups in SARA, resins and asphaltenes, are too big to significantly dissolve. For toxicity calculations however, a different standard that considers even trace dissolution may be more appropriate (French-McCay et al., 2016).

A commonly employed measure in dissolution calculations is the mass partition coefficient, K_{ow}, defined as the saturation concentration in octanol (serving as oil proxy) compared to the mass concentration in water. The larger the number, the less of the particular hydrocarbon escapes from the slick and dissolves into the surrounding water. As the numbers for all hydrocarbons tend to be large (little solubility), it is often the logarithm of K_{ow} that is reported.

Stevens, Thibodeaux, Overton, Valsaraj, and Walker (2017) created an oil droplet dissolution model of only two components, a dissolvable component, X_{diss}, and an inert component. The total volume of the oil droplet then is the volume sum

$$V_{drop} = V_{diss} + V_{inert} = V_{drop}(X_{diss} - (1 - X_{diss})) \tag{2.30}$$

Dissolution is estimated as

$$\frac{d}{dt}\left(V_{drop}X_{diss}\right) = -\,k_w X_{diss} A_{drop} \tag{2.31}$$

The area of the droplet A_{drop} is given by the assumed droplet size distribution with the challenge being estimation of the oil to water mass transfer coefficient k_w. Simplifying approximations for k_w include assuming a first-order relationship with $\langle K_{ow} \rangle$, representing an average of the mass partition coefficients of the soluble aromatics, and letting the controlling transfer factor be the diffusion to the droplet surface of the soluble fraction.

2.6.3 Photooxidation

The combination of hydrocarbons with oxygen is called oxidation. The resulting modified hydrocarbons may have a critical impact on toxicity and on many of the previous described weathering processes. Despite its importance in understanding spill behavior, photooxidation is often not included in spill weathering forecasts and comparatively little relevant research has been done to construct an appropriate submodel. Instead, published studies concentrate on reporting empirical results for specific oils (Ganjali, Niknafs, & Khosravi, 2007). Saturates are resistant to photooxidation while aromatic compounds are more sensitive. This includes PAHs (Plata, Sharpless, & Reddy, 2008). It is likely that photooxidation plays a key role in "skinning" of surface slicks for certain oils (Garret, Pickering, Haith, & Prince, 1998). These skins then probably retard evaporation. When combined with tar ball formation, this can yield what responders call "m&ms," where clumps of oil are hard on the outside and much more fluid on the inside.

2.7 Discussion and caveat

The author and others (Lehr, Simecek-Beatty, & Fingas, 2019) have predicted that the next decade will see significant advances in forecasting spill weathering behavior as newly published research identifies more appropriate oil and environmental characterization and hints at how to attack complex processes such as emulsification by using artificial intelligence techniques. Of course, this presumes adequate funding and direction to generate a sufficiently large oil database and conduct necessary experiments for the computational methods employing artificial intelligence to perform properly. However, forecasting the activities of funding agencies is a task well outside the pay-grade of the author. This chapter hopefully represents, with the restrictions previously mentioned, an accurate picture of the present state of spill weathering modeling. Nevertheless, it is inevitable in a work of this scope that errors of commission and omission, while unintentional, are to be found. Readers are advised to consult the originals of the cited publications before using any of the concepts and formulas discussed for their own research and other applications. Nothing in this work should be construed as either criticism or recommendation of any commercial product or service. This work did not receive any specific grant from funding agencies in the public, commercial, or not-for-profit sectors. The opinions and conclusions expressed in this article are those of the author and not necessarily those of any American government agency.

2.8 Notation (bracket shows frequently used units for dimensional terms)

A_{drop} = oil droplet area (m^2)
A_0 = initial slick area (m^2)
A_{slick} = slick surface area (m^2)
D_{eddy} = eddy diffusion coefficient (m^2/s)
f_{bw} = breaking wave fraction
f_{evap} = evaporated mass fraction
f_{max} = maximum water fraction of oil–water emulsion
f_w = water mass fraction of emulsion
H_s = significant wave height (m)
K_{ow} = mass partition coefficient
k_w = mass transfer coefficient (m/s)
L_{lang} = Langmuir cell surface separation distance (m)
N_{oil} = oil droplet number density (m^{-3})

N_s = sediment particle number density (m^{-3})
r_{a-r} = asphaltene–resin ratio
T_b = oil boiling point (°C or °K)
T_{ref} = reference temperature (°C or °K)
T_s = period of the significant wave (s)
T_{sl} = slope of distillation curve tangent (°C or °K)
T_w = water temperature (°C)
PP = pour point (°C or °K)
U_{10} = 10 m height surface wind (m/s)
X_{diss} = soluble fraction
u_* = friction velocity (m/s)
z_0 = roughness length (m)
δ_{oil} = oil droplet diameter (μm)
$\overline{\delta}_s$ = Sauter mean oil droplet diameter (μm)
ε = energy dissipation rate per unit mass (m^2/s^3)
ρ_{oil} = oil density (g/cm^3 or kg/m^3)
ρ_{ref} = reference density (g/cm^3 or kg/m^3)
ρ_w = water density (g/cm^3 or kg/m^3)
υ_{emul} = emulsion kinematic viscosity (cSt)
υ_{oil} = oil kinematic viscosity (cSt)
υ_w = water kinematic viscosity (cSt)
σ_{o-w} = oil–water interfacial tension (dynes/cm)
Δ_ρ = fractional density difference between water and oil
Θ = evaporative exposure

References

Ahlstrom, S. (1975). *A mathematical model for predicting the transport of oil slicks in marine waters*. Richland WA: Battelle Pacific Northwest Laboratories.

Aske, N., Kallevik, H., & Sjoblom, J. (2002). Water-in-crude oil emulsion stability studied by critical electric field measurements. Correlation to physic-chemical parameters and near-infrared spectroscopy. *Journal of Petroleum Science and Engineering*, *36*, 1–17.

ASTM (2007). Manual of Petroleum Measurement Standards, Chapter 11, Section 1, Addendum 1, ASTM D 1250–04.

Babinsky, E., & Sojka, P. (2002). Modeling drop size distributions. *Progress in Energy and Combustion Science*, *28*, 303–329.

Belore, R., & Buist, I. (1994). Sensitivity of oil fate model predictions to oil property inputs. *Proceedings of 17th Arctic and Marine Oilspill Program Technical Seminar*, 837–848.

Blaikely, D., Dietzel, G., Glass, A., & van Kleef, P. (1977). Slictrac—A computer simulation of offshore oil spills, cleanup, effect, and associated costs. In *Proceedings of the 1977 oil spill conference*, American Petroleum Institute, Washington D.C., pp. 44–53.

Brakstad, O., Lewis, A., & Beegle-Krause, C. (2018). A critical review of marine snow in the context of oil spills and oil spill dispersant treatment with focus on Deepwater Horizon oil-spill. *Marine Pollution Bulletin*, *135*, 346–356.

Daling, P., Molestad, M., Johansen, O., Lewis, A., & Rodal, J. (2003). Norwegian testing of emulsion properties at sea—the importance of oil type and release conditions. *Spill Science & Technology Bulletin*, *8*, 123–136.

Delvigne, G., & Sweeney, C. (1988). Natural dispersion of oil. *Oil and Chemical Pollution Journal*, *4*, 281–310.

Drennan, W., Taylor, P., & Yelland, M. (2005). Parameterizing the sea surface roughness. *Journal of Physical Oceanography*, *35*, 835–848.

Durgut, I., & Reed, M. (2017). Modeling spreading of oil slicks based on random walk methods and Voronoi diagrams. *Marine Pollution Bulletin*, *118*, 93–100.

Elliot, A., & Hurford, N. (1989). The influence of wind and wave shear on the spreading of a plume at sea. *Oil and Chemical Pollution*, *5*, 347–363.

Elliot, A., Hurford, N., & Penn, C. (1986). Shear diffusion and the spreading of oil slicks. *Marine Pollution Bulletin*, *17*, 308–313.

Elliott, A. (1986). Shear diffusion and the spread of oil in the surface layers of the North Sea. *Deutsche Hydrografische Zeitschrift*, *39*, 113–137.

ESTD (2019). Oil Base, http://www.etc-cte.ec.gc.ca/databases/oilproperties/. Accessed February, 2019.

Fay, J. (1971). Physical processes in the spread of oil on a water surface. In *Proc. of the joint conf. on prevention and control of oil spill*, American Petroleum Institute, Wash. D.C. pp. 463–467.

Fay, J. (2007). Spread of large LNG pools on the sea. *Journal of Hazardous Materials*, *140*, 541–551.

Fingas, M. (1996). The evaporation of oil spills: Prediction of equations using distillation data. *Spill Science & Technology Bulletin*, *3*, 191–192.

Fingas, M. (2013). Modeling oil and petroleum evaporation. *Journal of Petroleum Science Research*, *2*, 104–115.

Fingas, M. (2014). Water-in-oil emulsions: Formation and prediction. *Journal of Petroleum Science Research*, *3*, 38–47.

Fingas, M., & Fieldhouse, B. (2012). Studies on water-in-oil products from crude oils and petroleum products. *Marine Pollution Bulletin*, *64*, 272–283.

French-McCay, D., Li, Z., Horn, M., Crowley, D., Spaulding, M., Mendelsohn, D., & Turner, C. (2016). Modeling oil fate and subsurface exposure concentrations from the Deepwater Horizon oil spill. *Arctic and Marine Oil Spill Program Technical Seminar, Contamination and Response Section*, 1–8.

Galt, J., & Overstreet, R. (2009). Development of Spreading Algorithms for MARO, Genwest Systems tech. Note, Seattle, WA.

Ganjali, S., Niknafs, B., & Khosravi, M. (2007). Photo-oxidation of crude petroleum maltenic fraction in natural simulated conditions and structural elucidation of photoproducts. *Iranian Journal of Environmental Health Science & Engineering*, *4*, 37–42.

Garcia-Martinez, R., Mata, L., & Flores-Tovar, H. (1986). A correction to the Mackay oil spreading formulation. In *Proceedings of the nineteenth arctic and marine oil spill program technical seminar, environment* Canada, Ottawa, Ontario, pp. 1627–1635.

Garret, R., Pickering, I., Haith, C., & Prince, R. (1998). Photo-oxidation of crude oils. *Environmental Science & Technology*, *32*, 3719–3723.

Gemmrich, J., & Farmer, D. (2004). Near surface turbulence in the presence of breaking waves. *Journal of Physical Oceanography*, *34*, 1067–1086.

Geng, X., Boufadel, M., Ozgokmen, T., King, T., Lee, K., Lu, Y., & Zhou, L. (2016). Oil droplets transport due to irregular waves: Development of large-scale spreading coefficients. *Marine Pollution Bulletin*, *104*, 279–289.

Giwa, A., & Jimoh, A. (2010). Development of models for the spreading of crude oil. *Australian Journal of Technology*, *14*, 66–71.

Gjosteen, J., & Loset, S. (2004). Laboratory experiments on oil spreading in broken ice. *Cold Regions Science and Technology, 38*, 103–116.

Guo, W., Wang, Y., Xie, M., & Cui, Y. (2009). Modeling oil spill trajectory in costal waters based on fractional Brownian motion. *Marine Pollution Bulletin, 58*, 1339–1346.

Hanna, S., & Drivas, P. (1993). Modeling VOC emissions and air concentrations from the Exxon Valdez oil spill. *Air and Waste, 43*, 298–309.

Hill, P., Khelifa, A., & Lee, K. (2003). Time scale for oil droplet stabilization by mineral particles in turbulent suspensions. *Spill Science & Technology Bulletin, 8*, 73–82.

Holthuijsen, L., & Herbers, T. (1986). Statistics of breaking waves observed as whitecaps in the open sea. *Journal of Physical Oceanography, 16*, 290–297.

Hwang, P., & Slatten, M. (2008). Energy dissipation of wind-generated waves and whitecap coverage. *Journal of Geophysical Research, 113*, C02012.

Jeffery, P. (1973). Large scale experiments on the spreading of oil at sea and disappearance by natural factors. In *Proceedings of the joint conference on prevention and control of oil spills*, pp. 469–474.

Johansen, Ø., Reed, M., & Bodsberg, N. (2015). Natural dispersion revisited. *Marine Pollution Bulletin, 93*, 20–26.

Jokuty, P., Whiticar, S., Wang, Z., Fingas, M., Fieldhouse, B., Lambert, P., & Mullin, J. (1999). Properties of crude oils and oil products. Manuscript Report EE-165. Environmental Protection Service, Environment Canada, Ottawa, ON.

Karafyllidis, I. (1997). A model for the prediction of oil slick movement and spreading using cellular automata. *Environment International, 23*, 839–850.

Khelifa, A., & Gamble, R. (2006). Prediction of tar ball formation. In *Proceedings of the twenty-ninth AMOP technical seminar, environment* Canada, Ottawa, ON, pp. 79–90.

Korinenko, A., & Malinovsky, V. (2014). Field study of film spreading on a sea surface. *Oceanologia, 56*, 461–475.

Kotzakoulakis, K., & George, S. (2018). Predicting the weathering of fuel and oil spills: A diffusion-limited evaporation model. *Chemosphere, 190*, 442–453.

Kowalczuk, P., & Drzymala, J. (2016). Physical meaning of the Sauter mean diameter of spherical particulate matter. *Particulate Science and Technology, 34*, 645–647.

Lehr, W. (1996). Modeling the benzene inhalation hazard from spilled oil. *Spill Science & Technology Bulletin, 3*, 199–202.

Lehr, W. (2001). Review of modeling procedures for oil spill weathering behavior. In C. Brebbia (Ed.), *Oil spill modeling and processes*. Ashurst, Southampton, UK: WIT Press.

Lehr, W. (2017). Developing a new emulsification algorithm for spill response models. In *Proceedings of the fortieth AMOP technical seminar, environment and climate change* Canada, Ottawa, Canada, pp. 572–585.

Lehr, W., Fraga, R., Belen, M., & Cekirge, H. (1984). A new technique to estimate initial spill size using a modified Fay-type spreading formula. *Marine Pollution Bulletin, 15*, 326–329.

Lehr, W., & Simecek-Beatty, D. (2000). The relation of Langmuir Circulation processes to the standard oil spill spreading, dispersion, and transport algorithms. *Spill Science & Technology Bulletin, 6*, 247–253.

Lehr, W., Simecek-Beatty, D., Aliseda, A., & Boufadel, M. (2014). Review of recent studies on dispersed oil droplet distributions. In *Proceedings of the 37th AMOP technical seminar on environmental contamination and response*, pp. 1–8.

Lehr, W., Simecek-Beatty, D., & Fingas, M. (2019). Whither oil spill models in the next decade? In *Proceedings of the forty-second AMOP technical seminar, environment and climate change* Canada, Ottawa, Canada, pp. 453–472.

Li, M., & Garrett, C. (1993). Cell merging and the jet/downwelling ratio in Langmuir circulation. *Journal of Marine Research*, *51*, 737–769.

Li, Z., Spaulding, M., & French-McCay, D. (2017). An algorithm for modeling entrainment and naturally and chemically dispersed oil droplet size distribution under surface breaking wave conditions. *Marine Pollution Bulletin*, *119*, 145–152.

Liu, P., & Schwab, D. R. (2002). Has wind-wave modeling reached its limit? *Ocean Engineering*, *29*, 81–98.

Mackay, D., Buist, I., Mascarenhas, R., & Patterson, R. (1980). *Oil spills processes and models.* Canada: Environment Canada, Ottawa.

Mackay, D., & Matsugu, R. (1973). Evaporation rates of liquid hydrocarbon spills on land and water. *Canadian Journal of Chemical Engineering*, *51*, 434–439.

MacKay, D., & McAuliffe, C. (1989). Fate of hydrocarbons discharged at sea. *Oil and Chemical Pollution*, *5*, 1–20.

Mackay, D., Shiu, W., Hossain, W., Stiver, K., McCurdy, D., Paterson, D., & Tebeau, P. (1982). *Development and calibration of an oil spill behavior model.* Washington D.C: US Coast Guard, Report No. CG-D-27–83.

MacKay, D., Stiver, W., & Tebeau, P., 1983. Testing of crude oils for petroleum products for environmental purposes. In *Proceedings of the international oil spill conference*, American Petroleum Institute, Washington DC, pp. 331–337.

Masuko, T., & Magill, J. (1988). A comprehensive expression for temperature dependence of liquid viscosity. *Journal of the Society of Rheology, Japan*, *16*, 22–26.

McLean, J., & Kilpatrick, P. (1997). Effects of asphaltene solvency on stability of water-in-crude-oil emulsions. *Journal of Colloid and Interface Science*, *189*, 242–253.

Monahan, E. (1971). Oceanic whitecaps. *Journal of Physical Oceanography*, *1*, 139–144.

Mullins, O., et al. (2012). Advances in asphaltene science and the Yen-Mullins model. *Energy & Fuels*, *26*, 3986–4003.

Nissanka, I., & Yapa, P. (2018). Calculation of oil droplet size distribution in ocean oil spills: A review. *Marine Pollution Bulletin*, *135*, 723–734.

Onyelucheya, O., Osaka, E., Onyelucheya, C., & Kamen, F. (2014). Modeling the viscosity behavior of weathered and unweathered bonny light crude oils. *International Journal of Engineering and Technology Research*, *2*, 26–28.

Overstreet, R., Lewandowski, A., Lehr, W., Jones, R., Simecek-Beatty, D., & Calhoun, D. (1995). Sensitivity analysis in oil spill models: Case study using ADIOS. In *Proceedings of the 1995 international oil spill conference*, American Petroleum Institute, Washington D.C. pp. 898–900.

Pajouhandeh, A., Kavousi, A., Schaffie, M., & Ranjbar, M. (2016). Toward a mechanistic understanding of rheological behaviour of water-in-oil emulsion: Roles of nano-particles, water volume fraction and aging time. *South African Journal of Chemistry*, *69*, 113–123.

Pal, R., & Rhodes, P. (1989). Viscosity relationships for emulsions. *The Journal of Rheology*, *33*, 1021–1045.

Payne, J., Kirstein,B., Clayton, J., Clary, C., Redding,R., Mcnabb, D., & Farmer, G. (1987). Integration of suspended particulate matter and oil transportation study, Report No. MMS 87-0083, United States Minerals Management Service, Anchorage AK.

Phillips, W. (2002). Langmuir circulation beneath growing or decaying surface waves. *Journal of Fluid Mechanics*, *469*, 317–342.

Plata, D., Sharpless, C., & Reddy, C. (2008). Photochemical degradation of polycyclic aromatic hydrocarbons in oil films. *Environmental Science & Technology*, *42*, 2432–2438.

Rahuma, K. (2016). Evaluation of oil viscosity performance using several empirical correlations for some Libyan crude oils. *American Journal of Engineering and Technology Management, 1*, 55–58.

Reynolds, O. (1886). On the theory of lubrication and its application to Mr. Beauchamp tower's experiments, including an experimental determination of the viscosity of olive oil. *Philosophical Transactions of the Royal Society of London, 177*, 157–234.

Riazi, M. (2005). *Characterization and properties of petroleum fractions.* West Conshohocken PA: ASTM.

Riazi, M., & Edalat, M. (1996). Prediction of the rate of oil removal from seawater by evaporation and dissolution. *Journal of Petroleum Science and Engineering, 16*, 291–300.

Schmidt, P. (1986). *Fuel oil manual* (4th Ed.). Woodbine New Jersey: Industrial Press.

Schramm, L. (1992). *Emulsions fundamentals and applications in the petroleum industry.* Washington D.C.: American Chemical Society.

Shemden, O. (1972). Wind generated current and phase speed of wind waves. *Journal of Physical Oceanography, 2*, 281–297.

Simecek-Beatty, D., & Lehr, W. (2017). Extended oil spill spreading with Langmuir circulation. *Marine Pollution Bulletin, 122*, 226–235.

Smith, C., & MacIntyre, G. (1971). Initial aging of fuel oil films of sea water. In *Proceedings of international oil spill conference*, pp. 457–461.

Society of Petroleum Engineers SPE. (1990). *The properties of petroleum fluids* (2nd ed.). Tulsa OK: Penwell Publishing.

Spaulding, M. (2017). State of the art review and future directions in oil spill modeling. *Marine Pollution Bulletin, 115*, 7–19.

Speight, J. (2005). *Handbook of petroleum product analysis* (2nd ed.). NY: John Wiley and Sons.

Stevens, C., Thibodeaux, L., Overton, E., Valsaraj, K., & Walker, N. (2017). Dissolution and heavy residue sinking of subsurface oil droplets: Binary component mixture dissolution theory and model-oil experiments. *Journal of Environmental Engineering, 143*(10).

Stiver, W., & Mackay, D. (1984). Studies on the evaporation regulation mechanisms of crude oil and petroleum products. *Environmental Science & Technology, 18*, 834–840.

Sutherland, P., & Melville, W. (2015). Field measurements of surface and near-surface turbulence in the presence of breaking waves. *Journal of Physical Oceanography, 45*, 943–965.

Terry, E., Donelan, M., Agraval, M., Drennan, Y., Kahma, W., Williams, K., . . . Kitaigorodskii, S. (1996). Estimates of kinetic energy dissipation under breaking waves. *Journal of Physical Oceanography, 26*, 792–807.

Thorpe, S. (2004). Langmuir circulation. *Annual Review of Fluid Mechanics, 36*, 55–79.

Venkatesh, S. (1988). The oil spill behavior model of the Canadian Atmospheric Environmental Service. Part 1: Theory and model evaluation. *Atmosphere-Ocean, 26*, 93–108.

Venkatesh, S. (1990). Modeling the behavior of oil spills in ice-infested waters. *Atmosphere-Ocean, 28*, 303–329.

Williams G., Hann, R., & James, W. (1975). Predicting the fate of oil in the marine environment. In *Proceedings of the joint conference on prevention and control of oil spills*, American Petroleum Institute, Washington D.C., pp. 567–572.

Xia, L., Lu, S., & Cao, G. (2014). Stability and demulsification of emulsions stabilized by asphaltenes or resins. *Journal of Colloid and Interface Science, 271*, 504–506.

Xie, H., Yapa, P., & Nakata, K. (2007). Modeling emulsification after an oil spill in the sea. *The Journal of Marine Systems, 68*, 489–506.

Yang, W., Gao, Y., & Casey, J. (2008). *Determination of trace elements in crude oils and fuel oils: A comprehensive review and new data, in solution chemistry: Advances in research and applications*. NY: Nova Science Publishers.

Yapa, P., & Weerasuriya, S. (1997). Spreading of oil spilled under floating broken ice. *Journal of Hydraulic Engineering*, 676–683, ASCE, August.

Yetilmezsoy, K., Fingas, M., & Fieldhouse, B. (2012). Modeling water-in-oil emulsion formation using fuzzy logic. *Journal of Multiple-Valued Logic and Soft Computing*, *18*, 329–353.

Zatsepa, S., Ivchenko, A., Korotenko, K., Solbakov, V., & Stanovoy, V. (2018). The role of wind waves in oil spill natural dispersion in the sea. *Oceanology*, *58*, 517–524.

Zedel, L., & Farmer, D. (1991). Organized structures in subsurface bubble clouds: Langmuir circulation in the open ocean. *Journal of Geophysical Research*, *96*, 8889–8900.

Zeinstra-Helfrich, M., Koops, W., & Murk, A. (2016). How oil properties and layer thickness determine the entrainment of spilled surface oil. *Marine Pollution Bulletin*, *110*, 184–193.

Zhao, D., & Toba, Y. (2001). Dependence of whitecap coverage on wind and wind-wave Properties. *Journal of Physical Oceanography*, *57*, 603–616.

Zhao, L., Boufadel, M., Geng, X., Lee, K., King, T., Robinson, B., & Fitzpatrick, F. (2016). A-drop, a predictive model for the formation of oil particle aggregates (OPAs). *Marine Pollution Bulletin*, *106*, 245–259.

Zhao, L., Boufadel, M., Socolofsky, S., Adams, E., King, T., & Lee, K. (2014). Evolution of droplets in subsea oil and gas blowouts: Development and validation of the numerical model VDROP-J. *Marine Pollution Bulletin*, *83*, 58–69.

Zheng, L., Kobayashi, T., & Yapa. P. (1995). Application of linked-list approach to pollutant transport models. Water Resources Engineering, pp. 1668–1672.

Chapter 3

Horizontal transport in oil-spill modeling

Rodrigo Duran[1,2], Tor Nordam[3,4], Mattia Serra[5,6] and Christopher H. Barker[7]

[1]National Energy Technology Laboratory, United States Department of Energy, Albany OR, United States, [2]Theiss Research, La Jolla, CA, United States, [3]SINTEF Ocean, Trondheim, Norway, [4]Department of Physics, Norwegian University of Science and Technology, Trondheim, Norway, [5]School of Engineering and Applied Sciences, Harvard University, Cambridge, MA, United States, [6]Department of Physics, University of California San Diego, La Jolla, CA, United States, [7]Emergency Response Division, Office of Response and Restoration, National Oceanic and Atmospheric Administration, Seattle, WA, United States

3.1 Introduction

Successfully forecasting the movement of oil during an oil spill, and knowing oil's current location, are the key ingredients needed to solve one of the most decisive problems during emergency response operations: where is the oil heading? Oil's current and future locations are critical for emergency planning and response, including recovery and containment. Reconstructing past spills through hindcasts is also needed for modeling improvements, environmental impact assessments, and forensics. The challenge of successfully forecasting or hindcasting an oil spill is a considerable one. There is no guarantee that a simulation will be successful, and deviations between simulated transport and observations are the norm rather than the exception. This chapter describes some of the main reasons why accurately simulating trajectories in the ocean is complicated and presents recent progress along different fronts that help remediate some of the problems. The focus is on hindcasts and forecasts, characterized by the need to replicate or anticipate observed oil transport.

One of the problems is related to local deficiencies in the velocity that propagate during integration when computing trajectories, these errors often grow exponentially due to the unstable nature of ocean circulation. For this problem, we present modern techniques called Objective Eulerian Coherent Structures (OECS; Serra & Haller, 2016) that can bypass errors in the

Marine Hydrocarbon Spill Assessments. DOI: https://doi.org/10.1016/B978-0-12-819354-9.00004-1

velocity, identifying instantaneous attracting regions that influence transport exceptionally, and that are computed without the need to integrate the velocity. This method allows trajectory forecasting without future information and can also be applied to ocean-model forecasts to limit the effect of discrepancies between the forecasted flow and reality.

The other type of problem we examine is related to a velocity assumed to simulate oil's motion, but that is missing some of the physical processes driving observed motion. The focus is in the upper centimeters of the ocean where atmospheric influence is strongest. The solution is to parameterize missing physics when necessary. Studies in the last few years have improved our understanding of the velocity within a fine surface layer, providing information that is somewhat at odds with common practice in oil-spill modeling. As an example, we show how progress in our understanding can explain the transport of oil during the 2003 Point Wells spill in the Salish Sea, a spill that had remained unexplained for 15 years. Also discussed are recent studies showing that parameterized near-surface processes are often responsible for oil beaching. Finally, we suggest the potential for surface convergence driven by subduction at the sea surface as an indicator of regions of interest for oil recovery.

A brief review of the basics of oil-spill modeling is given in Section 3.2, Section 3.3 overviews transport in the ocean, the unstable nature of ocean currents, and a description of velocity products typically available to simulate oil transport. Section 3.4 is about motion in the upper layer of the ocean, describing recent progress due to numerical simulations and unique at-sea experiments. Section 3.5 describes some of the novel tools that may help overcome velocity errors that would result in erroneous trajectories and exemplify their use by revisiting transport observed during the Deepwater Horizon accident in May 2010. We conclude in Section 3.6 with conclusions and an outlook of what progress can be expected, and how the techniques presented here fit into that picture.

3.2 The physics, the mathematics, and the numerics

Oil-spill modeling is often a multidisciplinary endeavor; it is not uncommon for biologists, chemists, physicists, geographers, mathematicians, oceanographers, engineers, and computer scientists to work together. It is therefore helpful to begin clarifying, somewhat informally, the basic physics, mathematics, and numerical solutions used to simulate the transport in a fluid that ultimately result in an oil-spill simulation. The physical mechanisms that drive motion in the ocean are described in later sections.

The physical approximations used to simulate transport forced by a vast variety of oceanographic processes are well known. Consequently, the mathematical equations are also well known (e.g., studied in most introductory partial differential equation courses). The mathematics of oil transport boils down to solving the advection–diffusion equation, also known as the dispersion–diffusion equation. In fluid dynamics and physical oceanography, the effects of advection and diffusion are often referred to as stirring and mixing, respectively. In oil-spill modeling, these equations are solved in a Lagrangian framework, that is, by computing the trajectories of individual elements ("particles"). However, it is illustrative to introduce the equations of mathematical physics in Eulerian terms, that is, as a function of fixed space, and return to the Lagrangian formulation when discussing the numerical solution.

Stirring within a fluid causes a tracer to deform into streaks and swirls while the along-path concentration of the tracer does not change. Effectively, the tracer is redistributed with the velocity field while preserving its concentration; such behavior is called conservative. Assuming the velocity divergence is negligible, the following evolution equation for a tracer C is satisfied:

$$\frac{\partial C}{\partial t} + \mathbf{u} \cdot \nabla C = 0 \tag{3.1}$$

where $\mathbf{u}$ is the two-dimensional velocity of the fluid. This equation is known as the advection equation or dispersion equation. Some scientists may use the word convection instead of advection, although oceanographers often reserve the term convection for a different type of physics (vertical motion related to buoyancy changes). The advection equation describes how a concentration evolves as the velocity field acts upon the gradient of the concentration, causing a redistribution of the tracer.

Mixing refers to the diffusion of a tracer with concentration C, typically simulated with the diffusion equation:

$$\frac{\partial C}{\partial t} = \nabla \cdot (\kappa \nabla C). \tag{3.2}$$

If κ is a molecular diffusion coefficient, then (3.2) represents the mixing of a tracer due to molecular collisions. By itself, this is an inefficient method of mixing a fluid. In the ocean, however, an eddy diffusion coefficient is used for κ, several orders of magnitude larger than the molecular coefficient that is characteristic of the fluid. From a physics point of view, the large coefficient means that the diffusion equation is modeling the mixing of concentration due to the collision of water parcels, not molecules. This is an ad hoc way of parameterizing small-scale stirring and overturning of water parcels as they undergo turbulent motions. Ironically, the physics simulated by Eq. (3.2) is well understood, yet the ocean physics that it parameterizes includes a variety of processes that are difficult to even measure (e.g., Moum & Rippeth, 2009). It is a fortunate turn of events that results from using a well-understood equation such as (3.2) are adequate for many purposes, including oil-spill modeling.

To simulate transport in the ocean, both stirring and diffusion are often used simultaneously: the evolution equation for the transport of a tracer is then the advection–diffusion (or dispersion–diffusion) equation:

$$\frac{\partial C}{\partial t} + \mathbf{u} \cdot \nabla C = \nabla \cdot (\kappa \nabla C). \tag{3.3}$$

In the case of oil-spill modeling, an additional source term can be added to represent the addition of oil as it is spilling into the ocean; a sink term can also be included to represent the removal of oil. This chapter is concerned with the transport of oil, and we will not consider sources or sinks. Salmon (1998) presents a discussion (his Section 14) on stirring and mixing, describing their individual and simultaneous effects on tracer variability. Stirring, diffusion, and their interplay are also discussed in Section 7.3 of Tennekes and Lumley (1972).

The eddy diffusion coefficient, also known as the turbulent diffusion coefficient, is an unknown that needs to be determined. The production of turbulence, and therefore the magnitude of the eddy diffusion coefficient, depends on many factors including the spatial structure of seawater's density, the spatial structure of the velocity field, heating or cooling of water parcels, and Earth's rotation. Turbulence in the ocean is often produced by instabilities that range in length from centimeters to hundreds of kilometers—the interested reader is referred to the free book on ocean instabilities by Smyth and Carpenter (2019). Due to a large number of instability types, the large range of spatial and temporal scales at which they occur and interact, and their often anisotropic nature, determining an adequate turbulent diffusion coefficient is a difficult problem.

Numerical ocean models require accurate mixing of momentum, heat, and salinity to produce a good simulation; they use sophisticated turbulence closure models that are computationally intensive and that require considerably more information than what is typically available during oil-spill simulations. Fortunately, the need for mixing parameterizations in oil-spill modeling is fundamentally different than in numerical ocean modeling and is not nearly as consequential. In oil-spill modeling, the diffusion equation is used to simulate the small-scale spreading of oil caused by oceanographic processes that are not resolved by the velocity in Eq. (3.1).

In oil-spill modeling, the most efficient way to determine the eddy diffusion coefficient is to choose a constant coefficient that matches the observed spread of oil. This is a strategy used for hindcasts and forecasts where the main objective for the simulation of diffusion is matching the spread of oil as observed through overflights, ships, and satellites. However, during response forecasts, the diffusion coefficient is chosen to err on the high side, so that the simulations are unlikely to underestimate the extent of impacted locations. For example, standard practice for the Office of Response and Restoration at the National Oceanic and Atmospheric Administration

(NOAA), is to use a random walk with a constant diffusion coefficient as described. Despite its rather crude and ad hoc nature, it is often important to simulate diffusive processes in oil-spill modeling because (1) it is needed to match observed oil spreading about the trajectories resulting from Eq. (3.1), (2) it provides a least-regret conservative estimate for the spreading and impact of oil during response forecasts, and (3) for simulations without wind, it provides one way for oil to beach, which is otherwise not available from most ocean-current velocity products (e.g., ocean models set to zero the velocity normal to the coast near the coastline, although it may cause beaching if there is a mismatch between the velocity product and the model coastline, or due to numerical instabilities). Beaching due to diffusion is most appropriate in the surf zone, the effects of which are not typically simulated in ocean circulation models. Other (more realistic) mechanisms that may drive oil beaching are discussed in Section 3.4.

There are types of oil-spill simulations that may need an automated method of determining an eddy diffusion coefficient. For example, ensemble-type modeling uses a large number of oil-spill simulations to evaluate the environmental impact of an oil spill in a statistical sense; such simulations may choose to include diffusive processes. The solution for oil-spill simulations that compute an eddy diffusivity as part of the problem (instead of choosing it to match observed spread) is described in Appendix A. Further discussion on stochastic parameterizations of diffusion and their implementation can be found in the technical documentation for NOAA's GNOME model (Zelenke, O'Connor, Barker, Beegle-Krause, & Eclipse, 2012) and Duran (2016).

The easiest way to numerically solve Eq. (3.3) for oil transport simulations is to separately solve (3.1) and (3.2) and then add the motion induced by each process to obtain the oil's movement. Solutions to (3.1) and (3.2) are often found separately in Lagrangian terms, that is, by computing oil-parcel trajectories, rather than in Eulerian terms where the equations are solved over a fixed numerical grid. Because the advection part of transport preserves concentrations, the simulation of advection reduces to integrating the velocity field to obtain trajectories given an initial time and position. Thus, the typical approach to simulate the advective part of oil's trajectory is the solution $\mathbf{x}(t)$ to the equation:

$$\frac{\mathrm{d}\mathbf{x}}{\mathrm{d}t} = \mathbf{u}(\mathbf{x}(t), t) \quad \mathbf{x}(0) = \mathbf{x}_0. \tag{3.4}$$

Eq. (3.4) can be solved with regular numerical methods for ordinary differential equations, although there are some specific considerations due to the discrete nature of the velocity data being integrated (e.g., Nordam & Duran, 2020; Nordam et al., 2018).

The diffusion part is typically modeled as a random walk, numerically simulating the random motions induced by the modeled collisions with

Eq. (3.2). The resulting motion is added to the solution of (3.4) at each time step. Typically, the precision needed for **u** in (3.4) is more consequential than the spread of oil modeled with diffusion. Because of this, in this chapter, we will focus on the advective part of oil's transport. In this approach, once the advection part is satisfactory, further simulations will add diffusion to the advective part. We note that formally, the interplay between advection and diffusion is more complicated than often appreciated; the effect of diffusion in the context of Lagrangian transport is currently being researched (Haller, Karrasch, & Kogelbauer, 2018).

The Eulerian representation of fluid flow and its associated numerical representation could, in principle, be used for oil-spill modeling. In practice, however, it is much more efficient to simulate oil transport using the Lagrangian representation. Among the problems inherent to the Eulerian representation is the need to set up a numerical grid for each domain. Also, the Eulerian representation is more computationally intensive because it requires solving the advection–diffusion equation at each point on the grid, regardless of whether there is oil there or not, while in the Lagrangian approach trajectories are integrated only for existing oil parcels. In the Eulerian approach, the advection term in (3.3) can be problematic for numerical methods (e.g., Durran, 2010, Sections 3.3, 3.4, 3.5.1, 3.5.2, 5.10), while the Lagrangian approach only requires integrating an ordinary differential equation that is, for the most part, straightforward and highly accurate. Finally, the Lagrangian approach works well with a velocity field saved at a series of discrete times; this is convenient because it readily allows additional experiments using the same velocity field. A comparison of advection results from Eulerian and Lagrangian formulations can be found in Wagner, Rühs, Schwarzkopf, Koszalka, and Biastoch (2019).

Oil-spill models can account for other processes related to the fate of oil separately. For example, oil droplets breaking into smaller-size droplets, oil dissolution at depth, or oil evaporation at the sea surface. These processes can affect the oil's buoyancy. As oil-spill and blowout models increase in complexity, the effects of vertical motion will be included. Oil's vertical motion may be induced by ambient conditions such as waves, or by oil's buoyancy (some oils are denser than water, some are less dense), droplet size, and weathering. Further details on some of the processes causing vertical motion can be found in Nordam, Skancke, Duran, and Barker (2021).

In this chapter, we assume that oil's vertical location is known, whether varying or fixed. This is a valid approach because (1) for some spills, the oil floating at the surface is of primary concern, and thus modeling horizontal trajectories alone may be enough to get satisfactory results and (2) when vertical motion is important, vertical and horizontal components of a parcel's trajectory are computed separately. This is because the mechanisms forcing horizontal motion tend to be different from mechanisms forcing vertical motion, and the resulting trajectory components can be added to give an updated location for the next integration step. Whenever the vertical dimension is included, the problem of determining oil's horizontal motion is more complicated. The vertical position of oil must first be determined to be able to sample the correct horizontal velocity, additionally, the velocity **u** driving horizontal motion will now need to be accurate at different vertical levels.

For this chapter's discussion, horizontal transport will be defined as the motion of oil at a constant depth, whether at the surface or deeper down within the water column. Horizontal motion in this chapter also means motion along surfaces of constant density (lateral motion in oceanographic terminology) as long as it does not involve abrupt vertical displacements, for example, where constant density surfaces rise to intersect the ocean surface. Some comments will be made regarding horizontal transport of oil at density fronts in Section 3.4.3. Ocean currents are driven by a variety of physical processes that, for oil-spill modeling, can be conveniently classified by their vertical location within the water column. The range of physical processes driving motion can then be narrowed down according to the vertical location of oil.

3.3 Overview of oil transport in the ocean

In practice, transport of oil in the ocean is successfully simulated as the movement of parcels moving with the velocity **u** of ocean currents, as in Eq. (3.4), this is demonstrated, for example, in Jones et al. (2016). Near the surface, additional movement induced by wind and waves may help drive motion. In general, the vertical location of oil is consequential because the effect of wind and wave changes abruptly in the upper meters of the ocean, as will be discussed in Section 3.4, and because ocean currents also tend to change with depth.

Over the last several decades, we have come to understand the ocean as a turbulent fluid in perpetual motion, rich in temporal variability. There are different types of turbulence in the ocean and a vast variety of instabilities—processes that can trigger oscillations with speeds considerably larger than the mean flow. Some of these oscillations (e.g., eddies) often result in hyperbolic points in the velocity field. Intuitively, hyperbolic points in the Eulerian velocity field suggest that initially close trajectories are likely to undergo exponential separation. While this basic intuitive idea turns out to be correct, the relation between the velocity at each point and the trajectories traversing this time-dependent velocity is not as intuitive, requiring a careful mathematical treatment to uncover, as discussed in Section 3.5.

As an example to illustrate chaotic behavior, we use Hybrid Coordinate Ocean Model-Gulf of Mexico (HyCOM-GoM), an ocean model that is likely to be used by oil-spill modelers in the Gulf of Mexico, to advect two groups of four trajectories initiated less than 5 km apart. Within each group, four trajectories are initiated within 300–500 m of each other. The two groups of

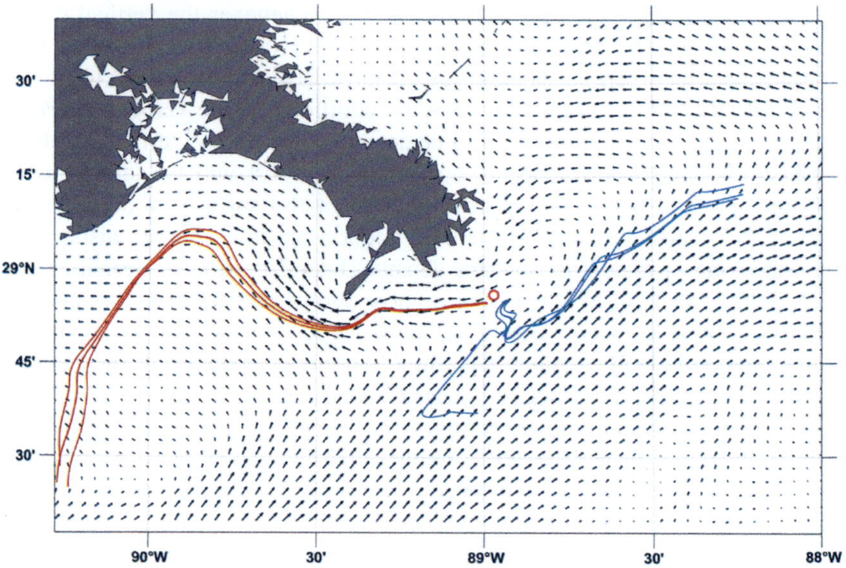

FIGURE 3.1 Sea-surface velocity (black vectors) near the Mississippi delta in the Gulf of Mexico from a Hybrid Coordinate Ocean Model-Gulf of Mexico operational simulation on March 14, 2016. Four trajectories (orange) are released at noon March 9, 2016, just southwest of Taylor well (red circle). Another four trajectories (blue) are released just southeast of Taylor well at the same time. A hyperbolic behavior separates initially close (<5 km) blue and orange trajectories more than 200 km over 5 days, some moving northeast, some moving southwest. The blue trajectories were initially 300–500 m apart, yet three remain close (<5 km), and one separates about 100 km.

trajectories undergo exponential separation as they move with realistic ocean currents, ending 200 km apart after 5 days (Fig. 3.1). These trajectories are initiated close to the Taylor Energy well that continued for many years to spill oil from the seafloor starting in 2004 (Sun et al., 2018), representing a realistic example of the uncertainty that an oil-spill modeler might face. Among the group of four trajectories initiated to the southeast of Taylor well, one trajectory separates 100 km from the other three trajectories, despite being initially 300–500 m away, further showing the chaotic nature of trajectories in the ocean. This example shows that for a time-dependent flow, the interaction of trajectories with a hyperbolic point in the velocity is complex.

Exponential separation of initially close trajectories (i.e., a sign of chaotic behavior) is an important part of why predicting trajectories in the ocean is difficult. Small errors in the velocity field or in the location at which trajectories are initiated are likely to result in large errors over short periods of hours to days. This is a problem inherent to ocean currents, it cannot be corrected with higher-order integration when computing trajectories. We return to this type of problem in Section 3.5.

A different type of problem occurs when a trajectory is deficient because the velocity is not representative of all the processes driving motion. In this case, it is sometimes possible to parameterize missing physics to complement the velocity. Oil does not necessarily remain at the sea surface. The depth at which oil is located is particularly important when parameterizing missing physics, as driving mechanisms change drastically even within tens of centimeters in the upper ocean, as discussed in Section 3.4.

3.3.1 Sources of velocity for oil-spill modeling

In this section, we provide an overview of some common sources of ocean current data for oil spill modeling. Two products are from remotely sensed measurements and are therefore limited to producing a sea-surface velocity up to the current time. Ocean models produce ocean currents from numerical simulations and can provide a forecast into the future, as well as a complete velocity field, both horizontally and vertically throughout the water column.

3.3.1.1 Numerical ocean models

The equations governing geophysical fluid dynamics—fluid dynamics on a rotating sphere—can be discretized and solved numerically. The equations themselves are very complicated, and their numerical solution is further complicated because motion at different spatial scales, from thousands of kilometers to centimeters, interact with each other in fundamental ways, yet computers are not powerful enough to simulate all such scales. Also, simulations are necessarily initiated from imperfect initial and boundary conditions, and geophysical flows tend to be chaotic. Notwithstanding, ocean models are surprisingly accurate in portraying a variety of physical processes in the ocean and are often used as the source for the velocity $\mathbf{u}$ needed to solve Eq. (3.4).

When using an ocean model, the vertical resolution should be a concern even when simulating oil transport exclusively at the sea surface. This is because the model's output for a surface velocity will be in reality a representation of the vertically sheared currents over the height of the top grid cell of the model, not a representation of the velocity at the very surface. The model output that is closest to the surface is a depth-averaged value, where averaging takes place over the thickness of the model's upper vertical level. Naturally, coarse resolutions result in greater smoothing and therefore, a less realistic representation of the surface velocity. In addition, even if a model uses a reactively thin surface layer, it usually does not include smaller-scale processes at the surface, notably wind waves.

Producing accurate velocity products with models is also complicated because hyperbolic trajectories are often linked to ocean instabilities that are not completely understood and are difficult to accurately simulate. For example, ocean eddies can have a profound effect on Lagrangian transport, and a numerical simulation may develop an eddy that does not exist in the ocean. Even if a model accurately simulates an eddy, small displacements in the eddy location relative to the correct position of the eddy in the ocean can result in large trajectory errors. Thus, ocean models often do not represent the oceanic structures that are most influential on trajectories with enough accuracy. This is true even when the numerical model assimilates a variety of ocean measurements in an attempt to replicate the real ocean. Observations used by data-assimilating models include satellite products such as sea-surface height (SSH), temperature, and salinity and in situ observations from oceanographic buoys, drifters, gliders, and other autonomous platforms. A recent overview of progress and challenges in ocean modeling can be found in Fox-Kemper et al. (2019).

3.3.1.2 High-frequency radars

High-frequency (HF) radars can measure sea-surface currents remotely near the coastline (<200 km) with resolutions typically 500 m to 6 km, and up to hourly in time. Velocity from HF radar is an exponentially weighted vertical average, with a decay scale that is proportional to the wavelength of the radar signal (e.g., Röhrs, Sperrevik, Christensen, Broström, & Breivik, 2015). Due to the resolution, which is unable to determine small-scale structures, and processing errors, trajectories from drifters designed to sample similar ocean currents as those measured by HF radar, differ from trajectories computed from HF radar velocity. Carefully calibrated radars at resolutions higher than about 1.5 km and 3 hours can replicate drifter trajectories with a separation rate of about a few kilometers over a day (Kirincich, De Paolo, & Terrill, 2012; Rypina, Kirincich, Limeburner, & Udovydchenkov, 2014). However, the quality of HF radar processing and HF radar resolution varies. Improving HF radar velocity products to better represent coastal currents is an ongoing endeavor (e.g., Kirincich, Emery, Washburn, & Flament, 2019). Other limitations include accuracy that varies with position relative to the antennae and gaps between stations and very near the coastline.

HF radars are an important part of operational models that assimilate the surface velocity to minimize the model's error. HF radar can measure currents near the coast, making them an excellent complement to altimetry that can only produce a geostrophic velocity (Section 3.3.1.3) further from the coast.

There are methods to improve HF radar data for Lagrangian purposes. For example, to produce trajectories from HF radar that are closer to drifter trajectories, the Eulerian velocity may be corrected using trajectory data (e.g., Berta et al., 2014). A downside of this approach is that it requires deploying drifters and allowing them to drift for some time before the corrections can be applied.

HF radar is increasingly available in the United States and around the world (Roarty et al., 2019); a review on HF radar can be found in Paduan and Washburn (2013).

3.3.1.3 Velocity products from satellites

Satellites with altimeters are able to measure SSH with enough accuracy that a geostrophic velocity proportional to the SSH gradient can be computed:

$$u_g(x,y,t) = -\frac{g}{f}\frac{\partial\eta(x,y,t)}{\partial y} \tag{3.5}$$

$$v_g(x,y,t) = +\frac{g}{f}\frac{\partial\eta(x,y,t)}{\partial x} \tag{3.6}$$

where (u_g, v_g) are respectively the east and north components of the geostrophic surface velocity, η is the SSH, g is the acceleration of gravity and $f = 2\Omega\sin\theta$ is the Coriolis parameter, $\Omega = 7.29 \times 10^{-5}\text{s}^{-1}$ is the rotation of the Earth, and θ is the latitude.

Velocity from altimetry has been shown to give good results for Lagrangian transport applications and may give superior results than numerical models that assimilate the same altimetry data. We cite a few studies as examples where using altimetric velocity for Lagrangian transport applications has been shown to be a good choice.

Ohlmann, Niiler, Fox, and Leben (2001) compare surface velocity from drifters and altimetry and find very good agreement in the Gulf of Mexico deeper than the 2000 m isobath and good agreement between the 200 and 2000 m isobaths. They find that these correlations depend on the length scale over which the differentiation in Eqs. (3.5) and (3.6) is computed, with best results at $\partial x, \partial y \sim 125$ km. A combined observational and modeling study in northern Norway found that the accuracy of trajectories calculated from satellite products was comparable to ocean model data, and in some cases better (Dagestad & Röhrs, 2019); the superior results they report from a free-running model are not surprising in coastal areas where geophysical flows are more predictable (e.g., Kim, Samelson, & Snyder, 2011) and altimetry measurements are less reliable.

Sudre, Maes, and Garçon (2013) show that globally the velocity from ARGO floats correlates well with the velocity they produce mainly from altimetry, except near the Equator for the meridional component. They also show excellent correlations between their altimetry-based velocity and the velocity from drifters in the Indian Ocean. Sudre et al. (2013) also show that their velocity product is capable of explaining Lagrangian transport visualized through satellite-sensed chlorophyll during an iron-release fertilization experiment. Olascoaga et al. (2013) show a very good correspondence between altimetric hyperbolic Lagrangian Coherent Structures (LCS) and satellite-observed transport (chlorophyll). Jacobs et al. (2014) then compare this transport to LCS from operational ocean models Navy Coastal Ocean Model (NCOM) and HyCOM. They find that altimetry gives accurate transport patterns, while the two models show a fictitious transport barrier that is crossed by offshore chlorophyll advection (see their Figs. 3.1 and 3.2); they propose a modification to the data assimilation scheme as a correction. NCOM and HyCOM are the former and current models used by the US Navy for their Global Ocean Forecast System that assimilates a variety of observations through the Navy Coupled Ocean Data Assimilation (NCODA) system. The results of Olascoaga et al. (2013) and Jacobs et al. (2014) are for July 2012, here we will analyze a similar transport pattern during the

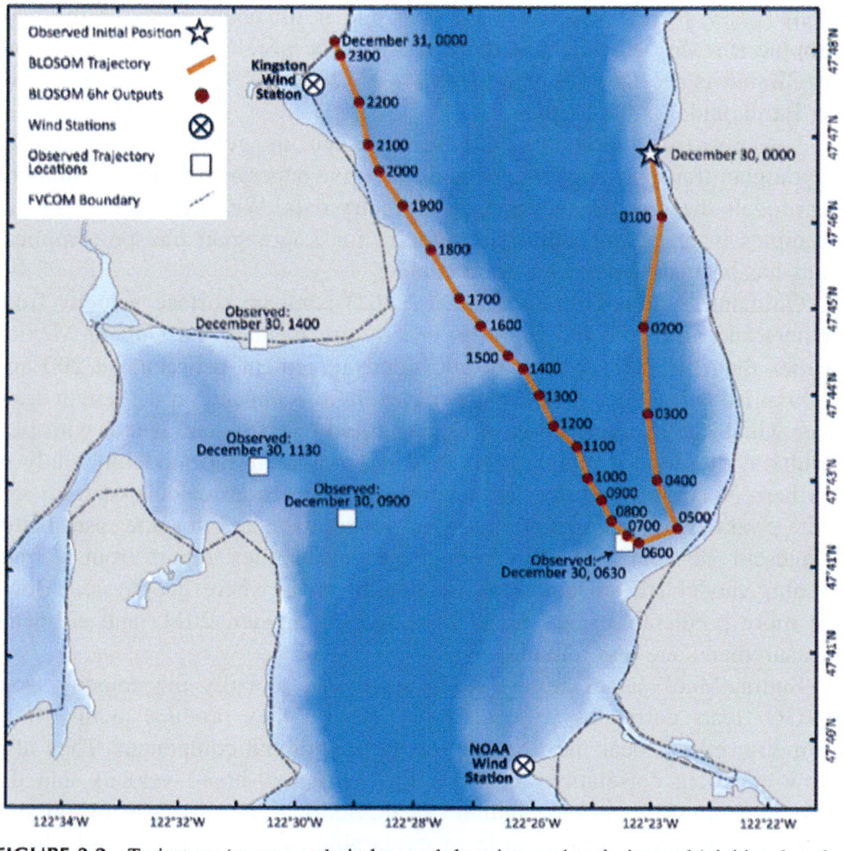

FIGURE 3.2 Trajectory (orange, red circles mark locations at hourly intervals) initiated at the time and location of the Point Wells 2003 spill in the Salish sea (white star) forced only with 6% of the wind measured by the NOAA wind station (white circle with black cross near bottom). Locations and times where oil was observed are marked with white squares and text.

Deepwater Horizon accident in May 2010 in Section 3.5.2. Liu, Weisberg, Vignudelli, and Mitchum (2014) found that different altimetry velocity products perform similarly and that trajectories simulated from altimetry perform better than from data-assimilative models. Berta et al. (2015) compare satellite-tracked drifters to synthetic trajectories from altimetry finding "satisfactory average results"; they also show how blending drifter data into the altimetric velocity considerably improves trajectory hindcasts and restores missing physics that cannot be explained by Ekman superposition. Beron-Vera, Wang, Olascoaga, Goni, and Haller (2013) and Beron-Vera, Olascoaga, Wang, Triñanes, and Pérez-Brunius (2018) show the relevance of Lagrangian coherence associated with eddies detected objectively from altimetry. Essink, Hormann, Centurioni, and Mahadevan (2019) found that trajectories advected with altimetry do well in replicating the main transport patterns observed with drifters, and even though they find that a variety of statistics from altimetry trajectories do not closely resemble those from observed trajectories, we note that the concern for oil-spill modelers is identifying prevailing oil movement. Another limitation of satellite-derived products is the low frequency of satellite passes over a given region—a synoptic view cannot be instantly obtained.

Work toward remotely sensed velocity products of higher resolution is currently underway (e.g., Chelton et al., 2019).

3.4 Transport in the upper layer of the ocean

One of the challenges oil-spill modelers face is that some of the physical processes driving oil's motion near the ocean surface may not be represented in available velocity products. In this section, we discuss how motion in the upper layer of the ocean may be strongly influenced by wind drift (windage) and Stokes drift from waves. As we will see, motion in the upper centimeter of the ocean can be significantly different than in the upper meter. Velocity from HF radar and ocean models do not typically include Stokes drift or windage. See Röhrs et al. (2015) for a discussion on Stokes drift in HF radar ocean currents. Velocity from altimetry does not include Stokes drift, windage, or Ekman transport, although the latter is sometimes added from additional satellite measurements. Because it is difficult to obtain a velocity that is representative of the upper centimeter, it may be necessary to parameterize certain types of physics if the trajectories of interest are in the order 1 cm upper layer of the ocean (or smaller: a typical oil "slick" may be on the order of microns thick). We emphasize that near the ocean's surface, the vertical location of oil makes a big difference; even when oil has a surface expression, large amounts of oil may circulate beneath the upper centimeter. The vertical location of oil may be determined by oil's density (it can be heavier than water) or due to a dynamic balance between entrainment, vertical mixing, and rise due to buoyancy (Nordam et al., 2021).

Only recently have adequate observations resulted in insight into the movement within the upper layer of the ocean. It is therefore timely to review how this information is relevant for oil spill modelers, as it suggests possibilities that were not typically considered in the past. We include the Point Wells spill in the Salish Sea as a recent example where this type of physics was used to explain the oil's trajectory after remaining a mystery since 2003. Cross-shelf transport is crucial for oil-spill modeling because it is needed for oil to beach, and beaching is one of the more consequential events during oil spills. A large fraction of ocean currents is in approximate geostrophic balance, strongly constraining flow in its ability to cross isobaths (Brink, 2016). Consequently, Lagrangian transport near the coast tends to move parallel to the coastline (more precisely along geostrophic contours) and is limited in its ability to move perpendicular to the coast (LaCasce, 2008). The processes capable of causing cross-shelf transport are eddying activity, ageostrophic processes such as Ekman transport, Stokes drift, and windage. We also present recent evidence that windage and Stokes drift are important because they are effective in driving large-scale beaching.

3.4.1 Windage

For some time now, it has been noted that as the wind increased in magnitude, the effect it had on motion near the sea surface increased. Recently, Lodise, Özgökmen, Griffa, and Berta (2019) used data from one of the largest Lagrangian experiments to date to show that undrogued drifters sampling the upper 5 cm of the ocean move with a velocity that is 3.4%–6.0% of the wind under strong wind (12–20 m/s) conditions, with a deflection to the right of wind direction increasing from 5 to 55 degrees, as wind increased from 12 to 20 m/s. For drogued drifters sampling the upper 60 cm of the ocean, the angle of deflection increased with increasing wind (12–20 m/s) from 30 to 85 degrees, and windage ranged between 2.3% and 4.1% of wind speed. In those experiments, an additional velocity component from Stokes drift was found to be about 1.2%–1.6% of the wind for undrogued drifters and about 0.5%–1.2% of the wind for drogued drifters, with a deflection to the left from wind's direction of about 5 degrees. Overall, windage and Stokes drift accounted for about 70% of the total velocity of drogued drifters, and about 80% of the velocity of undrogued drifters. Laxague et al. (2018) use a variety of instruments including different drifters to measure currents in the upper meters of the ocean with an emphasis on the upper centimeters. They find that the velocity in the upper centimeter, about 60 cm/s, is twice the velocity averaged over the upper meter, and four times the velocity averaged over the upper 10 m. Röhrs and Christensen (2015) use two types of drifters: an undrogued drifter to sample the surface layer and a drogued drifter to sample the upper 70 cm. They find that the upper layer is influenced by wind, while subsurface motion has a stronger link to ocean dynamics, the result being that the surface response to wind forcing is distinct from the response 70 cm below. Androulidakis et al. (2018) also show examples of how drogued drifters have significantly different trajectories than undrogued drifters, with the latter heavily influenced by wind.

Often the direction of motion induced by wind is not the same as the wind direction. However, the angle of deflection can vary with several factors including the ocean's stratification, the buoyancy of the particle, latitude, and the magnitude of the wind, making it difficult to predict the direction of windage. Another complication with oil spills is that the windage of the oil changes over time as the oil weathers and is transformed. Similar to the strategy where the eddy diffusion coefficient is adjusted to match the observed spread of oil, it might be necessary to use observations when possible, to adjust the windage coefficient and the angle of deflection to match the observed motion. Some discussion on the diverse range of deflection angles that have been measured at sea can be found in Duran (2016) and windage for different objects can be found in Breivik, Allen, Maisondieu, and Roth (2011) and Maximenko, Hafner, Kamachi, and MacFadyen (2018).

3.4.1.1 The Point Wells oil spill

On midnight of December 30, 2003, almost 5000 gallons (about 110 barrels) of fuel spilled into the Puget Sound after a tank barge accidentally overtopped near Richmond Beach in Shoreline, Washington. Helicopter overflights early in the morning observed that the surface expression of the spill drifted south, yet oil-spill models forecasted northward movement. The temporal and spatial extent of the spill was small enough (about a day and a 15 km trajectory) for overflights to suffice supporting response efforts. However, the reason why typical oil-spill model forcing, such as ocean currents and 1%–3% of the wind, could not explain the surface-oil trajectory remained a mystery. This was particularly puzzling in an enclosed sea where predictable tides are an important component of the circulation.

Published work where windage reaches 6% of the wind is unusual in the oil spill modeling community, although not unprecedented. In Duran, Romeo, et al. (2018), it was conjectured that oil's motion during the Point Wells oil spill was driven by a combination of 6% of the wind with a 9-degree deflection to the right of wind's direction and ocean currents. This hypothesis explained the trajectory of the spill that had previously been a mystery, although the use of such a high windage coefficient was unusual. It was not until a year later that measurements of motion in a fine upper layer of the ocean were published by Lodise et al. (2019), documenting motion dominated by windage at 6% of the wind speed.

Hindcasts of the Point Wells spill advecting oil only with wind from a meteorological station in the vicinity of the spill were suggestive for two reasons: (1) excellent agreement with the observed trajectory in the first 6 hours and (2) it forced oil toward the south, an observed-trajectory feature that had been difficult to emulate with ocean currents and typical windage (Fig. 3.2). When 6% of the wind from a meteorological station and ocean currents from an ocean model that replicated tides with high skill were combined to force the trajectory computation, the resulting trajectory matched the correct locations at the correct times throughout the spill, finally beaching at the correct location in the afternoon of December 30, 2003. A full account of the numerical experiments can be found in Duran, Romeo, et al. (2018).

3.4.2 Stokes drift

Stokes drift is a net drift in the direction of wave propagation caused by the asymmetrical orbital motion of particles near the surface induced by passing waves. Some authors convey the idea that Stokes drift is canceled in the mean due to the Coriolis effect. However, there is a large body of evidence suggesting that cancellation in the near-surface is negligible in the presence of turbulence induced by wind stress, which is the typical condition in the ocean (e.g., Clarke & Van Gorder, 2018). Stoke's drift was an important driver of oil during the Deepwater Horizon: it was responsible for the observed beaching patterns and it is believed to have avoided oil being entrained by the Loop Current (Carratelli, Dentale, & Reale, 2011; Le Hénaff et al., 2012; Weisberg, Lianyuan, & Liu, 2017). This is consistent with other studies reporting that Stokes drift can exceed the Eulerian mean in the cross-shelf direction (e.g., Monismith & Fong, 2004).

Stokes drift is mainly driven by high-frequency waves, that is, waves forced by local wind rather than remote swell (Clarke & Van Gorder, 2018; D'Asaro, 2014). Using many years of hourly concurrent wind and directional wave spectra from buoys in the Gulf of Mexico and the Pacific, Clarke and Van Gorder (2018) derived a simple formula with which Stokes drift can be parameterized directly from local wind, with good accuracy (within about 1 cm/s from the average Stokes drift) for common wind speeds (between 1 and 50 m/s):

$$u_{\mathrm{Stokes}} = 4.4u_* \ln\left(0.0074U_{10}/u_*\right) \tag{3.7}$$

where u_{Stokes} is the magnitude of Stokes drift, U_{10} is the wind speed 10 m above sea level and $u_* = \sqrt{\tau/\rho_0}$ is the frictional velocity, the square root of wind-stress magnitude divided by a reference seawater density. The direction of Stokes drift is given by the unit vector in the direction of the wind. This is good news because local wind data is often available, whether from meteorological stations or operational models. Further good news is that this result holds even in the presence of swell.

Both Clarke and Van Gorder (2018) and Onink, Wichmann, Delandmeter, and van Sebille (2019) note that an additional contribution to transport at the surface, also in the direction of the wind, may be necessary due to wave breaking. It is also possible that swell may induce surface transport near the coastline, as waves become increasingly nonlinear due to interactions with the bottom. Deeper down within the water column, Stokes drift from internal waves at the pycnocline has also been observed to be an effective driver of oil transport (Shanks, 1987).

3.4.3 Horizontal organization induced by vertical motion

How tracers respond in the upper ocean when they sample velocity structures with influential vertical motion along their path is an active topic of research. It is receiving considerable attention as new observational tools and experiments allow measuring smaller-scale processes, while higher numerical-model resolutions become accessible. Considerable progress was made when such technological advances coincided with funding that became available following the Deepwater Horizon accident.

Identification of submesoscale structures, such as fronts and Langmuir cells, can be important because actionable countermeasures while responding to an oil spill require oil to reach a certain thickness. Intense convergence of oil along water-mass boundaries may, therefore, create an ideal location for mitigation strategies when conditions are right (Gula, Molemaker, & McWilliams, 2014). It has often been observed during oil spills that oil collects in windrows formed by Langmuir circulation as well as at convergences associated with fresh water at river mouths. Water-mass subduction forecasting and detection are therefore suggested as an aid to identifying regions of thick oil. It should also be cautioned that using divergence as a diagnostic to identify regions of accumulation can lead to false positives and negatives (Serra et al., 2020). Clustering may happen in a region of positive Eulerian velocity divergence; we present an example in Section 3.5.1. Another potential caveat is the effect of strong wind acting directly on buoyant oil. The experiment in Romero et al. (2019) suggests that wind order 10 m/s does not impede strong vertical motion at fronts, although the tracer in their experiment was neutrally buoyant, so may not reflect the behavior of a positively buoyant tracer such as oil.

Types of vertical motion that are known to affect the horizontal distribution of oil are related to ocean fronts, filaments, and Langmuir circulation. When two water masses meet, a front is formed along the boundary between the two. Whether one water type sinks under the other because it is heavier, or because of cabbeling, fronts in the ocean tend to be accompanied by water subduction. A vertical circulation due to similar reasons also forms along the boundaries of filaments (McWilliams, 2017). Thus, a downward vertical velocity is induced at the boundaries of water masses, which implies convergence in the horizontal plane. Frontal regions are characterized by relative vorticity and negative divergence that can be several times greater than planetary vorticity. This can have a profound local effect on transport, collapsing floating material to essentially a point. An example of drifters originally spanning a width of about 10 km, collapsing to 60 m, can be found in D'Asaro et al. (2018). A comparison between two- and three-dimensional circulation at a scale of about 100 m, illustrating the distribution of a neutrally buoyant tracer due to vertical motion, can be found in Romero et al. (2019), showing the tracer sinking relatively rapidly. If the tracer were buoyant, as often is oil, an agglomeration of tracer could be expected at the surface. Androulidakis et al. (2018) found that the front-induced circulation dominated the trajectories of undrogued drifters, even under considerable wind, although wind may modulate their speed along the front. Wind is also one of the factors determining the location of the front. Fronts are also of interest because they may serve as horizontal transport barriers (Androulidakis et al., 2018).

Langmuir circulation in its most basic form results from the interaction of Stokes drift induced by surface waves, and the vertical shear induced by the turbulent transfer of momentum from wind to the upper ocean (Sullivan, Romero, Mcwilliams, & Kendall Melville, 2012; Thorpe, 2004). The book by Bühler (2014) describes how a mean flow is induced by an instability (Craik-Leibovich instability), a mechanism that turns out to be robust and therefore explains why Langmuir cells are ubiquitous in the ocean. As with the circulation associated with fronts, Langmuir circulation also has an important vertical component, that likewise concentrates oil into bands within minutes to hours, typically at scales of meters, to hundreds of meters (Chang et al., 2019; D'Asaro, 2000; Simecek-Beatty & Lehr, 2017). Convergence due to frontal circulation may dominate convergence due to Langmuir circulation (Romero et al., 2019). Changes in the vertical location of oil droplets (e.g., McWilliams & Sullivan, 2000) induced by Langmuir circulation enhances the dispersion of oil by subjecting droplets to different ocean currents, as determined by vertical shear (Thorpe, 2004). Also, at least sometimes, Langmuir circulation may be a more important part of ocean dynamics than previously thought (D'Asaro, 2014). For example, the newest ocean climate models—designed to study climate change—now parameterize the effect of Langmuir mixing at the ocean's surface. Evidence that Langmuir circulation may induce a large-scale coastal circulation can be found in Kukulka, Plueddemann, and Sullivan (2012).

In the larger picture, the spatial scales of intense subduction structures (hundreds of meters to a few kilometers) imply that they are likely to be embedded within larger (> 50–100 km) mesoscale structures that advect the smaller structures and therefore determine their location (Androulidakis et al., 2018; D'Asaro et al., 2018; Jacobs et al., 2014; McWilliams, 2019). This suggests that the horizontal motion might still be dominated by larger-scale features, although the local organization might be strongly influenced by the smaller scales. However, separating the length scales of oceanographic processes driving motion is not a trivial endeavor (e.g., Beron-Vera & LaCasce, 2016; Essink et al., 2019).

Because regions of subduction (fronts, filaments, and Langmuir cells) are ubiquitous in the ocean, it is suggested that effective oil-spill planning and response should study how to incorporate the associated material clustering in near real time. Satellite and other remotely sensed data, such as sea-surface temperature sensed by an airplane or an UAV (Unmanned Aerial Vehicle, or drone), may be accurate and relatively inexpensive means of identifying regions of subduction during response operations, complementing the information available from HF radar and numerical ocean models. Sea-surface velocity measured from shipboard X-Band radar seems to be a promising way to identify the strength of the velocity divergence (e.g., Lund et al., 2018), although the reach of radar measurements shown in studies so far (less than 10 km) may be small for an appropriate sampling of the more relevant Lagrangian quantity of along-path divergence. Further research will be needed to understand the interaction between confluence unrelated to divergence, wind, and buoyant material accumulation related to vertical motion.

3.5 Modern Lagrangian tools

The regions where the separation of initially close trajectories and the attraction of initially separate trajectories occur are regions of special interest when studying horizontal motion. It is these regions that have an exceptional influence on the movement of nearby parcels and thereby play a leading role in organizing the flow into identifiable and predictable patterns. In a two-dimensional flow, these regions are hyperbolic lines, and in a three-dimensional flow, they are surfaces. A rigorous approach to detecting these regions has been developed by identifying regions with maximal normal attraction, typically referred to as hyperbolic LCS (Farazmand & Haller, 2012; Haller, 2015).

LCS theory builds on the concept of a flow map, a function that maps every initial ($t = t_0$) position within a domain of interest $\mathbf{x}_0 \in U$, to its current ($t = t_1$) position $\mathbf{x}(t_1)$; that is, $\mathbf{F}_{t_0}^{t_1}(\mathbf{x}_0) := \mathbf{x}(t_1 : \mathbf{x}_0, t_0)$. The Jacobian of the flow map $D\mathbf{F}_{t_0}^{t_1}$ is given by

$$D\mathbf{F}_{t_0}^{t_1}(\mathbf{x}_0) = \begin{pmatrix} \dfrac{\partial x}{\partial x_0} & \dfrac{\partial x}{\partial y_0} \\ \dfrac{\partial y}{\partial x_0} & \dfrac{\partial y}{\partial y_0} \end{pmatrix}. \tag{3.8}$$

Informally, the Jacobian of the flow map (3.8) can be used to map trajectory perturbations from one time to another, and this linear approximation can then be used to optimize quantities of interest. For example, normal attraction of nearby fluid parcels along a trajectory over a time interval can be maximized with respect to perturbations of the initial-time normal vector. This is the strategy used to find hyperbolic LCS, trajectories characterized by maximal normal attraction, or repulsion. Working out the math for this optimization problem—a formal account of which can be found in Haller (2015) and references therein—the Cauchy–Green (CG) strain tensor arises naturally. The CG tensor is defined as

$$\mathbf{C}_{t_0}^{t_1}(\mathbf{x}_0) := \left[D\mathbf{F}_{t_0}^{t_1}(\mathbf{x}_0)\right]^{\mathrm{T}} D\mathbf{F}_{t_0}^{t_1}(\mathbf{x}_0). \tag{3.9}$$

In particular, the eigenvalues $0 < \lambda_1(\boldsymbol{x}_0) < \lambda_2(\boldsymbol{x}_0)$ and normalized eigenvectors $\hat{\xi}_1(\boldsymbol{x}_0) \perp \hat{\xi}_2(\boldsymbol{x}_0)$ of (3.9) are used to set up ordinary differential equations from which hyperbolic, elliptic, and parabolic LCS can be found (Haller, 2015). Thus, the CG tensor is central in LCS theory. Note that to obtain the CG tensor, one must integrate the velocity; we return to the CG tensor in Section 3.5.1.

The mathematical formality behind LCS has proven a versatile approach to understanding Lagrangian motion. Hyperbolic LCS will accurately identify how fluid will deform (i.e., along attracting hyperbolic LCS), anticipating the more influential transport patterns. However, the final results ultimately depend on the accuracy of the velocity field, which is what induces Lagrangian transport in the first place. If a velocity field is relatively accurate while having localized errors, trajectories traversing the time and location of errors in the velocity are likely to give wrong results relative to observed trajectories. LCS will be negatively affected by those velocity errors as well, correctly identifying transport induced by the velocity, yet remaining incorrect relative to observed transport. The computation of trajectories propagates localized velocity errors, often resulting in trajectories that are incorrect relative to observations when hindcasting or forecasting the transport of oil. The need to integrate the velocity field limits the suitability of observational velocity data sets and of ocean models that assimilate such data, for Lagrangian transport purposes.

Velocity products in our time can be relatively accurate thanks to accurate measurements over wide areas, with satellite altimetry being particularly relevant because of good global coverage and because it captures what is often an important part of the velocity at the sea surface (Section 3.3.1.3). However, the coarse temporal and spatial resolution of altimetry means that the resulting velocity will almost certainly have deficiencies. Ocean models assimilating data inherit these deficiencies and have shortcomings of their own. Given the chaotic sensitivity of trajectories, it has therefore been a natural development to try to bypass the sensitivity resulting from localized velocity errors.

Another complication with hyperbolic LCS from an applied point of view is that there is a timescale T involved in the computation. When computing (3.8), a choice must be made for the initial time t_0 and the final time $t_1 = t_0 + T$. The choice for T and sometimes even the choice for t_0 are often not clear a priori, forcing subjective choices. Since LCS are material lines moving with turbulent flow, these choices can result in big differences. In Section 3.5.1, we describe a way to bypass sensitivity to the velocity field and the need to choose T.

3.5.1 Eulerian Coherent Structures

The fundamental equation translating from the Eulerian and Lagrangian characterizations of fluid flow is (3.4), an equivalence between the velocity of a parcel traversing a trajectory $\mathbf{x}(t)$, and the Eulerian velocity $\mathbf{u}$, at the parcel's time and location. Based on this instantaneous correspondence between Lagrangian and Eulerian descriptions of fluid flow, it is natural to search for an Eulerian counterpart to the CG tensor, seeking to describe fluid deformation near the instantaneous limit. As mentioned in Section 3.5, the CG tensor is central to finding hyperbolic LCS, trajectories that maximize normal attraction, thereby maximizing the influence on nearby water parcels and thus organizing flow. Serra and Haller (2016) developed OECS, including attracting hyperbolic OECS, they showed that the Taylor expansion of the CG tensor with respect to time is given in terms of the strain-rate tensor $\mathbf{S}$:

$$\mathbf{C}_{t_0}^{t_1}(\mathbf{x}_0) = \mathbf{I} + 2\mathbf{S}(\mathbf{x}_0, t_0)(t_1 - t_0) + O\left(|t_1 - t_0|^2\right).$$

This means that for time close enough to t_0, Lagrangian deformation is approximated by the Eulerian strain-rate tensor, the ijth entry of which is given by $\frac{1}{2}\left(\partial u_i/\partial u_j + \partial u_j/\partial x_i\right)$. The strain-rate tensor has eigenvalues s_1, s_2 with corresponding eigenvectors $\mathbf{e}_1$ and $\mathbf{e}_2$. Attracting hyperbolic OECS are tangent to $\mathbf{e}_2$, their cores given by minima in the eigenvalue s_1, which is the rate of change of the length of the normal eigenvector $\mathbf{e}_1$ due to the deformation induced by the flow; equivalently, s_1 is the strength of attraction normal to $\mathbf{e}_2$. Negative values of s_1 mean that there is attraction normal to $\mathbf{e}_2$, the more negative the stronger the attraction. Thus, Serra and Haller (2016) extended the theory of LCS from finite time to their instantaneous limit in terms of an Eulerian quantity, where there is no longer a need to integrate the velocity field, bypassing the attendant sensitivity.

The strain-rate tensor is objective; that is, the results from computing Eulerian Coherent Structures are frame invariant (Haller, 2015; Serra & Haller, 2016). This means that changes of reference frames characterized by time-dependent rotations and translations will not affect the results. This is important because nonobjective methods might give different results under coordinate transformations; for example, Eulerian velocity hyperbolic points are not Galilean invariant (Serra & Haller, 2016).

Serra et al. (2020) use attracting hyperbolic OECS, which they call TRAPs (TRansient Attracting Profiles), to demonstrate that, in a series of experiments with satellite-tracked drifters and Search & Rescue Training Manikins, TRAPs organize flow and perform better than trajectory computations in predicting drifter locations. They use a carefully calibrated HF radar velocity and a data-assimilating model similar to what the US Coast Guard would use for search and rescue operations. Similar to hyperbolic LCS, TRAPs are lines in a two-dimensional flow; they have a core that is where normal attraction is maximal, with attraction strength decaying along the rest of the TRAP. We present an example of satellite-tracked drifters converging to TRAPs computed from HF radar velocity in Martha's Vineyard in Massachusetts (Fig. 3.3). In this example, there is confluence of drifters at a TRAP where the Eulerian horizontal velocity divergence is positive. A description of these data, how TRAPs organize Lagrangian motion, and how TRAPs can be used for search and rescue operations, can be found in Serra et al. (2020). In Section 3.5.2, we present an example where TRAPs can predict the movement of oil at least 8 days in advance, while trajectories diverge from the observed transport due to a likely erroneous hyperbolic point in the velocity.

3.5.2 Revisiting the Deepwater Horizon with modern tools

The difficulty of simulating Lagrangian transport can be easily experienced by trying to replicate observed trajectories. During the Deepwater Horizon, at least six different ocean models were used in an attempt to forecast the location of oil to provide critical information for response and planning (Liu,

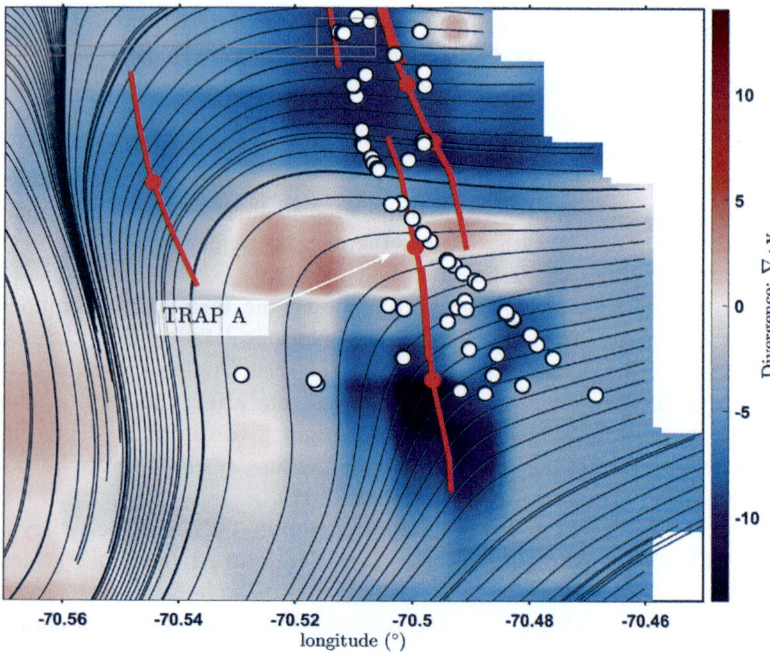

FIGURE 3.3 TRansient Attracting Profile (TRAPs) (red lines) and their cores (red circles) computed from high-frequency radar velocity off Martha's Vineyard, MA, plotted over the velocity divergence (color contours; day^{-1}) with satellite-tracked drifters (white dots) converging to TRAPs. Black lines are streamlines. TRAP A is in a region of positive horizontal velocity divergence. TRAPs remain invisible to divergence fields and streamlines.

Weisberg, Hu, & Zheng, 2011). However, there was enough intermodel variability that ensemble averaging was recommended to produce a forecast that was more likely to occur. Even then, forecasts were limited to 2 days due to forecast error growth. We note that these were ideal conditions as often there are not that many ocean models available for ensemble averaging.

Here we present a different method that seeks to bypass the sensitivity that causes error growth, using TRAPs that are computed from instantaneous snapshots of an Eulerian velocity. We show that an analysis combining TRAPs and LCS is enough to (1) accurately forecast the observed movement of oil at least 8 days in advance and (2) understand why the simulated Lagrangian transport does not conform to observations. In this example, forecasts only depend on a previous-day single velocity snapshot, and the LCS are computed from only past information to complement the information obtained from TRAPs and pinpoint the source of error in the velocity field, and its Lagrangian manifestation.

A variety of velocity products from altimetry are available, some of them including an Ekman component. The product that we use here is a daily velocity by GEKCO2 (Geostrophic and EKman Current Observatory; Sudre et al., 2013), from satellite altimetry and wind. We confirm our results using a daily instantaneous velocity from HyCOM Global at about 9 km resolution in the Gulf of Mexico, the current US Navy Operational model (Burnett, Harper, Preller, Jacobs, & LaCroix, 2014) that assimilates a variety of observations using the Navy Coupled Ocean Data Assimilation (NCODA) and that is forced with the NAVy Global Environmental Model (NAVGEM). Transport simulated with HyCOM Gulf of Mexico, which has a similar setup as HyCOM Global but at a 4 km resolution, also produces similar simulated transport as the other velocity products used in this experiment.

We analyze daily forecasts of the movement of oil during the Deepwater Horizon accident by comparing TRAPs to the observed outline of oil and the advection of oil obtained by integrating the velocity field, that is, computing trajectories between May 11 and 17, 2010, that are initiated at the location of oil observed on May 10, 2010. TRAPs are computed from snapshots of the velocity on previous days, thus providing forecasts within this hindcast exercise.

The forecast on May 10 shows a weak TRAP (near 28.2N, 88.3W) suggesting slight oil movement toward the southwest, coinciding with the outline of oil observed on May 11, and with trajectories computed between May 10 and 11, although the simulated oil and the TRAP are slightly offset from the observed oil (Fig. 3.4). The strength of attraction of the TRAP is low (about 0.3 day^{-1}), accurately forecasting slight oil movement. The only other TRAP core in contact with the observed oil (near 29.2N, 88W) is the core of a TRAP almost entirely contained within the observed oil on May 11 and, therefore, it cannot be expected to cause a significant rearrangement of oil outside of the observed oil outline. TRAPs in the southeast section of

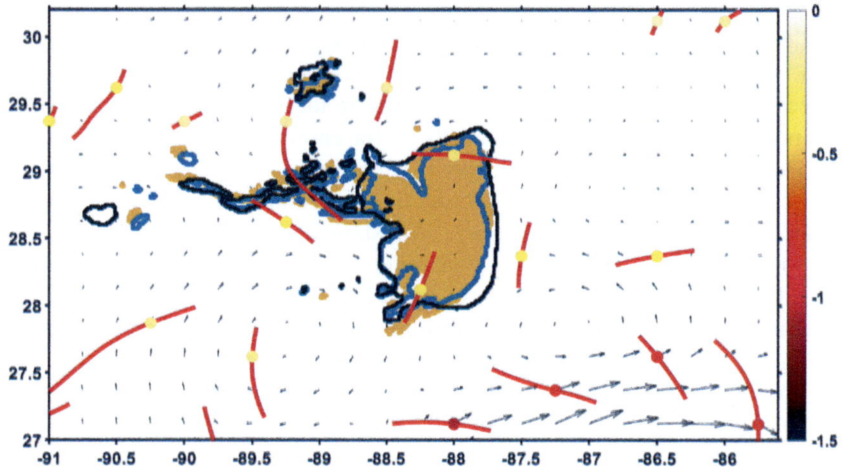

FIGURE 3.4 The blue line is the outline of oil as observed from satellites on May 10, 2010, used as initial conditions for trajectory computations. In orange are the final positions (May 11) of trajectories initiated within the blue outline on May 10, computed by integrating Geostrophic and EKman Current Observatory (GEKCO2) velocity. The black line is the outline of oil as observed on May 11, 2010. Black vectors are the velocity from GEKCO2 on May 10, and the red lines are TRansient Attracting Profile (TRAP) computed with the velocity on May 10, TRAP cores (colored circles) are colored according to attraction strength (day^{-1}; color scale on the right).

Fig. 3.4 have higher strengths of attraction of about 1 day^{-1} but are still relatively far from the oil.

By May 13, oil trajectories continued along the path toward the southwest, the weak TRAP computed from the May 12 velocity accurately forecasting that path (Fig. 3.5). This will be the last day this weakly attracting TRAP appears near 28.2N, 88.3W. The TRAPs with strongest attraction computed with the May 10 velocity remain when TRAPs are computed with the May 12 velocity (Figs. 3.4 and 3.5), these are the TRAPs in the southeast of our domain, with attraction three to four times stronger than TRAPs directly interacting with oil.

By May 15, observed oil has aligned with one of the strongest TRAPs, the one that has remained near 27.4N and 87.25W since May 10. Meanwhile, the simulated oil trajectory continues its original path toward the southwest, by now clearly diverging from the observed path (Fig. 3.6). The weak TRAP originally indicating the path toward the southwest (near 28.2N, 88.3W in Figs. 3.4 and 3.5) is no longer present in the May 14 velocity and will not be seen again during the rest of our analysis.

By May 17, observed oil has deformed toward the south then east, while the simulated oil trajectory has deformed toward the south then west, thus the observed and simulated trajectories are heading in opposite directions

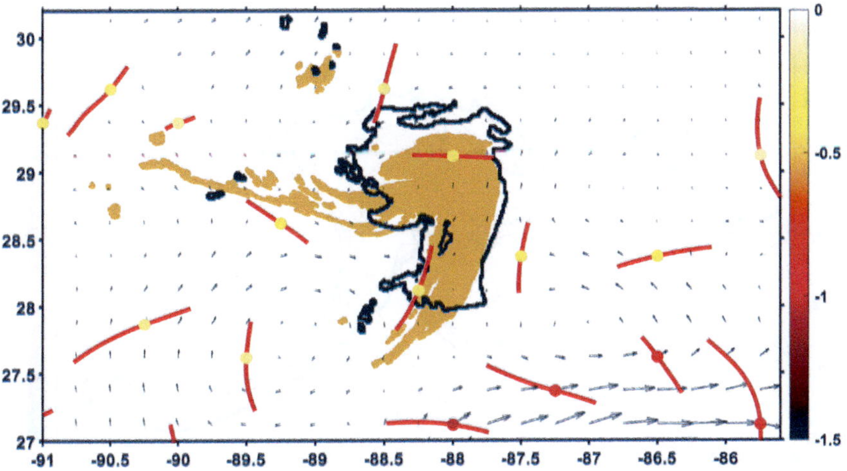

FIGURE 3.5 Same as in Fig. 3.4 but on May 13, 2010.

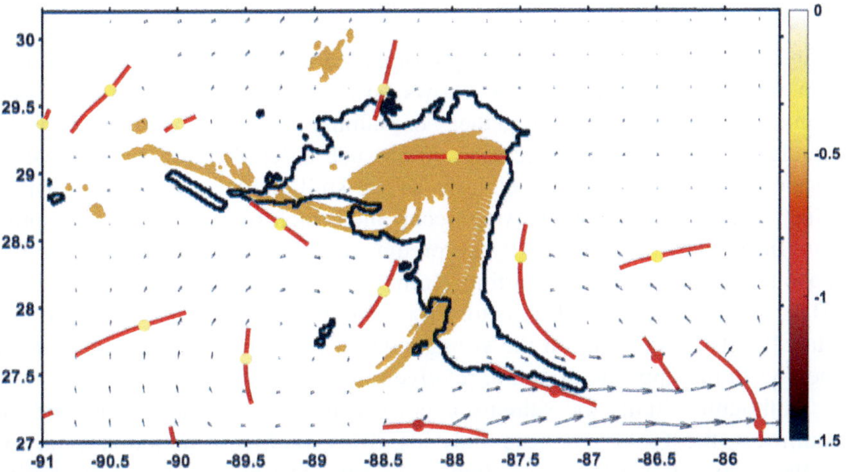

FIGURE 3.6 Same as in Fig. 3.4 but on May 15, 2010.

(Fig. 3.7). LCS illustrate a hyperbolic point near 28.4N and 87.7W where transport splits, the western part heading toward the south then west (simulated oil follows this LCS) and the eastern part moving south then east (observed oil follows this LCS). Thus, LCS show that the simulated tracer just barely missed the observed transport pattern that is accurately depicted by the LCS on the eastern side of the hyperbolic point. Note there is a TRAP above the LCS on the eastern side of the hyperbolic point near 28.4N and 87.7W, accurately selecting the altimetric LCS that agrees with the path of observed oil transport. Further confirmation comes from the strong TRAP

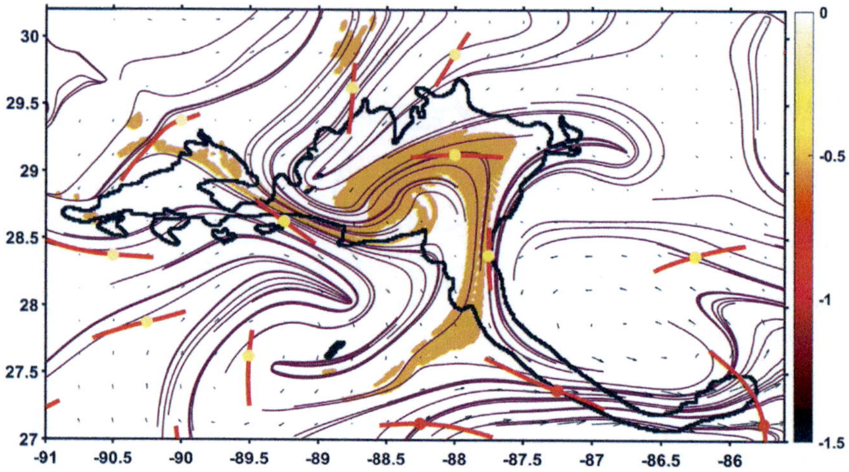

FIGURE 3.7 Same as in Fig. 3.4 but on May 17, 2010; purple lines are attracting Lagrangian Coherent Structures computed back in time between May 17 and May 10.

near 27.4N and 87.25W forecasting elongation of oil along the correct direction since at least May 10. This shows how TRAPs and LCS provide complementary information, together explaining the discrepancy between simulated and observed trajectories. With such a detailed Lagrangian characterization of the available velocity, an oil-spill modeler can then identify which patterns are most likely to occur and which patterns are likely spurious. In this example, a TRAP can predict the observed movement of oil at least 8 days in advance, starting from the velocity snapshot on May 10, 2010, while the simulated trajectory for oil initiated from the observed oil on May 10, 2010 is caught on the wrong side of a hyperbolic point and ends up moving in the opposite direction relative to observed transport.

Although initially the velocity is correct in inducing transport toward the southwest, the TRAP that accurately identifies southwest motion is weak and it disappears after a few days (Figs. 3.4–3.6). The hyperbolic point causing the divergence of simulated transport relative to observed transport (Fig. 3.7) therefore seems to originate from a disparity in the velocity arising from coarse temporal resolution and resulting in error accumulation during Lagrangian integration. The problem may be related to the period between passes of altimetry satellites being too long to capture changes in the ocean velocity on timescales of a few days and the attendant influence on trajectories when integrating the velocity. Fortunately, velocity from altimetry accurately captures the features that result in the main transport patterns; it is just that velocity integration is not an adequate tool to extract this information.

The above results are from the GEKCO2 velocity; very similar results are obtained using HyCOM Global (not shown). Simulated transport is also very similar when using HyCOM GoM (not shown). The computation of TRAPs for flows at high resolutions (about 4 km or less in the ocean) may require filtering and is a topic of current research. Global GEKCO2 velocity is available since 1993 to date minus 2 days.

3.6 Conclusion and outlook

Recent advances suggest that better results for oil-spill modelers are a reachable goal. In this chapter, we have shown examples of how a basic understanding of the physics driving motion in the sea and the use of novel Lagrangian and Eulerian Coherent Structures techniques can result in improved oil-spill modeling.

Recent progress in Coherent Structures techniques—computing attracting structures that shape material transport from an Eulerian snapshot—is a promising development for oil-spill modeling efforts. We have shown how TRAPs (or attracting OECSs) can bypass errors in the velocity that produce large errors in simulated trajectories. Olascoaga and Haller (2012) explored similar ideas by searching for the most persistent Lagrangian hyperbolic cores using 15-day integrations. Lagrangian integration over such a period acts as a filter, removing short-term variability and focusing on finite-time mesoscale features. Our results are consistent with theirs: a highly attractive hyperbolic core persists for over a week, accurately anticipating prominent fluid deformation. The advantages of TRAPs are that they do not require velocity integration, they can be computed from a single velocity snapshot, and they predict hyperbolic attraction cores whether persistent or ephemeral.

As described in Section 3.5.2, TRAPs were able to identify the correct transport patterns, while LCS and simulated trajectories were influenced by a deficient velocity. The erroneous simulated transport is initially correct as evidenced by the observed movement of oil, aptly identified in the velocity by a weak, ephemeral TRAP. However, as time advances, simulated trajectories become erroneous when integration causes velocity-error accumulation, while a strong persistent TRAP marked the correct region of oil confluence well in advance of observed deformation.

Higher resolution observations of sea-surface velocity and surface processes are expected to advance our understanding, ultimately resulting in improved velocity products. For example, high-resolution observations, theoretical modeling, and coupled ocean-atmosphere-wave numerical models can be expected to improve our understanding of the ocean's surface (Villas Bôas et al., 2019). Improvements in observations and understanding should translate to improved oil transport forecasts. As velocity products improve by including more of the physics relevant to simulating oil's movement, the techniques highlighted here will become more relevant. A basic understanding of ocean physics will continue to be needed to supplement velocity products lacking certain types of forcing, but also to understand new velocity products that will incorporate more types of physics than previously available.

Despite improvements in velocity products, the sensitivity of trajectory computations to small errors will likely continue to produce erroneous trajectories. This suggests that the novel techniques that seek to bypass the sensitivity inherent to trajectory computations are likely to become important tools for the oil-spill modeler. Thus, as an effort that is parallel to improving velocity products, progress in techniques bypassing the problems inherent to the unstable nature of ocean currents can be expected. Examples include OECS for instantaneous transport patterns, and climatological LCS for climatological transport patterns. The latter is an empirical approach developed in Duran, Beron-Vera, et al. (2018) where it was found that filtering the velocity by computing a climatology is surprisingly accurate for identifying recurrent Lagrangian transport patterns if the proper Lagrangian tools are used. Among their results, the transport pattern studied in Section 3.5.2 turns out to be recurrent, and therefore a pattern that is likely to be seen in May through August of any given year. A climatological approach should not replace forecasts, yet it does provide a valuable general understanding of persistent transport barriers, trajectories, regions of persistent attraction, and persistent isolation. Thus, Lagrangian climatologies in the sense of Duran, Beron-Vera, et al. (2018) complement the interpretation of forecasts while providing a broad understanding of Lagrangian motion in a region of interest. The climatological approach suggests that progress can be made by understanding the connection between the inherently time-dependent trajectories of an instantaneous ocean velocity and a low-pass filtered climatological velocity. An alternative approach to understanding uncertainty in oil-spill modeling is the use of ensemble simulations to create a surrogate model (Zhang et al., 2020). Future work might be able to bridge the Lagrangian climatology strategy of Duran, Beron-Vera, et al. (2018) with the ensemble-based surrogate model of Zhang et al. (2020).

As ocean observations and ocean models improve with new satellite products, an increasing number of HF radars, drifters and autonomous vehicles, and advances in data processing, assimilation, and numerical modeling, oil-spill modelers should be able to capitalize from the material presented here, achieving a higher rate of success in forecasting and hindcasting the movement of oil.

Accurately simulating Lagrangian transport in the ocean is of considerable societal interest for a variety of reasons including oil spills, the fate of other contaminants, fisheries, ocean ecology, search and rescue, tracing accidents or crimes back in time (forensic work), climate change, weather predictions, and more. Many countries have conducted oceanographic research for several decades now. Consequently, enough progress has been made to where simulating trajectories in the ocean often produces valuable information. For the needed progress to continue, we must understand the ocean's importance for society at large and that the relevance of oceanographic endeavor is increasing due to pressing issues including climate change, coastal development, population growth, and globalization.

Acknowledgments

The GEKCO2 product used in this study was developed and extracted by Joël Sudre et al. at LEGOS, France. GEKCO2 data can be requested at http://www.legos.obs-mip.fr/members/sudre/gekco_form. Funding for the development of HyCOM has been provided by the National Ocean Partnership Program and the Office of Naval Research. Data assimilative products using HyCOM are funded by the US Navy. Computer time was made available by the DoD High Performance Computing Modernization Program. The output is publicly available at https://hycom.org. Deepwater Horizon oil products from NOAA are available at http://www.ssd.noaa.gov/PS/MPS/deepwater.html. RD would like to thank M. J. Olascoaga for helpful conversations.

The work of RD was performed in support of the US Department of Energy's Fossil Energy, Oil and Natural Gas Research Program. It was executed by NETL's Research and Innovation Center, including work performed by Leidos Research Support Team staff under the RSS contract 89243318CFE000003. M.S. acknowledges support from the Schmidt Science Fellowship and the Postdoc Mobility Fellowship from the Swiss National Foundation.

This work was funded by the Department of Energy, National Energy Technology Laboratory, an agency of the US Government, through a support contract with Leidos Research Support Team (LRST). Neither the US Government nor any agency thereof, any of their employees, LRST, any of their employees, makes any warranty, expressed or implied, or assumes any legal liability or responsibility for the accuracy, completeness, or usefulness of any information, apparatus, product, or process disclosed, or represents that its use would not infringe privately owned rights. Reference herein to any specific commercial product, process, or service by trade name, trademark, manufacturer, or otherwise, does not necessarily constitute or imply its endorsement, recommendation, or favoring by the US Government or any agency thereof. The views and opinions of authors expressed herein do not necessarily state or reflect those of the US Government or any agency thereof.

References

Androulidakis, Y., Kourafalou, V., Özgökmen, T., Garcia-Pineda, O., Lund, B., Le Hénaff, M., ... Horstmann, J. (2018). Influence of river-induced fronts on hydrocarbon transport: A multiplatform observational study. *Journal of Geophysical Research: Oceans*. Available from https://doi.org/10.1029/2017JC013514.

Beron-Vera, F. J., & LaCasce, J. H. (2016). Statistics of simulated and observed pair separations in the Gulf of Mexico. *Journal of Physical Oceanography*. Available from https://doi.org/10.1175/JPO-D-15-0127.1, arXiv:1505.03475.

Beron-Vera, F. J., Olascoaga, M. J., Wang, Y., Triñanes, J., & Pérez-Brunius, P. (2018). Enduring Lagrangian coherence of a Loop Current ring assessed using independent observations. *Scientific Reports*, *8*, 11275. Available from https://doi.org/10.1038/s41598-018-29582-5.

Beron-Vera, F. J., Wang, Y., Olascoaga, M. J., Goni, G. J., & Haller, G. (2013). Objective detection of oceanic eddies and the Agulhas Leakage. *Journal of Physical Oceanography*. Available from https://doi.org/10.1175/JPO-D-12-0171.1.

Berta, M., Bellomo, L., Magaldi, M. G., Griffa, A., Molcard, A., Marmain, J., ... Taillandier, V. (2014). Estimating Lagrangian transport blending drifters with HF radar data and models: Results from the TOSCA experiment in the Ligurian Current (North Western Mediterranean Sea). *Progress in Oceanography*. Available from https://doi.org/10.1016/j.pocean.2014.08.004.

Berta, M., Griffa, A., Magaldi, M. G., Özgökmen, T. M., Poje, A. C., Haza, A. C., & Josefina Olascoaga, M. (2015). Improved surface velocity and trajectory estimates in the Gulf of Mexico from Blended satellite altimetry and drifter data. *Journal of Atmospheric and Oceanic Technology*. Available from https://doi.org/10.1175/JTECH-D-14-00226.1.

Breivik, Ø., Allen, A. A., Maisondieu, C., & Roth, J. C. (2011). Wind-induced drift of objects at sea: The leeway field method. *Applied Ocean Research*, *33*, 100–109.

Brink, K. (2016). Cross-shelf exchange. *Annual Review of Marine Science*. Available from https://doi.org/10.1146/annurev-marine-010814-015717.

Bühler, O. (2014). Waves and mean flows. *Cambridge monographs on mechanics* (2nd ed.). Cambridge University Press. https://doi.org/10.1017/CBO9781107478701.

Burnett, W., Harper, S., Preller, R., Jacobs, G., & LaCroix, K. (2014). Overview of operational ocean forecasting in the US navy: Past, present, and future. *Oceanography*. Available from https://doi.org/10.5670/oceanog.2014.65.

Carratelli, E. P., Dentale, F., & Reale, F. (2011). *On the effects of wave-induced drift and dispersion in the Deepwater Horizon oil spill. Monitoring and modeling the Deepwater Horizon oil spill: A record breaking enterprise*. Wiley. https://doi.org/10.1029/2011GM001109.

Chang, H., Huntley, H. S., Kirwan, A. D., Carlson, D. F., Mensa, J. A., Mehta, S., ... Poje, A. C. (2019). Small-scale dispersion in the presence of Langmuir circulation. *Journal of Physical Oceanography*. Available from https://doi.org/10.1175/JPO-D-19-0107.1.

Chelton, D. B., Schlax, M. G., Samelson, R. M., Farrar, J. T., Molemaker, M. J., McWilliams, J. C., & Gula, J. (2019). Prospects for future satellite estimation of small-scale variability of ocean surface velocity and vorticity. *Progress in Oceanography*. Available from https://doi.org/10.1016/j.pocean.2018.10.012.

Clarke, A. J., & Van Gorder, S. (2018). The relationship of near-surface flow, Stokes drift and the wind stress. *Journal of Geophysical Research: Oceans*. Available from https://doi.org/10.1029/2018JC014102.

Csanady, G. (1973). *Turbulent diffusion in the environment*. Dordrecht, Holland: D. Reidel Publishing Company.

D'Asaro, E. (2000). Simple suggestions for including vertical physics in oil spill models. *Spill Science & Technology Bulletin*. Available from https://doi.org/10.1016/s1353-2561(01)00039-1.

D'Asaro, E. A. (2014). Turbulence in the upper-ocean mixed layer. *Annual Review of Marine Science*. Available from https://doi.org/10.1146/annurev-marine-010213-135138.

D'Asaro, E. A., Shcherbina, A. Y., Klymak, J. M., Molemaker, J., Novelli, G., Guigand, C. M., ... Özgökmen, T. M. (2018). Ocean convergence and the dispersion of flotsam. *Proceedings of the National Academy of Sciences of the United States of America*. Available from https://doi.org/10.1073/pnas.1718453115.

Dagestad, K.-F., & Röhrs, J. (2019). Prediction of ocean surface trajectories using satellite derived versus modeled ocean currents. *Remote Sensing of Environment*, *223*, 130–142.

Davidson, P. A. (2015). *Turbulence: An introduction for scientists and engineers* (2nd ed.). Oxford, UK: Oxford University Press.

Duran, R. (2016). *Sub-grid parameterizations for oceanic oil-spill simulations.* Technical Report EPAct Technical Report Series. United States Department of Energy, National Energy Technology Laboratory. Available from https://edx.netl.doe.gov/dataset/sub-grid-parameterizations-for-oceanic-oil-spill-simulations.

Duran, R., Beron-Vera, F. J., & Olascoaga, M. J. (2018). Extracting quasi-steady Lagrangian transport patterns from the ocean circulation: An application to the Gulf of Mexico. *Scientific Reports.* Available from https://doi.org/10.1038/s41598-018-23121-y.

Duran, R., Romeo, L., Whiting, J., Vielma, J., Rose, K., Bunn, A., & Bauer, J. (2018). Simulation of the 2003 Foss Barge-Point Wells oil spill: A comparison between BLOSOM and GNOME oil spill models. *Journal of Marine Science and Engineering.* Available from https://doi.org/10.3390/jmse6030104.

Durran, D. R. (2010). *Numerical methods for fluid dynamics—With applications to geophysics.* Springer. https://doi.org/10.1007/978-1-4419-6412-0.

Essink, S., Hormann, V., Centurioni, L. R., & Mahadevan, A. (2019). Can we detect submesoscale motions in drifter pair dispersion? *Journal of Physical Oceanography.* Available from https://doi.org/10.1175/jpo-d-18-0181.1.

Farazmand, M., & Haller, G. (2012). Computing Lagrangian Coherent Structures from their variational theory. *Chaos: An Interdisciplinary Journal of Nonlinear Science, 22*, 013128.

Fox-Kemper, B., Adcroft, A., Böning, C. W., Chassignet, E. P., Curchitser, E., Danabasoglu, G., ... Yeager, S. G. (2019). Challenges and prospects in ocean circulation models. *Frontiers in Marine Science, 6*, 013165. Available from https://doi.org/10.3389/fmars.2019.00065.

Gula, J., Molemaker, M. J., & McWilliams, J. C. (2014). Submesoscale cold filaments in the Gulf Stream. *Journal of Physical Oceanography, 44*, 2617–2643. Available from https://doi.org/10.1175/JPO-D-14-0029.1.

Haller, G. (2015). Langrangian Coherent Structures. *Annual Review of Fluid Mechanics, 47.* Available from https://doi.org/10.1146/annurev-fluid-010313-141322, 140906185740003. URL. Available from http://www.annualreviews.org/doi/abs/10.1146/annurev-fluid-010313-141322.

Haller, G., Karrasch, D., & Kogelbauer, F. (2018). Material barriers to diffusive and stochastic transport. *Proceedings of the National Academy of Sciences of the United States of America.* Available from https://doi.org/10.1073/pnas.1720177115, arXiv:1808.04787.

Jacobs, G. A., Bartels, B. P., Bogucki, D. J., Beron-Vera, F. J., Chen, S. S., Coelho, E. F., ... Wei, M. (2014). Data assimilation considerations for improved ocean predictability during the Gulf of Mexico Grand Lagrangian Deployment (GLAD). *Ocean Modelling.* Available from https://doi.org/10.1016/j.ocemod.2014.09.003.

Jones, C. E., Dagestad, K. F., Breivik, Ø., Holt, B., Röhrs, J., Christensen, K. H., ... Skrunes, S. (2016). Measurement and modeling of oil slick transport. *Journal of Geophysical Research: Oceans.* Available from https://doi.org/10.1002/2016JC012113.

Kim, S., Samelson, R. M., & Snyder, C. (2011). Toward an uncertainty budget for a coastal ocean model. *Monthly Weather Review.* Available from https://doi.org/10.1175/2010MWR3352.1.

Kirincich, A., Emery, B., Washburn, L., & Flament, P. (2019). Improving surface current resolution using direction finding algorithms for multiantenna high-frequency radars. *Journal of Atmospheric and Oceanic Technology.* Available from https://doi.org/10.1175/JTECH-D-19-0029.1.

Kirincich, A. R., De Paolo, T., & Terrill, E. (2012). Improving HF radar estimates of surface currents using signal quality metrics, with application to the MVCO high-resolution radar system. *Journal of Atmospheric and Oceanic Technology.* Available from https://doi.org/10.1175/JTECH-D-11-00160.1.

Kloeden, P. E., & Platen, E. (1992). *Numerical solution of stochastic differential equations.* Berlin: Springer-Verlag.

Kukulka, T., Plueddemann, A. J., & Sullivan, P. P. (2012). Nonlocal transport due to Langmuir circulation in a coastal ocean. *Journal of Geophysical Research: Oceans, 117*. Available from https://doi.org/10.1029/2012JC008340, https://agupubs.onlinelibrary.wiley.com/doi/abs/10.1029/2012JC008340.

LaCasce, J. H. (2008). Statistics from Lagrangian observations. *Progress in Oceanography, 77* (1), 1–29. Available from https://doi.org/10.1016/j.pocean.2008.02.002.

Lawrence, G. A., Ashley, K. I., Yonemitsu, N., & Ellis, J. R. (1995). Natural dispersion in a small lake. *Limnology and Oceanography, 40*, 1519–1526.

Laxague, N. J., Özgökmen, T. M., Haus, B. K., Novelli, G., Shcherbina, A., Sutherland, P., ... Molemaker, J. (2018). Observations of near-surface current shear help describe oceanic oil and plastic transport. *Geophysical Research Letters*. Available from https://doi.org/10.1002/2017GL075891.

Le Hénaff, M., Kourafalou, V. H., Paris, C. B., Helgers, J., Aman, Z. M., Hogan, P. J., & Srinivasan, A. (2012). Surface evolution of the Deepwater Horizon oil spill patch: Combined effects of circulation and wind-induced drift. *Environmental Science and Technology*. Available from https://doi.org/10.1021/es301570w.

Liu, Y., Weisberg, R. H., Hu, C., & Zheng, L. (2011). *Trajectory forecast as a rapid response to the Deepwater Horizon oil spill. Monitoring and modeling the Deepwater Horizon oil spill: A record breaking enterprise* (pp. 153–165). Wiley. Available from https://doi.org/10.1029/2011GM001121.

Liu, Y., Weisberg, R. H., Vignudelli, S., & Mitchum, G. T. (2014). Evaluation of altimetry-derived surface current products using Lagrangian drifter trajectories in the eastern Gulf of Mexico. *Journal of Geophysical Research: Oceans*. Available from https://doi.org/10.1002/2013JC009710.

Lodise, J., Özgökmen, T., Griffa, A., & Berta, M. (2019). Vertical structure of ocean surface currents under high winds from massive arrays of drifters. *Ocean Science*. Available from https://doi.org/10.5194/os-15-1627-2019.

Lund, B., Haus, B. K., Horstmann, J., Graber, H. C., Carrasco, R., Laxague, N. J., ... Özgökmen, T. M. (2018). Near-surface current mapping by shipboard marine X-band radar: A validation. *Journal of Atmospheric and Oceanic Technology*. Available from https://doi.org/10.1175/JTECH-D-17-0154.1.

Lynch, D. R., Greenberg, D. A., Bilgili, A., McGillicuddy, D. J., Jr, Manning, J. P., & Aretxabaleta, A. L. (2014). *Particles in the coastal ocean: Theory and applications*. Cambridge University Press.

Maximenko, N., Hafner, J., Kamachi, M., & MacFadyen, A. (2018). Numerical simulations of debris drift from the Great Japan Tsunami of 2011 and their verification with observational reports. *Marine Pollution Bulletin, 132*, 5–25.

McWilliams, J. C. (2017). Submesoscale surface fronts and filaments: Secondary circulation, buoyancy flux, and frontogenesis. *Journal of Fluid Mechanics, 823*, 391–432. Available from https://doi.org/10.1017/jfm.2017.294.

McWilliams, J. C. (2019). A survey of submesoscale currents. *Geoscience Letters*. Available from https://doi.org/10.1186/s40562-019-0133-3.

McWilliams, J. C., & Sullivan, P. P. (2000). Vertical mixing by Langmuir circulations. *Spill Science and Technology Bulletin*. Available from https://doi.org/10.1016/S1353-2561(01)00041-X.

Monismith, S. G., & Fong, D. A. (2004). A note on the potential transport of scalars and organisms by surface waves. *Limnology and Oceanography*. Available from https://doi.org/10.4319/lo.2004.49.4.1214.

Moum, J. N., & Rippeth, T. P. (2009). Do observations adequately resolve the natural variability of oceanic turbulence? *Journal of Marine Systems*. Available from https://doi.org/10.1016/j.jmarsys.2008.10.013.

Murthy, C. (1976). Horizontal diffusion characteristics in Lake Ontario. *Journal of Physical Oceanography*, *6*, 76–84.

Nordam, T., Brönner, U., Skancke, J., Nepstad, R., Rønningen, P., & Alver, M. (2018). Numerical integration and interpolation in marine pollutant transport modelling. In *Proceedings of the forty-first AMOP technical seminar*.

Nordam, T., & Duran, R. (2020). Numerical integrators for Lagrangian oceanography. *Geoscientific Model Development*, *13*, 5935–5957. Available from https://doi.org/10.5194/gmd-13-5935-2020.

Nordam, T., Skancke, J., Duran, R., & Barker, C. (2021). Vertical mixing in oil-spill modeling. *Marine hydrocarbon spill assessments*. Elsevier.

Ohlmann, J. C., Niiler, P. P., Fox, C. A., & Leben, R. R. (2001). Eddy energy and shelf interactions in the Gulf of Mexico. *Journal of Geophysical Research: Oceans*. Available from https://doi.org/10.1029/1999jc000162.

Okubo, A. (1971). Oceanic diffusion diagrams. *Deep Sea Research and Oceanographic Abstracts*, *18*, 789–802.

Okubo, A., & Levin, S. A. (2013). *Diffusion and ecological problems: Modern perspectives*. New York: Springer-Verlag.

Olascoaga, M. J., Beron-Vera, F. J., Haller, G., Triñanes, J., Iskandarani, M., Coelho, E. F., … Valle-Levinson, A. (2013). Drifter motion in the Gulf of Mexico constrained by altimetric Lagrangian Coherent Structures. *Geophysical Research Letters*. Available from https://doi.org/10.1002/2013GL058624.

Olascoaga, M. J., & Haller, G. (2012). Forecasting sudden changes in environmental pollution patterns. *Proceedings of the National Academy of Sciences of the United States of America*. Available from https://doi.org/10.1073/pnas.1118574109.

Onink, V., Wichmann, D., Delandmeter, P., & van Sebille, E. (2019). The role of Ekman currents, geostrophy, and Stokes drift in the accumulation of floating microplastic. *Journal of Geophysical Research: Oceans*. Available from https://doi.org/10.1029/2018JC014547.

Paduan, J. D., & Washburn, L. (2013). High-frequency radar observations of ocean surface currents. *Annual Review of Marine Science*, *5*, 115–136. Available from https://doi.org/10.1146/annurev-marine-121211-172315.

Richardson, L. F. (1926). Atmospheric diffusion shown on a distance-neighbour graph. *Proceedings of the Royal Society of London, Series A*, *110*, 709–737.

Richardson, L. F., & Stommel, H. (1948). Note on eddy diffusion in the sea. *Journal of Meteorology*, *5*, 238–240.

Roarty, H., Cook, T., Hazard, L., George, D., Harlan, J., Cosoli, S., … Grilli, S. (2019). The global high frequency radar network. *Frontiers in Marine Science, 6*, 164. Available from https://doi.org/10.3389/fmars.2019.00164, https://www.frontiersin.org/article/10.3389/fmars.2019.00164.

Röhrs, J., & Christensen, K. H. (2015). Drift in the uppermost part of the ocean. *Geophysical Research Letters*. Available from https://doi.org/10.1002/2015GL066733.

Röhrs, J., Sperrevik, A. K., Christensen, K. H., Broström, G., & Breivik, Ø. (2015). Comparison of HF radar measurements with Eulerian and Lagrangian surface currents. *Ocean Dynamics*. Available from https://doi.org/10.1007/s10236-015-0828-8.

Romero, L., Ohlmann, J. C., Pallàs-Sanz, E., Statom, N. M., Pérez-Brunius, P., & Maritorena, S. (2019). Coincident observations of dye and drifter relative dispersion over the inner shelf. *Journal of Physical Oceanography*. Available from https://doi.org/10.1175/jpo-d-19-0056.1.

Rypina, I. I., Kirincich, A. R., Limeburner, R., & Udovydchenkov, I. A. (2014). Eulerian and Lagrangian correspondence of high-frequency radar and surface drifter data: Effects of radar resolution and flow components. *Journal of Atmospheric and Oceanic Technology*. Available from https://doi.org/10.1175/JTECH-D-13-00146.1.

Salmon, R. (1998). *Lectures on geophysical fluid dynamics*. Oxford University.

Serra, M., & Haller, G. (2016). Objective Eulerian Coherent Structures. *Chaos*. Available from https://doi.org/10.1063/1.4951720, arXiv:1512.02112.

Serra, M., Sathe, P., Rypina, I., Kirincich, A., Ross, S. D., Lermusiaux, P., … Haller, G. (2020). Search and rescue at sea aided by hidden flow structures. *Nature Communications*. Available from https://doi.org/10.1038/s41467-020-16281-x, arXiv:1909.07828.

Shanks, A. L. (1987). The onshore transport of an oil spill by internal waves. *Science, 235*, 1198–1200. Available from https://doi.org/10.1126/science.235.4793.1198, https://science.sciencemag.org/content/235/4793/1198.

Simecek-Beatty, D., & Lehr, W. J. (2017). Extended oil spill spreading with Langmuir circulation. *Marine Pollution Bulletin*. Available from https://doi.org/10.1016/j.marpolbul.2017.06.047.

Smagorinsky, J. (1963). General circulation experiments with the primitive equations: I. The basic experiment. *Monthly Weather Review, 91*, 99–164.

Smyth, W. D., & Carpenter, J. R. (2019). *Instability in geophysical flows*. Cambridge University Press.

Spivakovskaya, D., Heemink, A. W., & Deleersnijder, E. (2007). Lagrangian modelling of multi-dimensional advection-diffusion with space-varying diffusivities: Theory and idealized test cases. *Ocean Dynamics, 57*, 189–203.

Sudre, J., Maes, C., & Garçon, V. (2013). On the global estimates of geostrophic and Ekman surface currents. *Limnology and Oceanography: Fluids and Environments*. Available from https://doi.org/10.1215/21573689-2071927.

Sullivan, P. P., Romero, L., Mcwilliams, J. C., & Kendall Melville, W. (2012). Transient evolution of Langmuir turbulence in ocean boundary layers driven by hurricane winds and waves. *Journal of Physical Oceanography*. Available from https://doi.org/10.1175/JPO-D-12-025.1.

Sun, S., Hu, C., Garcia-Pineda, O., Kourafalou, V., Le Hénaff, M., & Androulidakis, Y. (2018). Remote sensing assessment of oil spills near a damaged platform in the Gulf of Mexico. *Marine Pollution Bulletin, 136*, 141–151. Available from https://doi.org/10.1016/j.marpolbul.2018.09.004, URL. Available from http://www.sciencedirect.com/science/article/pii/S0025326X18306441.

Tennekes, H., & Lumley, J. L. (1972). *A first course in turbulence*. MIT Press.

Thorpe, S. A. (2004). Langmuir circulation. *Annual Review of Fluid Mechanics, 36*, 55–79. Available from https://doi.org/10.1146/annurev.fluid.36.052203.071431.

Villas Bôas, A. B., Ardhuin, F., Ayet, A., Bourassa, M. A., Brandt, P., Chapron, B., … van Sebille, E. (2019). Integrated observations of global surface winds, currents, and waves: Requirements and challenges for the next decade. *Frontiers in Marine Science*. Available from https://doi.org/10.3389/fmars.2019.00425.

Wagner, P., Rühs, S., Schwarzkopf, F. U., Koszalka, I. M., & Biastoch, A. (2019). Can Lagrangian tracking simulate tracer spreading in a high-resolution ocean general circulation model? *Journal of Physical Oceanography*. Available from https://doi.org/10.1175/JPO-D-18-0152.1.

Weisberg, R. H., Lianyuan, Z., & Liu, Y. (2017). On the movement of Deepwater Horizon Oil to northern Gulf beaches. *Ocean Modelling*. Available from https://doi.org/10.1016/j.ocemod.2017.02.002.

Zelenke, B., O'Connor, C., Barker, C., Beegle-Krause, C., & Eclipse, L.E. (2012). *General NOAA Operational Modeling Environment (GNOME)*. Technical Documentation, NOAA Technical Memorandum NOS OR&R 40. Technical Report October NOAA.

Zhang, R., Wingo, P., Duran, R., Rose, K., Bauer, J., & Ghanem, R. (2020). *Environmental economics and uncertainty: Review and a machine learning outlook*. Available from https://oxfordre.com/environmentalscience/view/10.1093/acrefore/9780199389414.001.0001/acrefore-9780199389414-e-572.

Appendix A: Automated oil-spill simulations

To simulate an oil spill with advection and diffusion but without the need to choose an eddy diffusion coefficient, the advection–diffusion equation is solved in Lagrangian terms including an automated method to determine an eddy diffusion coefficient κ. Mathematically, and considering two-dimensional horizontal transport, this amounts to solving a stochastic differential equation (SDE), given by

$$d\mathbf{X} = (\mathbf{u} + \nabla\kappa)\mathrm{d}t + \sqrt{2\kappa}\mathrm{d}\mathbf{W}(t), \tag{3.A1}$$

where κ is assumed to be a scalar function of space and time, and where the random variable $\mathbf{W}(t)$ is a two-dimensional Wiener process (see, e.g., Kloeden & Platen, 1992, p. 28, 70). Here, we have assumed that the diffusivity is isotropic; that is, it is the same in both horizontal directions. For details of the anisotropic case, the interested reader is referred to Spivakovskaya, Heemink, and Deleersnijder (2007). If we solve this equation for a sufficiently large number of particles, the density of particles will evolve in time in the same way as the concentration, C, described by Eq. (3.3) (Lynch et al., 2014, pp. 122–126).

The diffusion part is typically modeled as a random walk, by numerically solving Eq. (3.A1), with $\mathbf{u} = 0$ if advection is separately accounted for. The simplest numerical scheme for SDEs is the Euler–Maruyama scheme (Kloeden & Platen, 1992, p. 305), which in our case (with $\mathbf{u} = 0$) is

$$\mathbf{X}_{n+1} = \mathbf{X}_n + (\nabla\cdot\kappa)\Delta t + \sqrt{2\kappa}\Delta\mathbf{W}_n. \tag{3.A2}$$

Here, $\mathbf{X}_n$ is the position of a particle at time t_n, Δt is the time step, and $\Delta\mathbf{W}_n$ are the increments of the two-dimensional Wiener process. That is, $\Delta\mathbf{W}_n$ is a vector with two independent, identically distributed random components, with zero mean, and variance Δt. If the diffusivity is spatially variable, accounting for its gradient in Eq. (3.A2) avoids nonphysical transport away from regions of high diffusivity (Lynch et al., 2014, p. 125). However, this problem is usually more important for vertical transport, as diffusivity gradients are usually both sharper and more persistent in the vertical (Nordam, Skancke, Duran, & Barker, 2021).

The diffusivity can be estimated by different means and is sometimes provided by an ocean model, but it will usually include uncertainty and errors. It is also important to remember that the eddy diffusivity does not directly correspond to any physical, measurable quantity in nature. Rather, it is a parameterization of the combined effect of unresolved eddy motion (subgrid stirring), and molecular diffusivity. Note that since the eddy diffusivity is intended to compensate for unresolved features in the ocean model, the eddy diffusivity will be higher for low-resolution models, and smaller for high-resolution models. A simple scheme suggested by Smagorinsky (1963) scales the eddy diffusivity with the square of the model grid cell size, which may be a useful rule-of-thumb.

Another option is to use a time-dependent diffusivity, which increases with the time since the release. The rationale for this approach is found in observations. In the ocean, eddies exist at a wide range of spatial scales, from the largest basin-scale gyres, down to the Kolmogorov length scale of millimeters or less. The effect of these eddies on a patch of dissolved tracer depends on the size of the eddy, relative to the size of the patch. Eddies that are much larger than the patch will only advect it, with little or no change to its shape. Eddies that are much smaller than the patch will only serve to deform its surface, without changing its overall shape. Eddies that are of the same size as the patch, on the other hand, will significantly change its shape, by stretching out filaments in different directions, thus increasing the overall size of the patch.

A small patch will initially be most affected by small eddies, but as it grows in size, it will be affected by increasingly large eddies. By constructing an argument based on the typical turnover time of eddies of different sizes, it is possible to arrive at an expression for how fast the size of the patch will grow in time. Following the argument of Davidson (2015, pp. 257–258), we let R be the mean radius of an initially small and spherical patch, and let the typical speed of an eddy of size r be $v_r \sim (\varepsilon r)^{1/3}$, where ε is the turbulent kinetic energy dissipation rate per unit mass. Since the patch is mainly affected by eddies of its own size, we get

$$\frac{\mathrm{d}R}{\mathrm{d}t} \sim v_R \sim (\varepsilon R)^{1/3}. \tag{3.A3}$$

Rewriting this expression by using that $\frac{\mathrm{d}}{\mathrm{d}t}R^2 = 2R\frac{\mathrm{d}R}{\mathrm{d}t}$, we get

$$\frac{\mathrm{d}R^2}{\mathrm{d}t} \sim \varepsilon^{1/3} R^{4/3}, \tag{3.A4}$$

which is known as Richardson's four-thirds law (Richardson, 1926). This expression is only valid for $\eta \ll R \ll \ell$, where η is the scale of the smallest eddies (the Kolmogorov scale), and ℓ is the scale of the largest eddies (Davidson, 2015, p. 258). A further limitation in our case is that on large scales, the ocean is essentially two-dimensional. We will return to this point.

Since R^2 is proportional to the variance of a patch of tracer, we see that the rate of increase of the variance is size-dependent, and thus time-dependent, when a patch is subject to turbulent mixing. This is contrary to the case in Fickian diffusion described by Eq. (3.2), where the variance grows linearly with time, proportional to the diffusivity:

$$\frac{\mathrm{d}R^2}{\mathrm{d}t} \sim \kappa. \tag{3.A5}$$

From the above, we can derive a time-dependent "effective diffusivity," $\kappa_{\text{eff}}(t)$, for a patch subject to turbulent mixing. Integrating Eq. (3.A3), we find that $r \sim \varepsilon^{1/2} t^{3/2}$, and inserting this into Eq. (3.A4) we find

$$\kappa_{\text{eff}}(t) \sim \varepsilon t^2. \tag{3.A6}$$

Hence, we find that the variance of a patch subject to turbulent mixing is proportional to t^3, since it grows at a rate proportional to t^2.

Early experimental investigation of the above results include observations of balloons released into the atmosphere (Richardson, 1926), and bits of parsnip thrown into a loch by Richardson's cabin in Scotland (Richardson & Stommel, 1948). Okubo (1971) published a summary of several dye release experiments, covering spatial scales from 100 to 10 and time scales from hours to several weeks. When plotting variance as a function of time (Fig. 1 in Okubo, 1971), he found $R^2 \sim t^{2.3}$, and when plotting effective diffusivity as a function of spatial scale, he found $\kappa_{\text{eff}} \sim R^{1.1}$. These results were later expanded with more observational data, by Murthy (1976) and Lawrence, Ashley, Yonemitsu, and Ellis (1995), still showing approximately the same trends.

If we for the moment accept sloppy notation with respect to units, an explicit expression for the time-dependent apparent horizontal diffusivity, κ_a, may be obtained from Eq. (3) in Okubo (1971),

$$\sigma_{rc}^2 = 0.0108 \cdot t^{2.34}, \tag{3.A7}$$

where σ_{rc}^2 is measured in cm^2 and t is in seconds. Combining this with the relation $\kappa_a = \sigma_{rc}/4t$, we get

$$\kappa_a = 0.0027 \cdot t^{1.34}, \tag{3.A8}$$

where κ_a is given in units cm/s^2. These observation-based results do not agree with the theoretical considerations summarized in Eq. (3.A4). However, it is clear that the ocean cannot be considered to be three-dimensional when considering a patch of size 100 m or above, released in the mixed layer. Hence, the theoretical results cannot be expected to hold exactly.

Based on the discussion above, it might seem reasonable to use a time-dependent diffusivity in an oil spill model. However, it is important to remember that the effective diffusivity is intended to mimic the mixing due to eddies in the ocean currents. If high-resolution current data is used as input to the modeling, more of those eddies will already be represented in the data and need not be accounted for in the time-dependent diffusivity. Hence, the diffusivity should in some sense be matched to the resolution of the ocean current data.

If the horizontal resolution of the current data is Δx, then any patch of tracer with $R \ll \Delta x$ will only be advected along the currents, without changing its shape significantly. Hence, it makes sense to apply a time-dependent diffusivity to small patches. However, once the patch grows in size such that $R > \Delta x$, differential advection by eddies represented in the current data will lead the patch to spread out further. Applying an additional time-dependent diffusivity to such a patch will then lead to too much diffusion.

In practice, it is easier to truncate the effective diffusivity based on time, rather than spatial scales. It is a simple matter to keep track of the "age" of numerical particles and use a time-dependent diffusivity in the random walk for each particle. Future work will be needed to determine when to truncate the time-dependent diffusivity and to quantify the difference of time-dependent diffusivity instead of a constant one in practical applications. Further reading can be found in Csanady (1973, Chapter IV), Okubo and Levin (2013, Chapter 2), and Lynch et al. (2014, Chapter 4).

Chapter 4

Vertical mixing in oil spill modeling

Tor Nordam[1,2], Jørgen Skancke[1], Rodrigo Duran[3,4] and Christopher H. Barker[5]
[1]SINTEF Ocean, Trondheim, Norway, [2]Department of Physics, Norwegian University of Science and Technology, Trondheim, Norway, [3]National Energy Technology Laboratory, United States Department of Energy, Albany OR, United States, [4]Theiss Research, La Jolla, CA, United States, [5]Emergency Response Division, Office of Response and Restoration, National Oceanic and Atmospheric Administration, Seattle, WA, United States

This chapter is intended to give a comprehensive overview of the processes, which relate to the vertical movement of oil spilled in the ocean, namely turbulent mixing, buoyancy, and entrainment. These processes will be explained in terms of their physics, approaches to numerical modeling, and the historical background on how the research field has developed.

The aim of this chapter is that the reader should be capable of formulating a reasonable "one-dimensional" oil spill model.

4.1 Introduction

An essential aspect of oil spill modeling is to capture the different processes that influence fate and behavior of oil on the ocean surface and oil in the water column. Surface oil is exposed to the atmosphere, wind, and waves, and undergoes surface spreading, evaporation, emulsification, and entrainment due to breaking waves. Submerged oil, on the other hand, experiences to a greater degree dissolution, microbial biodegradation, and three-dimensional dispersion. In terms of interaction, surface oil may foul birds and surface-interacting mammals and may be washed ashore to cover coastline habitats. Subsurface oil in dissolved form represents an exposure risk to marine life. This is especially the case for early life stages like eggs and larvae. Oil in droplet form may for example cause toxicity by adhering to the surface of fish eggs (Hansen et al., 2018).

In addition to the distinction between surface and submerged oil, the vertical distribution of oil within the water column has a major impact on horizontal transport, due to current shear. The importance of vertical mixing for

Marine Hydrocarbon Spill Assessments. DOI: https://doi.org/10.1016/B978-0-12-819354-9.00002-8

horizontal transport has been known for a long time. Both Bowden (1965) and Okubo (1968) suggest that Bowles, Burns, Hudswell, and Whipple (1958) were the first to use the term "shear effect" in relation to mixing and transport in the sea, in a paper on dilution of radioactive waste water. Okubo (1968) states that the shear effect is "the dispersion of a vertical column of fluid due to the variation of velocity with depth combined with vertical diffusion."

Surface wind can create strong vertical gradients of current speed and direction in the upper meters of the water column (Fernandez, Vesecky, & Teague, 1996; Laxague et al., 2018), thus leading to dispersion of oil submerged at different depths. Vertical density gradients can also act to separate vertical layers, allowing them to move in different horizontal directions or at different speeds, contributing to current shear.

In the context of oil spill modeling, Johansen (1982) provides an early discussion of the importance of vertical distribution for determining horizontal transport. He describes the continuous exchange of oil between the surface and the subsurface due to breaking waves and surfacing, discussing the importance of rise speed and the droplet size distribution produced by natural entrainment. He also explicitly formulates a one-dimensional Eulerian model for the vertical transport of oil droplets, based on the advection–diffusion equation, and uses this model to discuss the implications for the drift and weathering of surface oil.

Another early work treating an oil spill as a three-dimensional process is that of Elliott (1986). This paper describes the elongation of an oil slick in the direction of the wind, attributing the effect to vertical current shear, combined with the continuous exchange of oil between the surface and the subsurface. The downward process is driven by turbulent mixing from waves and the upward process by the buoyancy of the oil droplets. Elliott (1986) also formulated a three-dimensional random-walk-based Lagrangian particle model for an oil slick. While that model allowed sufficiently large oil droplets to remain at the surface, as their rise due to buoyancy would always dominate the random displacement due to diffusion, it did not include a slick formation process or a separate state for surface oil.

In modern oil spill models, surfaced oil is usually assumed to form continuous patches, while submerged oil is in the form of individual droplets of different sizes. More advanced models also include a nonbuoyant dissolved oil fraction. The mass exchange between the surface and subsurface compartments is a function of the state of the oil and the state of the wave field, the latter of which can be parameterized from wind speed and fetch length, or modeled by a wave model (coupled to the ocean model, or separate). To calculate entrainment, an oil spill model must predict the mass of oil entrained per area and time, over what depth that oil should be distributed, and the droplet size distribution of the entrained oil. Surfacing of oil, on the other hand, is found as a balance between the vertical rise of droplets and turbulent mixing. Vertical transport brings droplets toward the surface, while turbulent mixing tends to distribute oil droplets over a certain depth. Strong turbulent mixing will therefore reduce the amount of oil surfacing by reducing the concentration of oil droplets in the near-surface water layer.

4.2 Vertical mixing in the ocean

In this section, we give a description of the mechanisms behind vertical mixing in the ocean. Some oceanographic background is given, though for further information the interested reader may refer to, e.g., *Introduction to ocean turbulence* (Thorpe, 2007), and the classic work *Turbulent diffusion in the environment* (Csanady, 1973, see particularly Chapters III, V, and VI).

4.2.1 Turbulent diffusion

Molecular diffusion is a fundamental physical process, caused by the random motion of molecules in gases and liquids, occurring even in completely stagnant conditions. The effect of this random motion is to reduce gradients in concentration, leading eventually to an even distribution of, for example, dissolved chemicals in water. The rate of molecular diffusion depends on temperature and the relative size and properties of the molecules involved, but this is in all cases a relatively slow process. As an example, Lee, Lee, Lee, and Tseng (2004) did experiments on the diffusion of ink in water and found that a droplet of ink took about 1 minute to spread to a radius of 1 cm, in water at room temperature.

In contrast, we would expect a droplet of ink to be evenly distributed in a glass of water within seconds, if the water was stirred. This latter process is often called turbulent diffusion, or turbulent mixing, and is akin to what happens in the ocean, where the origins of the turbulent mixing can for example be breaking waves, current shear, bottom friction, overturning, etc.

Despite the name, turbulent diffusion is not a pure diffusion process, but rather a combination of an advection process and molecular diffusion (see, e.g., Thorpe (2005, pp. 20–21) for a particularly clear description). The crucial point is that turbulence causes stirring at a wide range of spatial scales, dramatically increasing the area of interface between regions of high and low concentrations. Fick's law (see, e.g., Csanady, 1973, p. 4) states that the diffusive flux of a substance (i.e., amount of substance transported per area per time) is given by

$$j_D = -K\frac{\mathrm{d}}{\mathrm{d}x}C(x), \tag{4.1}$$

where K is the diffusion parameter and C is the concentration of a substance. If we consider two initially separated volumes of water, with different concentrations of some substance, then it is clear that mixing will be faster if the area of the interface between the two volumes increases. This is precisely what turbulent mixing achieves, and the effect can be quite dramatic,

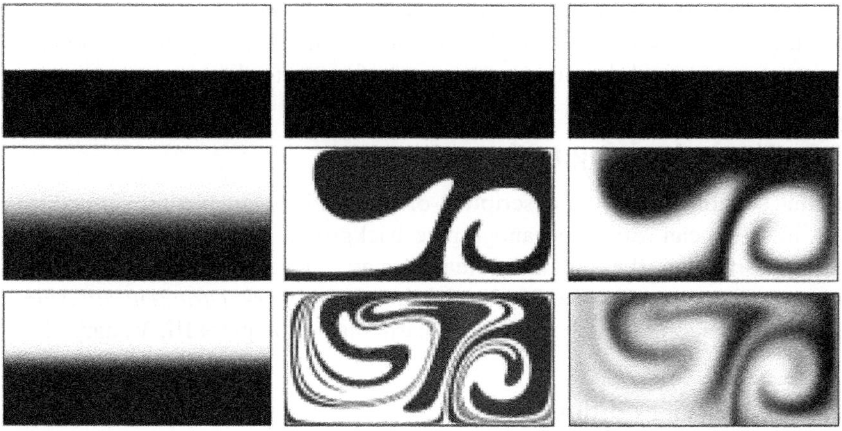

FIGURE 4.1 Mixing of a passive tracer, initially located in the bottom half of the domain. In the left column, Fickian diffusion is applied; in the middle column, advection by a double gyre; and in the right column, both advection and diffusion.

increasing the effective mixing by many orders of magnitude (see, e.g., Tennekes & Lumley, 1972, pp. 8–10).

An illustration of this has been made in Fig. 4.1, where a tracer initially located in the bottom half of a closed domain, is moved first with diffusion only (Fig. 4.1, left column), then advected by a double gyre (middle column), and finally both diffused and advected (right column). While this is only intended as a schematic illustration, it shows how the combined effect of advection by a gyre and Fickian diffusion leads to a faster mixing than diffusion alone due to an increased interface between regions of high and low concentration.

While "turbulent mixing" is in reality a combination of stirring by turbulence and molecular diffusion, it will in almost all cases be impractical to model it as such. In particular, eddies in the ocean exist at all scales, from the largest ocean gyres with a scale of several thousand kilometers, to the smallest turbulent eddies at the Kolmogorov scale (Davidson, 2015, p. 25), on the order of 1 mm or less. Most numerical ocean models currently have a horizontal resolution somewhere between tens of meters and tens of kilometers, and a vertical resolution ranging from meters to hundreds of meters. Any eddies smaller than the resolution of the model cannot be resolved, and therefore their contribution to the mixing must be parameterized, as the so-called eddy diffusivity. When talking about diffusion in the context of oil-trajectory modeling, and indeed for the rest of this chapter, one typically refers to the eddy diffusivity.

4.2.2 Origins of vertical mixing in the ocean

There are many sources of turbulent kinetic energy (TKE) leading to vertical mixing in the ocean. The most obvious (and spectacular) may be breaking waves, which contribute to mixing in the upper part of the water column. Furthermore, turbulent motion may be caused as currents flow across the seabed, in narrow straits, or due to shear flow between two fluid layers. In cold or windy conditions, evaporation or cooling at the surface will increase the density of the water, and if the water at the surface becomes denser than the underlying water, overturning will occur, leading to vertical mixing. In very cold conditions, sea ice will form. During the freezing process, the salinity of the ice is reduced through rejection of brine. This cold, high-salinity water will have a density higher than the water below, again leading to overturning and mixing of water masses.

In all cases, the stratification of the water column has a strong influence on the vertical diffusivity. Near the surface, there will usually be a layer of uniform density, called the surface mixed layer, or just the mixed layer. The thickness, or depth, of the mixed layer will vary with geographical location, time of year, and it will be influenced by local factors such as wind, air temperature, solar radiation, rainfall, and freshwater input from rivers.

While eddy diffusivity is typically high in the interior of the mixed layer, it drops both close to the surface and toward the bottom of the mixed layer. Toward the surface, the mixing efficiency (which the eddy diffusivity describes), is limited because the surface limits the size of turbulent eddies (Craig & Banner, 1994). This is a version of the so-called mixing length argument (see, e.g., Davidson, 2015, pp. 112–113), stating that the mixing depends not only on the TKE but also on the size of eddies.

At the bottom of the mixed layer, the density increases, either due to an increase in salinity, a decrease in temperature, or a combination of both. This region of increasing density is called the pycnocline. Stable stratification, where a layer of light water overlays denser water, will tend to prevent mixing across the density gradient, due to the energy required to lift the dense water against gravity. For this reason, vertical diffusivity can be quite high throughout the mixed layer, and then drop, sometimes by several orders of magnitude, at the pycnocline. The interested reader is referred to Gräwe, Deleersnijder, Shah, and Heemink (2012) for further discussion of this particular case, in the context of Lagrangian particle modeling.

As a side note, one might want to ask if the motion of the sea is actually turbulent. If we make the approximation that the water in the mixed layer behaves as an isolated slab of water, experiencing friction forces from the wind at the top, and from the deep water at the pycnocline, then the Reynolds number (Thorpe, 2005, p. 23) for this flow is

$$\mathrm{Re} = \frac{\Delta u L}{\nu}, \tag{4.2}$$

where Δu is the difference between the speed at the top and the bottom of the mixed layer, L is the thickness of the mixed layer, and ν is the kinematic viscosity of water, which is approximately $1.4 \times 10^{-6}\ \mathrm{m/s^2}$ for seawater at

10°C. Turbulent flow is commonly said to occur at Reynolds numbers above approximately 4000. If we choose for example a mixed layer thickness of $L = 10$ m, we see that any velocity difference of more than about 0.5×10^{-3} m/s will give turbulent flow.

The above discussion of turbulent flow assumes that the density truly is constant throughout the mixed layer. In regions where the density increases slowly with depth, a relevant parameter to consider is the Richardson number,

$$\mathrm{Ri} = \frac{g}{\rho_w} \frac{\frac{\partial \rho_w}{\partial z}}{\left(\frac{\partial u}{\partial z}\right)^2}, \tag{4.3}$$

which is a dimensionless number related to the ratio between the stabilizing forces of stratification and the destabilizing forces of current shear. If $\mathrm{Ri} \gg 1$, then the shear forces are not strong enough to break down the density gradient and cause vertical mixing.

4.2.3 Modeling ocean turbulence

The vertical eddy diffusivity can be described through models of different complexity, including simple parameterizations based on fitting simplified models against experiments, and more complex models that try to solve dynamic equations for transport and dissipation of TKE.

The simplest model for vertical turbulent mixing would be to simulate a vertical advection–diffusion process using a constant eddy diffusivity. However, in light of the discussion in Section 4.2.2, it should be clear that this will in many cases be too simple. In particular, the diffusivity in the mixed layer can often be several orders of magnitude higher than the diffusivity at greater depths.

Hence, the second simplest approach might be to model the diffusivity profile as a step function, with a high value in the mixed layer, and a lower value below the pycnocline. However, note that care must be taken to avoid numerical artifacts when using a step-function diffusivity. See Sections 4.7.2 and 4.7.4 for examples and additional discussion of this topic.

Another approach is to use a simplified continuous model for the vertical diffusivity. One such model, used in some oil spill modeling studies (Nordam et al., 2018; Skognes & Johansen, 2004), is due to Ichiye (1967), who suggested the following relation for the vertical eddy diffusivity as a function of depth (z positive downward):

$$K(z) = 0.028 \frac{H_s^2}{T_p} \mathrm{e}^{-2kz}, \tag{4.4}$$

where H_s is the significant wave height, T_p is the peak wave period, and k is the wave number. This relation takes mixing due to waves into account, but ignores the limiting effects of stratification, and does not feature reduced diffusivity toward the surface.

A third option is to obtain eddy diffusivity from an ocean model. All or most ocean models calculate eddy diffusivity, potentially taking into account waves, density stratification, current shear, and more complex processes such as Langmuir circulation (Thorpe, 2005, pp. 251–255). Several different approaches at different degrees of complexity exist. So-called turbulence closure schemes are a research field in themselves, and we will not go into detail on the schemes themselves here. The interested reader is referred to Davidson (2015, p. 27) and Haidvogel and Beckmann (1999, Chapter 5).

Relevant in the context of oil spill modeling is that many ocean models provide eddy diffusivity as output on the same formats as the ocean current data. A three-dimensional oil spill model will probably already be using currents from an ocean model, and the advantage of using eddy diffusivity from the same model is then that the fields are dynamically consistent. However, it is important to be aware that the eddy diffusivity in an Eulerian ocean model also serves the additional purpose of suppressing numerical instabilities that can occur in advection-dominated problems. Hence, it is possible that the eddy diffusivity from an ocean model may be somewhat higher than it should be and therefore unsuited for direct use in a Lagrangian transport model. Nevertheless, diffusivity from an ocean model would be expected to take the important effects of stratification into account.

Finally, it is worth mentioning the existence of separate, standalone models for vertical ocean turbulence. The most well-known of these is probably the General Ocean Turbulence Model (GOTM; Umlauf, Burchard, & Bolding, 2005).[1] This is an open-source one-dimensional water column model that can be set up to model a range of different cases, with different forces as input, and using different turbulence closure schemes, such as Mellor–Yamada (Mellor & Yamada, 1982), k–ε (Launder & Spalding, 1983), and k–ω (Wilcox, 2008). In an oil spill modeling context, using a one-dimensional turbulence model is not as convenient as using eddy diffusivity from an ocean model but might be an option for a localized area.

4.2.4 Wave modeling

As previously mentioned, breaking waves are a source of turbulent mixing in the ocean. In the context of oil spill modeling, however, breaking waves are perhaps even more important as the mechanism by which an oil slick at the surface is broken up into droplets and submerged in the water column. We will return to this point in Section 4.3, but here we will mention some approaches to obtain wave data for use in an oil spill model.

1. See also http://www.gotm.net.

As with turbulence, there exist different approaches to obtaining wave data, at different levels of complexity. Advanced wave models that calculate the entire wave spectrum exist and may be run coupled to an ocean model (or atmosphere-ocean model), such that the waves affect the calculation of the current, and vice versa. An example of such a model is SWAN (Booij, Holthuijsen, & Ris, 1997), which may for example be coupled with the ROMS ocean model (see, e.g., Warner, Perlin, & Skyllingstad, 2008).

A simpler approach is to parameterize the wave state from the wind speed, usually given at an altitude of 10 m above sea level. In the following example, the significant wave height and the peak wave period, H_s and T_p, are derived from the JONSWAP spectrum and associated empirical relations (Carter, 1982). The sea state is assumed to be either fetch-limited or fully developed. Fetch-limited means that a steady state is reached, where the sea does not reach a fully developed state because the fetch (the distance over which the wind acts on the sea) is too short. This is relevant close to the coast in off-shore wind conditions. Fully developed, on the other hand, refers to the steady-state wave conditions that are found far away from the coast. In both cases, the wave state is assumed *not* to be time-limited; that is, the wind is assumed to have been constant for a sufficiently long time to allow a steady wave state to develop.

Here, H_s and T_p are given by

$$H_s = \frac{u_{10}^2}{g} H_c \sqrt{\frac{gL_f}{u_{10}^2}} \tag{4.5a}$$

$$T_p = \frac{u_{10}}{g} T_c \sqrt[3]{\frac{gL_f}{u_{10}^2}} \tag{4.5b}$$

in the fetch-limited case, and

$$H_s = \frac{u_{10}^2}{g} H_0 \tag{4.6a}$$

$$T_p = \frac{u_{10}}{g} T_0 \tag{4.6b}$$

in the fully developed case. Here, $H_0 = 0.243$, $H_c = 0.0016$, $T_0 = 8.134$, and $T_c = 0.286$ are dimensionless parameters, g is the gravitational acceleration, L_f is the fetch length, and u_{10} is the wind speed at 10 m above sea level.

4.3 Entrainment of surface oil

When a wave breaks on an oil slick, part of the oil in the breaking zone of the wave will be entrained into the water column in the form of droplets. The amount of entrained oil will increase with the height of wind-driven waves and therefore depends on the wind speed. To describe this in an oil spill model, it is necessary to formulate a model for the mass of oil entrained per unit of time for a given surface slick in a given wave field. Some of the earliest quantitative work on the entrainment of oil due to breaking waves is that of Delvigne and Sweeney (1988). In this classic paper, based on experiments in a turbulence tank and two different mesoscale wave flumes (0.43 and 4.3 m depth), they provided empirical relations for the three key parameters in surface oil entrainment:

(1) Droplet size distribution of the entrained oil
(2) Entrainment rate
(3) Intrusion depth

The relationships obtained by Delvigne and Sweeney (1988) were used for decades in oil spill models, with new models only starting to take hold nearly 30 years later. The empirical relationship for entrainment rate by Delvigne and Sweeney was formulated in a convoluted way, where the entrainment rate depends on the droplet size distribution, and the models lack theoretical support. These and other aspects of the Delvigne and Sweeney models have been criticized by others who have formulated alternative models in recent years (Johansen, Reed, & Bodsberg, 2015; Li, Spaulding, & French-McCay, 2017; Boufadel et al., 2020).

4.3.1 Droplet size distribution of entrained oil

After entrainment of a surface slick, smaller droplets take longer to resurface compared to larger droplets (see Section 4.4.1). For this reason, the droplet size distribution of entrained oil is an important factor that influences both the horizontal transport of the oil, and the lateral dispersion from current shear. To describe dispersion, it is therefore necessary to use an accurate droplet size model. Experimental evidence has shown that the size distribution of droplets in a breaking wave event depends on oil viscosity (Delvigne & Sweeney, 1988; Reed, Leirvik, Johansen, & Brørs, 2009), oil–water interfacial tension (Li, Miller, Wang, Koley, & Katz, 2017; Zeinstra-Helfrich, Koops, & Murk, 2016), oil film thickness (Zeinstra-Helfrich, Koops, Dijkstra, & Murk, 2015; Zeinstra-Helfrich et al., 2016), and energy in the breaking wave (Delvigne & Sweeney, 1988). Several published models exist that use these and other parameters to estimate a droplet size distribution (Delvigne & Sweeney, 1988; Johansen et al., 2015; Li, Spaulding, French McCay, Crowley, & Payne, 2017; Nissanka & Yapa, 2017; Reed et al., 2009; Zhao et al., 2014).

Two main droplet size model types can be distinguished. One type is an equilibrium description, where the model consists of an expression for a characteristic droplet size, such as the median, and a static droplet size distribution function, such as a Rosin–Rammler, log-normal, or power-law

where D_{ba} is the dissipated energy per surface area [J/m^2], S_{cov} is the sea surface area fraction covered by oil, F_{wc} is the sea surface area fraction hit by breaking waves per second [s^{-1}], d is the droplet size [m], and Δd is the width of the droplet size interval (centered on d). The prefactor C is an empirical constant that can include the effects of oil state, such as viscosity, interfacial tension, and density; in Delvigne and Sweeney (1988), only viscosity is included for different values of C.

It is not explicitly stated in the original work of Delvigne and Sweeney (1988) how this equation should be applied to calculate the total entrainment rate; it has however been interpreted in the literature (Li, Spaulding, French McCay, Crowley, et al., 2017). The equation gives the entrainment rate over a droplet size interval, which means that an integration over size intervals must be performed. This means that a lower and upper limit of the droplet size must be decided upon. It is likely that different interpretations of Eq. (4.12) exist, meaning that models using this equation differently will provide somewhat different results.

From a modeling point of view, a more elegant solution is to have an expression of the entrainment rate that is completely independent of the droplet size distribution, and heuristic concepts such as "permanently dispersed oil." Johansen (1982) describes such an approach, modeling the vertical transport (rise due to buoyancy, and vertical mixing due to eddy diffusivity) with the advection–diffusion equation. He points out that in order to represent a distribution of droplet sizes (and thus a distribution of rise speeds), one must solve the advection–diffusion equation for each size class.

In such a model, the entrainment rate describes the amount of oil that is submerged, and the droplet size distribution will describe how that oil is distributed across size classes. The vertical transport model will then determine the future development of those droplets, allowing large droplets to surface rapidly, while small droplets remain submerged for longer periods.

Recent formulations of the entrainment rate adhere to this principle. Both the two following examples describe the submersion of surface oil as a first-order decay process

$$\frac{\mathrm{d}Q_s}{\mathrm{d}t} = -\alpha Q_s, \tag{4.13}$$

where Q_s is the amount of oil at the surface, and α is the entrainment rate. Johansen et al. (2015) describe a simple model where the submersion of surface oil is related to the white-cap coverage fraction, f_{wc} and the mean wave period, T_m:

$$\alpha = P f_{wc}/T_m. \tag{4.14}$$

In Johansen et al. (2015), Eq. (4.14) was used as a standalone model to describe the development of oil mass on the surface. It was thus assumed that droplets larger than some limiting diameter would resurface directly, and P was then taken to be the fraction of droplets smaller than this limiting diameter. For use in a modeling system where a vertical transport model determines the fate of the droplets, we assume that oil is entrained at the full rate, setting $P = 1$. Then, the transport model will allow the larger droplets to surface quickly.

Li, Spaulding, French McCay, Crowley, et al. (2017) developed an empirical relation parameterizing the entrainment rate, Q, in terms of the dimensionless Weber and Ohnesorge numbers:

$$\frac{Q}{F_{bw}} = a\mathrm{We}^b\mathrm{Oh}^c. \tag{4.15}$$

Here, F_{bw} is the white-capping fraction per unit time [s^{-1}], the Weber number is $\mathrm{We} = d_o\rho_w g H_s/\sigma_{ow}$, where ρ_w is the density of water, and the Ohnesorge number is $\mathrm{Oh} = \mu_o/\sqrt{\rho_o\sigma_{ow}d_o}$. The length scale is the Rayleigh–Taylor instability maximum droplet diameter, given by

$$d_o = 4\left(\frac{\sigma_{ow}}{(\rho_w - \rho)g}\right)^{1/2}. \tag{4.16}$$

The values of the empirical parameters are $a = 4.604 \times 10^{-10}$, $b = 1.805$, and $c = -1.023$ (Li, Spaulding, French McCay, Crowley, et al., 2017).

4.3.3 Entrainment depth of oil due to breaking waves

The linear parameterization of intrusion depth and distribution of oil found by Delvigne and Sweeney (1988) is that after a wave breaking event, the oil is distributed evenly in the interval

$$(1.5 - 0.35)H_w < z < (1.5 + 0.35)H_w, \tag{4.17}$$

where H_w is the wave height. No more recent general model formulations for the intrusion depth have been published in the oil spill literature. However, in disagreement with this model, recent experiments in a wave tank gave intrusion depth centers of less than half the wave height (Li, Miller, et al., 2017). Other work that may be relevant in this context includes studies of bubble entrainment by breaking waves (see, e.g., Leifer & De Leeuw, 2006) and observations of vertical distributions of buoyant fish eggs (see, e.g., Röhrs et al., 2014).

4.4 Submerged oil

The vertical transport processes that affect submerged oil droplets are rise due to buoyancy (or sinking in some cases; see, e.g., King, Robinson, Boufadel, & Lee, 2014), turbulent mixing, and vertical advection by currents. Of these three, advection by vertical currents is probably the least important.

Hence, we will not discuss this further; other than to state that if current data with a vertical current component is available, it can be used to advect the oil, just like the horizontal components.

Vertical turbulent mixing has already been discussed in Section 4.2.3, and how to use the eddy diffusivity in an oil spill model will be discussed in Sections 4.5 and 4.6. Hence, the main content of this section will be the calculation of rise speeds for oil droplets.

4.4.1 Calculation of droplet rise speeds

It is commonly assumed that droplets, bubbles, sediment particles, etc. rise or sink at their terminal velocity. The terminal velocity is derived by starting from the observation that buoyancy exerts a constant force, F_b, on the submerged particle:

$$F_b = \frac{4}{3}\pi r^3 g(\rho_a - \rho_p) = \frac{4}{3}\pi r^3 \rho_a g', \tag{4.18}$$

where $g' = g\frac{\rho_a - \rho_p}{\rho_a}$ is the reduced gravity, with ρ_a and ρ_p the density of the ambient fluid and the moving particle, respectively. While the buoyancy is constant, the drag force, F_D, increases with the velocity and has direction opposite to the velocity:

$$F_D = -\frac{1}{2}\rho_a v^2 C_D A \cdot \frac{v}{|v|}, \tag{4.19}$$

where v is the velocity of the particle relative to the fluid, A is the cross-sectional area of the particle, and C_D is a drag coefficient. By setting the total force equal to 0, we get an equation that can be solved to find the terminal speed, v_b:

$$v_b = \sqrt{\frac{4\,dg'}{3\,C_D}}. \tag{4.20}$$

The drag coefficient, C_D, is not constant, but rather a function of the Reynolds number, which for a sphere is given by

$$\mathrm{Re} = \frac{vd}{\nu_a} = \frac{\rho_a v d}{\mu_a}. \tag{4.21}$$

Here, v is the speed of the sphere, d is the diameter of the sphere, ν_a and μ_a are the kinematic and dynamic viscosities of the surrounding fluid, and ρ_a is its density.

At low Reynolds numbers, $\mathrm{Re} \ll 1$, the drag coefficient is given by

$$C_D = \frac{24}{\mathrm{Re}}. \tag{4.22}$$

With this drag coefficient, the expression for the drag force becomes

$$F_D = -6\pi r \mu_a v. \tag{4.23}$$

Solving for the terminal speed, v_b, one obtains

$$v_b = \frac{2}{9}\frac{\rho_p - \rho_a}{\mu_a} g r^2. \tag{4.24}$$

Eq. (4.23) for the drag force is commonly known as Stokes' law, after George Gabriel Stokes (Stokes, 1856), although Eqs. (4.22) and (4.24) are also sometimes referred to as Stokes' law.

In the derivation of Stokes' law, an assumption was made that the flow around the spherical particle is laminar. At higher Reynolds numbers, the flow around the sphere is no longer laminar, and Stokes' law no longer holds. Various empirical formulas exist for the case of high Reynolds number. Clift et al. (1978) combined several previously published results and developed a piecewise parameterization of C_D as a function of the Reynolds number, which they call the Standard Drag Curve for the drag coefficient of a spherical particle (Clift et al., 1978, pp. 110−112). This parameterization is shown in Fig. 4.2, together with Stokes' law (Eq. (4.22)) and two other parameterizations.

As the highest range of Reynolds numbers in Fig. 4.2 is not relevant for oil droplets rising due to buoyancy, simpler expressions than the Standard Drag Curve have been suggested. Delnoij, Lammers, Kuipers, and van Swaaij (1997) suggested a parameterization of C_D given by

$$C_D = \begin{cases} \frac{24}{\text{Re}}(1 + 0.15\text{Re}^{0.687}) & \text{if} \quad \text{Re} < 1000 \\ 0.44 & \text{if} \quad \text{Re} \geq 1000 \end{cases}. \tag{4.25}$$

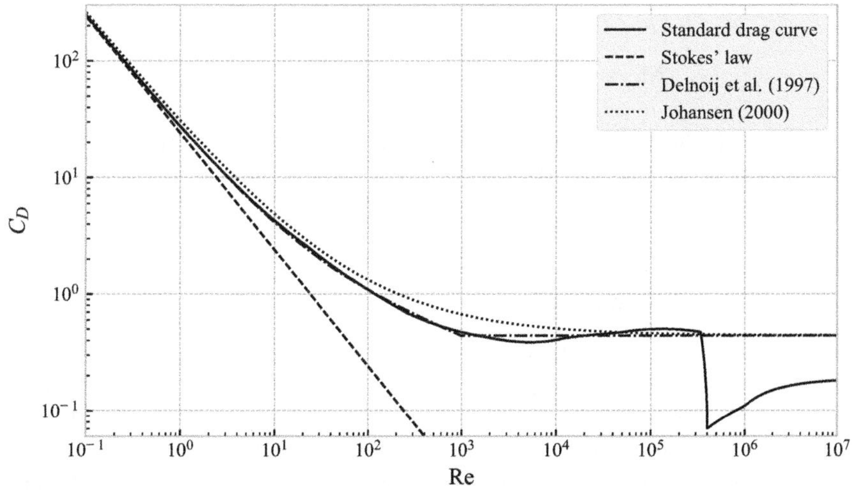

FIGURE 4.2 Parameterization of the Standard Drag Curve, due to Clift, Grace, and Weber (1978). Stokes' law, which is valid for Re ≪ 1, is shown as a dashed line.

Johansen (2000) proposed a variant of the above scheme where the terminal rise velocity is instead given by a harmonic transition between the high and low Reynolds number cases:

$$v_b = \left(\frac{1}{v_1} + \frac{1}{v_2}\right)^{-1}, \tag{4.26}$$

where v_1 and v_2 are calculated from Eq. (4.20), with a drag coefficient of $C_D = 24/\mathrm{Re}$ in v_1, and $C_D = 0.44$ in v_2. The parameterizations due to Delnoij et al. (1997) and Johansen (2000) are also shown in Fig. 4.2

All of the above expressions assume spherical particles. In reality, an oil droplet rising through water will deform to some degree, depending among other things on the volume, density, and oil–water interfacial tension. Work on this topic includes that of Bozzano and Dente (2001), which takes droplet deformation into account. They developed empirical relations for the drag coefficient and deformation of droplets and bubbles in terms of the Reynolds, Morton, and Eötwös numbers:

$$C_D = f\left(\frac{a}{R_0}\right)^2, \tag{4.27a}$$

where

$$f = \frac{48}{\mathrm{Re}}\left(\frac{1 + 12\mathrm{Mo}^{1/3}}{1 + 36\mathrm{Mo}^{1/3}}\right) + 0.9\frac{\mathrm{Eo}^{3/2}}{1.4(1 + 30\mathrm{Mo}^{1/6}) + \mathrm{Eo}^{3/2}}, \tag{4.27b}$$

$$\left(\frac{a}{R_0}\right)^2 = \frac{10(1 + 1.3\mathrm{Mo}^{1/6}) + 3.1\mathrm{Eo}}{10(1 + 1.3\mathrm{Mo}^{1/6}) + \mathrm{Eo}}, \tag{4.27c}$$

and the Morton and Eötwös numbers are given by

$$\mathrm{Mo} = \frac{g\mu_a^4(\rho_a - \rho_p)}{\rho_a^2\sigma^3}, \quad \mathrm{Eo} = \frac{g d_0^2(\rho_a - \rho_p)}{\sigma}. \tag{4.27d}$$

Here, σ is the interfacial tension between oil (or gas) and water, and d_0 is the equivalent diameter of the particle, that is, the diameter of a sphere with the same volume.

4.4.2 Role of dispersants

Oil dispersants are specifically designed surfactant chemicals intended to reduce the oil–water interfacial tension. Dispersants can be used as a countermeasure during oil spill response, with the objective of enhancing the dispersion of the spill by facilitating the creation of small oil droplets. This can be done subsea during a blowout, where the dispersants are injected directly into the oil stream, facilitating breakup of the oil into smaller droplets in the turbulent jet, or it can be done at the surface, where the treated oil will be broken up into small droplets when hit by breaking waves or other mechanical energy.

Looking at Eqs. (4.7) and (4.9), we see that reduced interfacial tension gives a larger Weber number, which in turn gives a smaller characteristic droplet size in natural dispersion, when everything else is kept constant. While droplet breakup in turbulent jets is outside the scope of this chapter, we can briefly mention that also in this case, the droplet size be may related to the Weber number, and reduced interfacial tension with everything else kept constant will lead to smaller droplets (Brandvik, Johansen, Leirvik, Farooq, & Daling, 2013; Johansen, Brandvik, & Farooq, 2013).

Small droplets produced by dispersant application will have a slower rise velocity, as discussed earlier in this chapter. As also mentioned, this will lead to the oil being spread over a larger volume of water, due to current shear and horizontal transport and mixing. Smaller droplets give rise to faster dissolution and faster biodegradation of the oil, potentially decreasing the overall lifetime of contamination after the spill. However, one should also be aware that applying dispersants does not remove the oil, even if it is less visible. Dispersant application can reduce the potential impact to sea birds, mammals, and shoreline habitats, but at the cost of potentially increasing the impact on marine life in the water column, as well as benthic species.

4.5 Eulerian model of vertical mixing

Calculating in the Eulerian picture means to consider concentrations at a set of points (or in a set of cells), and looking at how the concentrations in those points change with time. Solving a partial differential equation (PDE), such as the advection–diffusion equation, for a discrete grid of points, is an example of an Eulerian calculation. For different reasons, which will be described in more detail in Section 4.6, it is not very common to solve oil spill problems in the Eulerian picture. Nevertheless, some background information on the Eulerian picture is very useful, as this forms the starting point for the Lagrangian, particle based approach of oil spill modeling.

4.5.1 Advection–diffusion equation

The change in concentration, along the vertical dimension, of oil droplets that rise due to buoyancy and are mixed due to ocean turbulence, is commonly modeled as an advection–diffusion problem. Assuming the droplets to rise with a constant, terminal velocity, v_b, and that the spatially dependent diffusivity can be expressed as a function of depth and time, $K(z, t)$, the concentration of droplets as a function of space and time, $C(z,t)$, is described by

$$\frac{\partial C}{\partial t} = \frac{\partial}{\partial z}\left(K\frac{\partial C}{\partial z}\right) - v_b\frac{\partial C}{\partial z}. \tag{4.28}$$

If we have $v_b = 0$ and $K(z,t) > 0$, then this equation is simply the diffusion equation (also known as the heat equation). For simple geometries and initial conditions, analytical solutions are known for many cases, in particular if $K(z,t)$ is a constant (see e.g., Carslaw & Jaeger, 1959). With $v_b \neq 0$ and $K(z,t) = 0$, Eq. (4.28) becomes the advection equation, which describes the transport of a concentration profile, without diffusion. For the special case of constant v_b, the advection equation in one dimension describes a concentration profile that moves at a speed of v_b, without changing its shape.

For practical applications, it is usually not possible to find analytical solutions to Eq. (4.28). In such cases, a range of numerical solution techniques exist. For details, the interested reader is referred to the wide range of literature on the topic of numerical solutions of PDEs (see, e.g., Hundsdorfer, 2003; Pletcher, 2013; Versteeg, 2007).

4.5.2 Boundary conditions

In oil spill modeling, it is essential to distinguish between surface oil and submerged oil. Surface oil is not only distinguished by being located at zero depth, but also by the fact that surface oil is not subject to vertical mixing due to the turbulence in the water column. The idea behind this is that submerged oil is found in the form of droplets, which are surrounded by water and thus subject to turbulent motion. Surface oil, on the other hand, is present in the form of continuous patches of different sizes. In order to submerge oil from a patch or slick at the surface, some high-energy event, such as a breaking wave, is required to break the surface tension of the oil. The opposite process, that is, surfacing, is usually calculated from the buoyant rise speed of the oil droplets. Oil that reaches the surface due to buoyancy may leave the water column and merge with the surface slick. This makes oil behave differently than, for example, buoyant fish eggs, as these do not "get stuck" at the surface in the same way (Sundby & Kristiansen, 2015).

One way to model this behavior of oil is to assume that the concentration in the water column is described by the advection–diffusion equation, with a partially absorbing boundary at the surface. In particular, the boundary at the surface should enforce zero diffusive flux, while allowing the advective flux due to buoyancy to leave the water column through the boundary at the surface. The oil that has left the water column in this manner is then counted as part of the surface oil. The physical rationale for this choice of boundary conditions is that higher buoyancy (due to either larger droplets or less dense oil) does lead to faster surfacing, while higher diffusivity does not lead to faster surfacing.

The advective and diffusive fluxes are given by:

$$j_A(z,t) = wC(z,t), \tag{4.29a}$$

$$j_D(z,t) = -K(z,t)\frac{\partial C(z,t)}{\partial z}, \tag{4.29b}$$

where Eq. (4.29b) is commonly known as Fick's law (see, e.g., Csanady, 1973, p. 4). Hence, a no-diffusive-flux boundary condition at $z=0$ can be enforced by requiring

$$\frac{\partial C(z,t)}{\partial z}\Big|_{z=0} = 0. \tag{4.30}$$

Another option for modeling the surfacing of oil is to consider the surfacing process as a loss term in the PDE (also known as a sink), where oil that is sufficiently close to the surface is simply removed at a rate that would typically be calculated from the rise speed of the oil droplets. See Tkalich and Chan (2002) for an example of this approach. Note that the same rate of surfacing can be modeled in both approaches.

When considering a finite water depth, the boundary at the bottom should also be reflecting for the diffusion step. As long as the oil considered is positively buoyant, the advective flux through the bottom will necessarily remain zero. In advanced oil spill models, interaction with seabed sediments of different types through a turbid bottom layer may be included, where adhesion of oil to sediments is explicitly modeled. One may also wish to account for the possibility of sinking droplets of oils that are denser than water settling onto the sediment.

4.5.3 Source term for entrainment of oil

In the scheme described above, the concentration of submerged oil in the water column is described by the advection–diffusion equation (Eq. (4.28)). In this case, we may model the entrainment of oil by adding a reaction term to Eq. (4.28), which adds oil at certain depths. For example, if oil is entrained at rate $Q(t)$ (units mass per time), and distributed evenly across a depth interval ranging from $H_{\min}$ to $H_{\max}$, we have

$$\frac{\partial C}{\partial t} = \frac{\partial}{\partial z}\left(K(z)\frac{\partial C}{\partial z}\right) - v_b\frac{\partial C}{\partial z} + R(z,t), \tag{4.31a}$$

where

$$R(z,t) = \begin{cases} Q/L & \text{if} \quad H_{\min} < z < H_{\max} \\ 0 & \text{otherwise} \end{cases}, \tag{4.31b}$$

where $L = H_{\max} - H_{\min}$.

4.5.4 Modeling a droplet size distribution

An important point in oil spill modeling is the concept of a droplet size distribution, as discussed in Section 4.3.1. As oil is submerged due to breaking waves, a range of droplet sizes are produced, and these will have a different fate in the water column. Not only does the droplet size strongly influence the rise speed (see Section 4.4.1) but also dissolution and biodegradation are affected by the droplet size, due to the change in surface area relative to volume.

To capture the effect of droplet size on rise velocity in an Eulerian model, one needs to separate the submerged oil into discrete droplet size classes and solve one advection–diffusion equation for each size class. No exchange between droplet size classes is necessary for the submerged oil, but the source term (Eq. (4.31b)) must be modified such that the correct proportion of the submerged oil is inserted into each size class.

4.5.5 The well-mixed condition

The well-mixed condition (WMC), described by Thomson (1987), states that a passive (i.e., neutrally buoyant) tracer that is initially well mixed must remain well mixed while undergoing diffusion. This holds regardless of the shape of the diffusivity profile and provided of course that the tracer cannot escape through domain boundaries or similar. The WMC follows directly from the diffusion equation for a concentration, $C(z, t)$:

$$\frac{\partial C}{\partial t} = \frac{\partial}{\partial z}\left(K(z)\frac{\partial C}{\partial z}\right). \tag{4.32}$$

If $\partial_z C(z, t) = 0$ everywhere (including at the boundaries), then the right-hand side of Eq. (4.32) is 0, and hence there will be no change in concentration as time passes.

In Eulerian modeling of diffusion, it is fairly straightforward to ensure that the WMC is satisfied. In Lagrangian modeling, on the other hand, this is not always trivial. However, as stated by Thomson (1987), the WMC is a necessary (though not sufficient) condition for a Lagrangian scheme to be consistent with the diffusion equation. We will return to this point in Section 4.6.

4.6 Lagrangian modeling of vertical mixing

In oil spill modeling, the most common approach to simulating the transport and mixing of oil at sea is to represent the oil as numerical particles, also called Lagrangian elements (or sometimes "spillets"). These numerical particles move with the current, rise or sink according to their buoyancy, and move randomly to account for turbulent mixing. When a large number of Lagrangian elements is simulated, their distribution may be used to approximate the concentration field of a substance, such as oil.

In this section, we describe some of the theory behind the use of particles to model an advection–diffusion problem, and some conditions that must be satisfied in order for this approach to be equivalent to the Eulerian approach described above. We also give numerical schemes for the transport and mixing, the boundary conditions, and the entrainment.

The link between the diffusion equation and the distribution of a collection of randomly moving particles have been known for a long time. More than 100 years ago, Einstein (1905) showed that the random motion of Brownian particles (e.g., tiny pollen grains suspended in water) caused them to spread out in accordance with the diffusion equation on long time scales. A few years later, Langevin (1908) presented a differential equation for the movement of Brownian particles, based on Newton's second law with a stochastic force term.

Since then, the mathematical field of stochastic differential equations (SDEs) has been developed further and has put these results on a more solid theoretical foundation. In the following, we shall only use a few elements of the theory of SDEs, but references to further reading will be given where relevant.

4.6.1 Modeling vertical diffusion as a random walk

Diffusion in a Lagrangian model is described by a random walk, that is, a random displacement of particles at each timestep. More formally, a random walk is an example of an SDE, which is a differential equation that includes one or more random terms. A general one-dimensional SDE with one noise term is written

$$\mathrm{d}z = a(z,t)\,\mathrm{d}t + b(z,t)\,\mathrm{d}W_t, \tag{4.33}$$

where $a(z,t)$ is called the drift term, $b(z,t)$ is called the diffusion term or noise term, and $\mathrm{d}W_t$ are the random increments of a standard Wiener process, $W(t)$ (Kloeden & Platen, 1992, p. 40).

To solve this equation numerically, we first introduce a discrete time,

$$t_n = t_0 + n\Delta t, \tag{4.34}$$

and then we seek a scheme to calculate the next position, z_{n+1}, given the position, z_n, at time t_n. Numerous numerical schemes for SDEs exist, the simplest of which is the Euler–Maruyama scheme (Kloeden & Platen, 1992, p. 305; Maruyama, 1955). In this scheme, the iterative procedure for integrating Eq. (4.33) reads

$$z_{n+1} = z_n + a(z_n, t_n)\Delta t + b(z_n, t_n)\Delta W_n, \tag{4.35}$$

where z_n is the position at time t_n, and ΔW_n is a random number drawn from a Gaussian distribution with zero mean, $\langle \Delta W \rangle = 0$, and variance $\langle \Delta W^2 \rangle = \Delta t$.

For our purposes, it can be shown that if one solves the following SDE for a large number of particles,

$$dz = (w + K'(z))\,dt + \sqrt{2K(z)}\,dW(t), \tag{4.36}$$

then their distribution will develop according to the advection–diffusion equation (Eq. (4.28)), with advection w, and diffusivity $K(z)$. Additionally, in Eq. (4.36),

$$K'(z) = \left.\frac{\partial K}{\partial z}\right|_z. \tag{4.37}$$

See Appendix A for details on how to derive Eq. (4.36) from Eq. (4.28).

If we let the advection term be equal to the rise speed due to buoyancy, $w = v_b$, and discretize Eq. (4.36) with the Euler–Maruyama scheme, we obtain

$$z_{n+1} = z_n + (v_b + K'(z_n))\,\Delta t + \sqrt{2K(z_n)}\,\Delta W_n. \tag{4.38}$$

This equation is a discrete formulation of the transport equation for numerical particles. A similar expression may be used for the horizontal directions. Some variant of this equation is commonly seen in papers on numerical oil spill modeling. Note, though, that it is also quite common to see this equation without the term $K'(z_n)\,\Delta t$, in which case it is not consistent with the advection–diffusion equation (except in the special case where K is a constant). See Section 4.7.1 and Nordam, Nepstad, Litzler, and Röhrs (2019) for further details.

Just as for Ordinary Differential Equations (ODEs), a range of different numerical schemes exist for solving SDEs such as Eq. (4.36). For a review of several different schemes in the context of marine particle transport, see Gräwe (2011) and Gräwe et al. (2012). The interested reader should also refer to the general SDE literature such as the classic work by Kloeden and Platen (1992). See also Section 4.9.1.

Note that by describing the theory for vertical transport separately, we have implicitly assumed that the vertical motion can be treated independently of the horizontal motion, at least within a timestep. This is usually a fair assumption, as discussed in the next section. However, for a more general treatment, including iso- and diapycnal diffusivity (which leads to a nondiagonal diffusivity tensor, $\mathbf{K}$, if the isopycnals are not horizontal), see Spivakovskaya, Heemink, and Deleersnijder (2007b).

4.6.2 Vertical timestep

Regarding the choice of timestep, it will in many cases make sense to have a far shorter timestep for the vertical motion in an oil spill model, than for the horizontal motion. Among the reasons for this is that available ocean data usually have a far higher resolution in the vertical direction than in the horizontal and that diffusivity gradients tend to be both sharper and more persistent in the vertical direction. Hence, inaccuracies in the vertical transport step can lead to systematic errors in the vertical distribution of oil, which in turn can lead to errors in, for example, the prediction of surface signature. See Section 4.7 for some relevant examples.

Visser (1997) wrote down a criterion for the length of the timestep, which is based on the requirement that the diffusivity profile should be approximately linear over the typical length of a random step. If this criterion is satisfied, the WMC (see Section 4.5.5) should be reasonably well satisfied. He obtained

$$\Delta t \ll \min\left|\frac{1}{K''(z)}\right| \tag{4.39}$$

where the minimum is to be taken over the entire water column, and $K''(z)$ is the second derivative of $K(z)$ with respect to z. (Note that Visser did not take the absolute value, but this is clearly an omission since $K''(z)$ can be negative.) According to Gräwe et al. (2012, Section 3.4), it is commonly agreed that the timestep should be at least one order of magnitude smaller than the limit in Eq. (4.39).

It is worth noting that if $K''(z)$ is not finite everywhere, for example because $K(z)$ is a step function, or is a piecewise linear function with discontinuous first derivative, then the Visser timestep condition can never be satisfied. Fundamentally, this problem stems from the fact that the equivalence between the advection–diffusion equation and the random walk described by Eq. (4.36) requires both the drift and diffusion coefficients in Eq. (4.36) to be continuous. See Appendix A for further details.

4.6.3 Boundary conditions

As was discussed in Section 4.5.2, it is common in oil spill modeling to treat the boundary at the surface differently for diffusion and advection (advection here refers to the buoyant rise of droplets). In a Lagrangian model, this is straightforward to achieve by separating the advection term and the diffusion term in Eq. (4.36) into two separate steps. During each timestep, each particle is first displaced randomly due to diffusion, reflected from the surface or sea floor, moved upward due to buoyancy, and finally removed from the water column if the buoyancy brought it above the surface.

This scheme can be formulated as the following series of operations carried out for each particle, during each timestep, in order to update the position, z. We here consider a water column of finite depth H (depth positive downward) and a particle rising with a constant terminal speed v_b.

Step 1: Displace particle randomly

$$z \to z + K'(z)\Delta t + \sqrt{2K(z)}\Delta W. \tag{4.40a}$$

Step 2: Reflect from boundaries

$$z \rightarrow \begin{cases} -z & \text{if} \quad z<0 \\ 2H - z & \text{if} \quad z>H \\ z & \text{otherwise.} \end{cases} \tag{4.40b}$$

Step 3: Rise due to buoyancy

$$z \rightarrow z - v_b \Delta t. \tag{4.40c}$$

Step 4: Set depth to 0, if above surface

$$z \rightarrow \begin{cases} 0 & \text{if} \quad z \leq 0 \\ z & \text{otherwise.} \end{cases} \tag{4.40d}$$

A particle that reaches the surface in Step 4 is removed from the water column and considered "surfaced," corresponding to the droplet merging with a continuous surface slick. It will then take energy in the form of breaking waves to reintroduce surfaced oil into the water column. In that case, a fifth step is also carried out at each timestep:

Step 5: If a particle is considered surfaced, it is resuspended with probability $p = 1 - e^{-\Delta t/\tau}$, in which case it is assigned random droplet size and depth, drawn from suitable distributions.

In Step 5, Δt is the timestep, and the lifetime, $\tau = 1/\alpha$, is calculated from the entrainment rate, α (where α, with units time^{-1}, is the decay rate of the amount of surface oil; see Eq. 4.13). Note that steps 1 to 4 are applied to all particles in the water column (i.e., those particles that are not part of the surface slick), while step 5 is applied to all particles that *are* part of the slick.

The particle scheme described by steps 1 to 5 is equivalent to Eulerian modeling of the advection–diffusion equation with a Neumann boundary condition at the surface, enforcing zero diffusive flux, while allowing an advective flux (Nordam, Kristiansen, Nepstad, & Röhrs, 2019).

4.7 Some examples and pitfalls

In this section, we describe some example calculations and pay particular attention to some common mistakes that should be avoided.

4.7.1 Naïve random walk

In the case of constant diffusivity, K, the random walk given by Eq. (4.36) simplifies to

$$dz = v_b\, dt + \sqrt{2K}\, dW(t). \tag{4.41}$$

or discretized with Euler–Maruyama

$$z_{n+1} = z_n + v_b \Delta t + \sqrt{2K} \Delta W_n. \tag{4.42}$$

However, if K is a function of position, Eq. (4.41) is not consistent with the advection–diffusion equation and gives unphysical results where a net transport away from regions of high diffusivity is seen (Holloway, 1994; Hunter, Craig, & Phillips, 1993; Visser, 1997). The difference between the two schemes is the term $K'(z)\Delta t$ in Eq. (4.36), which is known as the pseudovelocity term (Lynch et al., 2014, p. 125).

In oil spill modeling, it seems fairly common to use the random walk scheme described by Eq. (4.42), even in combination with spatially variable diffusivity. In, for example, the plankton modeling community, the importance of using a consistent random walk appears to have been well known for two decades, with a particularly clear account of this issue being that of Visser (1997). In what follows, we will use the terminology of Visser and refer to Eq. (4.42) as the naïve random walk.

An investigation of this issue in the context of oil spill modeling is presented in Nordam, Nepstad, Litzler, and Röhrs (2019), where it is found that use of the naïve random walk scheme may lead to both over- and underprediction of the amount of surface oil, compared to the consistent random walk scheme (Eq. (4.36)). The difference depends on the nature of the diffusivity profile, as well as the relevant droplet size distribution.

An example is shown in Fig. 4.3, where initially evenly distributed neutrally buoyant tracers have been modeled with the naïve scheme (Eq. (4.42)), we observe that the initially constant concentration profile breaks down and the tracers start to accumulate in the regions of low diffusivity. While the

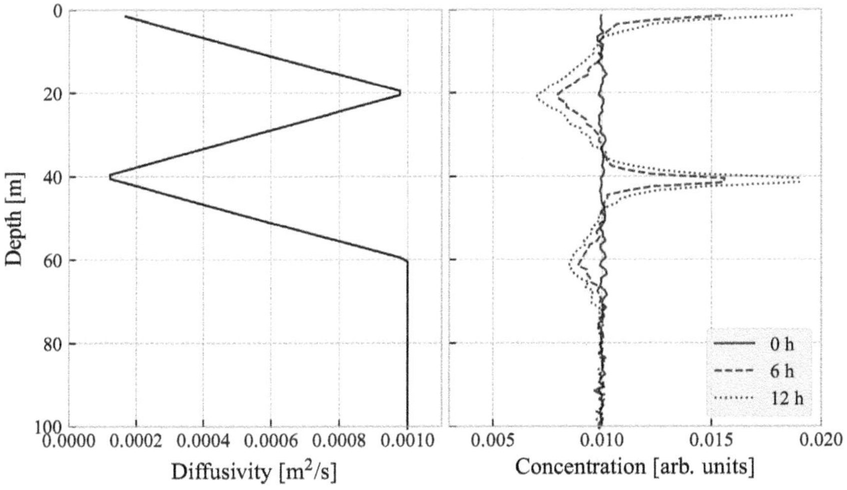

FIGURE 4.3 Concentration of initially well-mixed neutrally buoyant tracers, simulated with the naïve scheme (Eq. (4.42)), shown after different times. The number of particles is $N_p = 1{,}000{,}000$, the timestep is $\Delta t = 300$ s, and concentration is calculated by bin count in 100 bins of width 1 m each.

example here uses neutrally buoyant particles, it is clear that this effect can lead to errors in modeling, for example, the surfacing of small oil droplets.

4.7.2 Step-function diffusivity

Due to the difficulty of obtaining good data on the vertical diffusivity in the water column, simple schemes are sometimes used. For example, a step-function diffusivity profile may represent the well-known fact that diffusivity tends to be higher in the mixed layer and lower below the pycnocline. An example of such a step-function profile used in oil spill modeling is found in De Dominicis et al. (2016, Section 6.3):

$$K(z) = \begin{cases} 10^{-2}\text{m}^2\text{s}^{-1} & \text{if} \quad z < 30\text{ m} \\ 10^{-4}\text{m}^2\text{s}^{-1} & \text{otherwise} \end{cases}, \tag{4.43}$$

where depth is positive downward. The diffusivity profile is illustrated in the left panel of Fig. 4.4.

It is clear that with this diffusivity profile, the Visser timestep criterion (Eq. (4.39)) can never be satisfied, and thus we cannot expect the WMC to be satisfied, regardless of the timestep. The problem can be understood intuitively by realizing that particles that are in the high-diffusivity region, but close to the transition depth to low diffusivity, have a good probability to make a relatively large jump into the region of low diffusivity. Once there, however, it would take this particle a large number of steps to return to the

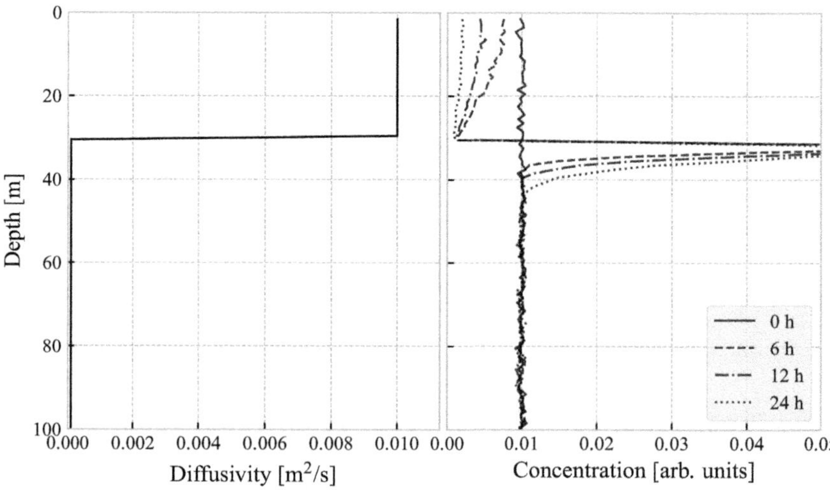

FIGURE 4.4 Concentration of initially well-mixed neutrally buoyant tracers, shown after different times. The number of particles is $N_p = 10{,}000{,}000$, the timestep is $\Delta t = 600$ s, and concentration is calculated by bin count in 100 bins of width 1 m each.

region of high diffusivity. The net result is that the particles tend to accumulate in the region of low diffusivity, in violation of the WMC.

The results of a numerical test of the WMC are shown in the right panel of Fig. 4.4. Neutrally buoyant particles have been initially evenly distributed across the water column, down to a depth of $H = 100$ m. A reflecting boundary condition has been used at the surface ($z = 0$), and at the bottom ($z = H$). Concentration profiles are shown for different times, and it is clear that the particle count in the high-diffusivity region is depleted, for the reason described above. As described in the discussion of the WMC in Section 4.5.5, the correct solution to the diffusion equation in this case is that the concentration should remain constant.

While this example uses neutrally buoyant tracer particles, it is clear that this behavior would also be a problem in an oil spill simulation. The effect of using this diffusivity profile is a net downward displacement of particles, which leads to reduced surfacing rates in an oil spill model, particularly for small droplets with slow rise speeds.

We note that with this diffusivity profile, we have $K'(z) = 0$ everywhere, except at $z = 30$ m, where the derivative of K is a Dirac delta-function. Hence, the naïve random walk (Eq. (4.42)) and the corrected random walk (Eq. (4.36)) are identical in this case, except in a single point, and the inclusion of the pseudovelocity term does not compensate for the spurious downward drift caused by the diffusivity profile.

We will now describe two approaches to avoid this problem. The first can be said to be a "workaround," that modifies the diffusivity function to make it into a smooth approximation of a step function, while the second approach uses a different numerical scheme to solve the SDE for diffusion.

The "workaround" to simulating this problem would be to replace the step-function diffusivity with a smooth sigmoid function with the same asymptotic values as the step function. In particular, the step function

$$K(z) = \begin{cases} K_0 & \text{if} \quad z < z_0 \\ K_1 & \text{otherwise} \end{cases}, \tag{4.44}$$

can be approximated as

$$K(z) = K_0 + \frac{K_1 - K_0}{1 + e^{-a(z-z_0)}}, \tag{4.45}$$

where the value of the parameter a determines the sharpness of the transition. By using a diffusivity profile given by Eq. (4.45), with large values of a, true step-function diffusivity can be approximated arbitrarily well. If this is done in combination with a timestep that satisfies the Visser criterion (Eq. (4.39)), one can make sure the WMC is satisfied. For the sigmoid diffusivity profile given by Eq. (4.45), the Visser timestep criterion becomes

$$\Delta t \ll \frac{\sqrt{3}}{18} \cdot \left| \frac{1}{a^2(K_0 - K_1)} \right|. \tag{4.46}$$

As mentioned in Section 4.6.2, the timestep should be kept at least an order of magnitude below this limit. This may make the timestep impractically short if a large value of a is chosen.

Another approach, which may be more efficient numerically, is to use the step-function diffusivity directly, but with a different numerical scheme for solving the SDE (Eq. (4.36)). Spivakovskaya, Heemink, and Deleersnijder (2007a) describe an alternative to the Euler–Maruyama scheme, which they call the backward Itô scheme. In this scheme, the position, z_{n+1}, of a particle at time t_{n+1} is given from its position z_n, at time t_n, by

$$\tilde{z}_n = z_n + \sqrt{2K(z_n)}\Delta W_n, \tag{4.47a}$$

$$z_{n+1} = z_n + v_b\,\Delta t + \sqrt{2K(\tilde{z}_n)}\Delta W_n, \tag{4.47b}$$

where ΔW_n is *the same realization* of a Gaussian random variable with zero mean and variance $\langle \Delta W_n^2 \rangle = \Delta t$ in both Eqs. (4.47a) and (4.47b). Hence, the net effect is to make a "trial step," to a position $\tilde{z}_n$, and then use the diffusivity at that point, $K(\tilde{z}_n)$, in the real step. With the backward Itô scheme, the WMC is satisfied for a step-function profile. However, the backward Itô scheme does not work as well as Euler–Maruyama for, for example, continuous, piecewise linear diffusivity functions. Experimentation is encouraged to verify that a chosen combination of diffusivity profile, numerical scheme, and timestep satisfies the WMC to acceptable accuracy.

4.7.3 Linearly interpolated diffusivity

Many ocean models provide eddy diffusivity as output, along with current, temperature, salinity, etc. If diffusivity is available, it can be used to drive the random walk, but care should be taken in the interpolation of the data. In particular, it is clear that the Visser timestep condition (Eq. (4.39)) can never be met if linear interpolation is used, as this will give a diffusivity profile that has piecewise constant first derivative, and hence a delta-function second derivative at each node in the interpolation.

As an example, we have carried out a test of the WMC for a piecewise linear diffusivity profile, as shown in the left panel of Fig. 4.5. While this profile is of course somewhat artificial, it has some realistic features, in that the diffusivity goes down toward the surface and has a minimum at some value representing the pycnocline (see, e.g., Gräwe et al., 2012 for a thorough discussion of the problem of a sharp pycnocline). A passive tracer represented by $N_p = 100{,}000{,}000$ particles was initially evenly distributed throughout the water column, down to a depth of 100 m. Reflecting boundaries were used at the bottom and surface.

In the right panel of Fig. 4.5, concentration profiles are shown for different time points. The results clearly indicate that there are deviations from constant concentration at the minima of the diffusivity profile, as well as below 60 m depth, where the diffusivity is constant. The degree to which this happens

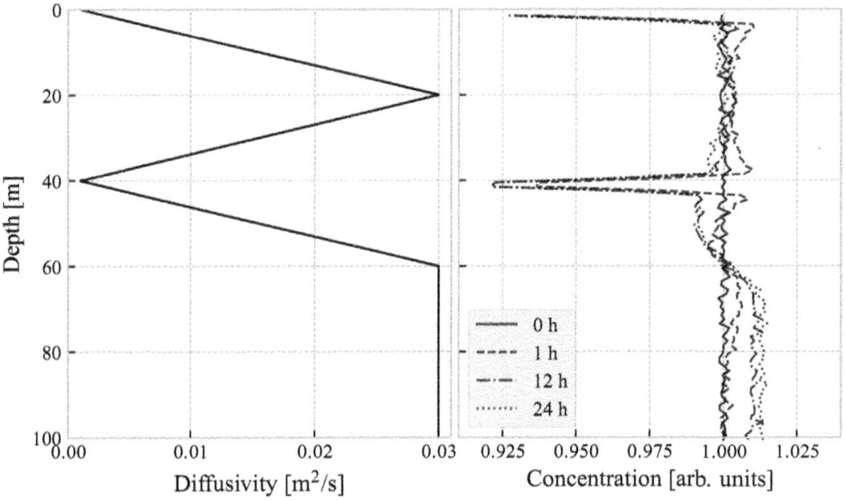

FIGURE 4.5 Concentration of initially well-mixed passive tracers, shown after different times. The number of particles is $N_p = 100{,}000{,}000$, the timestep is $\Delta t = 600$ s, and concentration is calculated by bin count in 100 bins of width 1 m each.

depends on the timestep, as well as the diffusivity profile, and a sufficiently short timestep will in practical applications remove the problem. Nevertheless, this demonstrates that unexpected things may happen if one uses linear interpolation of input data without checking that the WMC is satisfied.

4.7.4 Chemically dispersed oil in the mixed layer

The final case is included as an example of a situation where a one-dimensional oil spill model may be of practical use. We consider an idealized situation where oil has been treated with surface dispersants and dispersed into the water column by means of mechanical energy, either through waves, prop wash, water jetting, or other means. The question is then, for a given droplet size, how long may one expect the oil to remain submerged. If the oil stays submerged for a long time, the dispersant operation may be said to have been successful.

In this idealized case, we will consider a single droplet size and a sigmoid diffusivity profile giving a high diffusivity in the mixed layer, and a low diffusivity below the pycnocline (see Section 4.7.2). In particular we choose to use a diffusivity profile given by Eq. (4.45), with parameters $K_0 = 1 \times 10^{-4} \text{m}^2/\text{s}$, $K_1 = 1 \times 10^{-2} \text{m}^2/\text{s}$, $z_0 = 20$ m, and $a = 2\ \text{m}^{-1}$. The diffusivity profile is shown in the left panel of Fig. 4.6.

We consider two droplet sizes: 500 and 50 μm. Assuming an oil density of 0.95 kg/L, and using Eq. (4.20) to calculate the rise speed, we get respectively $v_b = 5.4\ \text{mm}/s$ and $v_b = 0.072\ \text{mm}/s$.

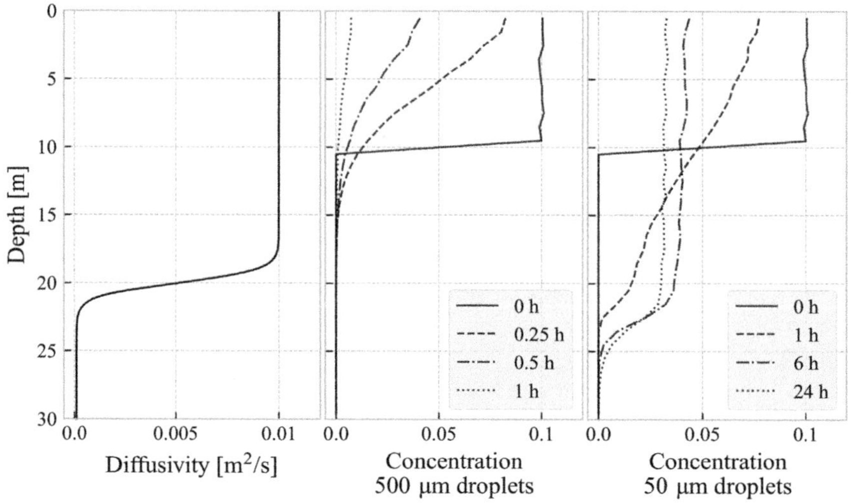

FIGURE 4.6 Left panel: Diffusivity as a function depth. Middle panel: Oil concentration, as a function of depth, for a droplet diameter of 500 μm. Right panel: The same, for a droplet diameter of 50 μm.

Before presenting simulation results, we will try to reason about what might be expected to happen. A useful quantity to consider here is the Péclet number,

$$\mathrm{Pe} = \frac{vH}{K}, \tag{4.48}$$

which gives the ratio between advective transport, and diffusive transport. Note that in our case, v is the rise speed of the droplets, H is the thickness of the mixed layer, and K is the (average) diffusivity in the mixed layer. If $\mathrm{Pe} \gg 1$, the transport is advection-dominated (advection here refers to the rise speed of the droplets), and if $\mathrm{Pe} \ll 1$, the transport is diffusion-dominated. With the parameters described above, we get $\mathrm{Pe} \approx 11$ for the large droplets, and $\mathrm{Pe} \approx 0.15$ for the small droplets.

Based on these considerations, we can begin to reason about the outcome of the dispersant operation, for the two droplet sizes we chose to look at. For the larger droplets, the vertical transport will be dominated by the rise speed. In the limit of zero diffusivity, the droplets will simply rise to the surface at their terminal velocity, v_b. If we assume an initial amount Q_0 of submerged oil, evenly distributed down to a depth L, then the amount of oil that remains submerged at time t is simply given by

$$Q(t) = Q_0\left(1 - t\frac{v_b}{L}\right), \quad 0 < t < L/v_b. \tag{4.49}$$

When $t = L/v_b$, all the oil droplets have had time to reach the surface, and there is no submerged oil remaining. While the diffusivity will never be zero in a real case, we will see later that Eq. (4.49) provides a reasonable approximation if $\mathrm{Pe} \gg 1$.

For the small droplets, transport is diffusion-dominated. Hence, they will be evenly distributed throughout the mixed layer, even if they were only initially entrained a short distance. Furthermore, the diffusivity in the mixed layer is sufficient to keep the remaining submerged droplets evenly distributed, even as the surfacing begins. We conclude that the fraction of submerged droplets that will surface during an interval Δt, is given by $v_b \Delta t / H$, where H is the thickness of the mixed layer. When a constant fraction resurfaces during an interval, we have a first-order decay process. If the initial amount of submerged oil is Q_0, then the remaining submerged oil is given by:

$$Q(t) = Q_0 \mathrm{e}^{-t/\tau}, \quad \tau = H/v_b. \tag{4.50}$$

Thus, we find that in addition to the difference in rise speed, there is also another difference that is relevant between advection-dominated transport (large droplets) and diffusion-dominated transport (small droplets), and that is the length scale. For large droplets, the entrainment depth is important, while for small droplets, the thickness of the mixed layer is important.

We will now look at some numerical simulation results. For both droplet sizes, we assume that the oil is initially evenly distributed down to a depth of $L = 10$ m. We run simulations using $N_p = 100{,}000$ particles. For the diffusivity profile described above, the Visser timestep limit (Eq. (4.39)) gives $\Delta t \ll 42$ s, and hence we choose $\Delta t = 2$ s.

In Fig. 4.6, the concentration of oil droplets is shown as a function of depth, for different times. We observe that the large droplets rise quickly to the surface and are not mixed any deeper than the initial depth of $L = 10$ m. For the small droplets, we observe that they fairly quickly mix down to the pycnocline, and that the concentration thereafter remains approximately constant with depth throughout the mixed layer.

Fig. 4.7 shows the remaining fraction of submerged oil as a function of time. Additionally, the idealized time developments given by Eqs. (4.49) and (4.50) are shown as a dashed lines.

The purpose of this example is to illustrate some special cases that may help provide some simple guidelines to reason about the outcome of a surface dispersant operation. In particular, we observe that if we assume $\mathrm{Pe} \gg 1$, then the time for the oil to surface is largely governed by the entrainment depth and the rise velocity. On the other hand, if $\mathrm{Pe} \ll 1$, then the time development is determined by the depth of the pycnocline and the rise velocity. The diffusivity does not appear in either case, other than in the estimation of Pe.

Finally, we note that it is of course not realistic to consider the entrainment and surfacing of oil as a purely one-dimensional problem over a period of several days, as in the right panel of Figs. 4.6 and 4.7. During this time, the oil will certainly be subject to horizontal advection and diffusion. This is of course precisely the goal of a surface dispersant operation, and such an operation will probably be said to be successful if the majority of the oil may be expected to remain submerged for several days.

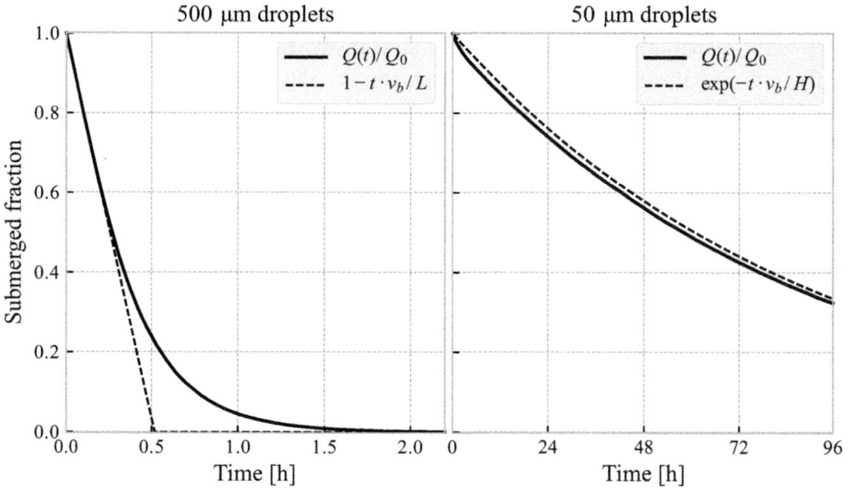

FIGURE 4.7 Left panel: Submerged fraction of oil, as a function of time, for a droplet diameter of 500 μm. The idealized time development given by Eq. (4.49) is shown as a dashed line. Right panel: The same, for a droplet diameter of 50 μm. The idealized time development given by Eq. (4.50) is shown as a dashed line.

4.8 Example cases

From the discussion in the preceding sections, it should be clear that the vertical movement of oil in the water column is an interplay between different effects. Entrainment moves oil from the surface, and into the water column. Buoyancy transports oil upward, and eventually to the surface, at a rate that is dependent on the droplet size distribution (which in turn depends on the conditions during entrainment).

Turbulent mixing tends to distribute the oil in the vertical. While this diffusion process does not itself have a preferred direction, the net effect can still be to move the center of mass of a concentration profile either up or down, due to the reflecting boundary at the surface, vertical variation in diffusivity, and dependence on initial conditions.

In breaking wave conditions, there will always be some entrainment of surface oil. Hence, some fraction of the oil will be submerged at any time, and some fraction will remain at the surface. The fraction at the surface will depend on wave conditions, vertical diffusivity in the subsurface, and the state of the oil (since droplet size distribution depends, among other things, on the viscosity of the oil). The surface fraction will change with time even if the wave conditions remain constant, as the oil weathers, and since the smaller droplets may remain submerged for a very long time.

It is clear that even though oil is typically buoyant, it is quite possible for the majority of the oil in a surface spill to be transported in the subsurface, in a state of dynamic equilibrium between entrainment and resurfacing. As discussed in the introduction, the vertical distribution of oil may have significant impact on horizontal transport, due to current shear effects. The aim of this section is to provide some examples of real oil spill scenarios where the vertical distribution of oil is of particular importance to the horizontal transport.

4.8.1 The 1993 *Braer* oil spill

On January 5, 1993, MV *Braer* ran aground within 100 m of the coast of Shetland (Reed et al., 1999) during a storm. It was carrying 85,000 tons of Gullfaks crude oil, which was released into the ocean over a period of several days (M. Spaulding, Kolluru, Anderson, & Howlett, 1994). During the event, model forecasts were made available but failed to accurately predict the movement of the oil (Turrell, 1994, 1995). Later, several hindcast modeling studies were made (see, e.g., Proctor, Elliot, & Flather, 1994; M. Spaulding et al., 1994; Turrell, 1994).

While the wind was mainly flowing toward the northeast, much of the oil moved toward the south, with oil found in the sediments up to 100 km to the south of the spill site (Proctor et al., 1994). One explanation would be that the relatively light Gullfaks crude dispersed as small droplets in the strong winds present during the spill, causing a large fraction of the spilled oil to be transported toward the south by the subsurface currents. An additional relevant mechanism is that of oil-mineral aggregation (which has not been discussed in this chapter), which may cause oil to sink when associated with high-density mineral particles.

4.8.2 The 2011 *Golden Trader* oil spill

On September 10, the bulk carrier MV *Golden Trader* collided with a fishing vessel off the north-west coast of Denmark. There were no casualties, but some bunker fuel was spilled from MV *Golden Trader*. The amount was later estimated at 150 tons. During the first 2 days after the spill, approximately 50 tons of oil were collected by Danish response vessels. After this, the wind picked up, and no further observations of oil were reported until September 15, when oil reached the Swedish shore. On September 16, it became clear that a significant amount of oil (estimated amount 25–30 tons) had beached (Transport Malta, 2012).

The distance from the release point to the site of the beaching is more than 250 km. Only approximately 15 km of the shoreline was heavily oiled (ITOPF, 2011). Combined with the fact that beaching appears to have occurred over a period of half a day or more, this indicates that the slick may have been elongated in the wind direction and relatively narrow in the cross-wind direction as discussed by Johansen (1982) and Elliott (1986).

To the best of our knowledge, no detailed hindcast of this incident has been published. Such a hindcast would however be an interesting exercise. It seems likely that a number of model processes will impact the arrival time of the oil and the site of the beaching, including droplet size distribution, vertical mixing, and possibly Stokes' drift (Broström, Drivdal, Carrasco, Christensen, & Mattsson, 2014).

4.9 Advanced topics and further reading

Historically, the mathematical and technical details of Lagrangian particle schemes have received limited attention in papers on oil spill modeling (see, e.g., Nordam, Nepstad, Litzler, & Röhrs, 2019 and references therein). A challenge is that the mathematical literature on SDEs is often very technical and not very accessible to nonspecialists. However, there exists a large body of work on the modeling of plankton, fish eggs, sediment particles, atmospheric dispersion, etc., where these schemes are treated more rigorously that what is commonly seen in the oil spill modeling literature. Much of this work is formulated in terms of familiar concepts from applied oceanography and may be more or less directly applied to the transport part of oil spill modeling.

In this section, we discuss some advanced topics and recommend some further reading for those who are interested in the details of these topics.

4.9.1 Higher-order stochastic differential equation solvers

Earlier, we used the Euler–Maruyama scheme to discretize Eq. (4.36), obtaining the following iterative scheme for particle positions:

$$z_{n+1} = z_n + (w + K'(z_n))\Delta t + \sqrt{2K(z_n)}\Delta W_n.$$

However, just like the Euler scheme is the simplest, and least accurate, solver for ordinary differential equations, so the Euler–Maruyama scheme is the simplest and least accurate SDE solver. Switching to higher-order schemes should in principle give improved accuracy at the same timestep or reduce computational effort by allowing a longer timestep to be used.

For SDE schemes, two types of convergence exist: weak and strong. Convergence in the weak sense means that for a large number of particles, the distribution of particles will converge toward the *true* distribution (which may or may not be known) as the timestep goes to zero. Technically, weak convergence is expressed in the following way: If, for a numerical SDE scheme, and for sufficiently short timesteps Δt, there exists a constant C, such that

$$|\langle f(z_N)\rangle - \langle f(z(t_N))\rangle| < C\Delta t^{\gamma}, \tag{4.51}$$

then the scheme is said to have order of convergence γ in the weak sense. Here, z_N is the numerical approximation at time t_N, and $z(t_N)$ is the true solution at the same time, and the angle brackets indicate ensemble average over many independent particles. The functions f are continuous functions that have polynomial growth and are at least $2(\gamma + 1)$ times differentiable. Since this class of functions includes all the integer powers of z, it follows that the moments of the distribution converge if the scheme converges in the weak sense. As any distribution is uniquely defined by its moments, this means that the modeled distribution converges to the true distribution.

Convergence in the strong sense is also called pathwise convergence. If, for a numerical SDE scheme, and for sufficiently short Δt, there exists a constant C, such that

$$\langle |z_N - z(t_N)| \rangle < C\Delta t^{\gamma}, \tag{4.52}$$

then the scheme is said to have order of convergence γ in the strong sense.

The Euler–Maruyama scheme has orders of convergence 1/2 in the strong sense and 1 in the weak sense. Higher-order schemes exist, but the complexity of the schemes grows fast as the order increases. An example of a higher order scheme is the first-order Milstein scheme, which has order of convergence 1, in both the strong and the weak sense (see, e.g., Kloeden & Platen, 1992, p. 345). Applied to our SDE for advection–diffusion problems (Eq. (4.36)), the first-order Milstein scheme yields

$$z_{n+1} = z_n + (w + K'(z_n))\Delta t + \sqrt{2K(z_n)}\Delta W_n + \frac{1}{2}K'(z_n)(\Delta t - \Delta W_n^2). \tag{4.53}$$

Gräwe et al. (2012) argue that in some cases, the Euler–Maruyama scheme is simply inadequate, even with very short timesteps. The example they give is that of a strong, sharp pycnocline where the diffusivity will drop almost to zero at the steepest point of the density gradient. In such a case, passive tracers should cross the pycnocline very slowly, a behavior that is modeled far more accurately by the first-order Milstein scheme, due to its higher order of convergence in the strong sense.

For a clear and readable presentation of a range of numerical SDE schemes, with a view to marine particle tracking applications, the interested reader is referred to Gräwe (2011) and Gräwe et al. (2012). Note however that some of the schemes have been found to contain small mistakes; hence, it is also advisable to consult other sources prior to implementation, for example, the classic work by Kloeden and Platen (1992).

4.9.2 Autocorrelated velocity or acceleration

Implementing a random walk scheme that makes random displacements at each timestep, with no correlation in time, makes the implicit assumption that a moving particle can instantly change its velocity. This may seem unreasonable. Furthermore, when very short timesteps are used we find that particle speed becomes arbitrarily large, since the average step-length is proportional to $\sqrt{2K\Delta t}$, and we have

$$\lim_{\Delta t \to 0} \frac{\sqrt{2K\Delta t}}{\Delta t} = \infty \tag{4.54}$$

for any positive K. Note, however, that while these points sound unreasonable from a physical point of view, there is no problem in using the random walk scheme with short timesteps. Eq. (4.36) was derived to be consistent with the advection–diffusion equation, and in the limit $\Delta t \to 0$, $N_p \to \infty$, the distribution of particles *will* converge to correct distribution, almost surely[2] (provided $K(z)$ and $K'(z)$ are sufficiently smooth functions).

In fact, the "infinite speed" of the particle is a feature that is built into the model from the start: The Wiener process, $W(t)$, whose increments appear in Eq. (4.36), has infinite total variation on any interval of nonzero length (Brzeźniak & Zastawniak, 1999, pp. 157–158). The apparent problem stems only from trying to extract a physically meaningful "speed" from a model that does not contain the speed of the particle as a variable.

Nevertheless, it might in some cases be desirable to have a more physically realistic random walk model. Recall that what we have been calling diffusivity is in reality a parameterization of mixing due to turbulence. If we consider a neutral tracer in a field of turbulent eddies, it is clear that the velocity at one instant will be at least somewhat correlated to the velocity a short time later. This behavior can also be captured in numerical modeling.

Lynch et al. (2014) describe a hierarchy of random walk models with different degrees of autocorrelation. The standard random walk that we have been considering so far is called AR0 in this hierarchy, as it has no autocorrelation in the displacement at each step. (Note that the position of a particle does of course have autocorrelation, as the position at time t_n depends on the position at time t_{n-1}.)

The next level of the hierarchy is called AR1, where the displacement at each step is related to the displacement at the previous step. In this scheme, there is autocorrelation not only in the position of a particle, but also in its velocity. This is in a way a more realistic model, as in reality, the movement of a particle from one instant to the next is correlated, with the decorrelation time being dependent on the turbulent fluctuations.

The original Langevin equation was formulated to describe Brownian motion, that is, the apparently random motion of small particles in fluids, caused by collisions with the molecules of the fluid (Langevin, 1908; Lemons & Gythiel, 1997), and reads

$$m\frac{d^2x}{dt^2} = -6\pi\mu a\frac{dx}{dt} + X. \tag{4.55}$$

2. The term "almost surely" is used in the technical sense meaning "with probability 1."

Here, m is the mass of the particle, $-6\pi\mu a \frac{\mathrm{d}x}{\mathrm{d}t}$ is the drag force from the bulk fluid, and X is a random force representing the collisions of molecules. Hence, this is simply Newton's second law, with a random component in the force.

Along the same lines, Lynch et al. (2014) write down a general equation for an AR1 scheme as

$$\frac{\mathrm{d}^2 z}{\mathrm{d}t^2} + \frac{1}{\tau}\frac{\mathrm{d}z}{\mathrm{d}t} = \eta, \tag{4.56}$$

where η is some random process acting as a forcing and τ is a timescale for decay of the velocity if no forcing is applied. Written as a pair of coupled first-order SDEs in standard notation, this becomes

$$\begin{aligned} \mathrm{d}v &= -v\frac{1}{\tau}\,\mathrm{d}t + \eta \mathrm{d}W_t, \\ \mathrm{d}z &= v\,\mathrm{d}t. \end{aligned} \tag{4.57}$$

In this model, there is a time correlation in the movement of the particle, since the velocity will only change by a small amount between timesteps. This is also called a "random flight." It should be noted that an AR1 scheme is fundamentally different from an AR0 scheme, in that it is *not* consistent with the diffusion equation. And that is of course the argument for using this schemes in the first place, since what we are trying to model is turbulent mixing, and not pure diffusion.

AR1 schemes have a long history of usage in dispersion models for the atmosphere (Thomson, 1987). However, comparisons between this and the AR0 scheme have shown small differences in the far field (Wilson & Yee, 2007). This implies that differences in the results of an oil spill simulation are unlikely to be substantial.

A review by M. L. Spaulding (2017) mentions regarding an AR1 scheme that "[u]se of this higher order model is possible if one has accurate estimates of the currents and dispersion," but does not elaborate further. Recent papers by Cui et al. (2018, 2020) solves the so-called Maxey–Riley equation, describing the inertia and drag forces on individual oil droplets due to turbulent motion of the surrounding waters, and compares the results to a regular random walk scheme. However, this work considers only small spatial scales underneath breaking waves, and the effects on larger scales are not investigated.

In conclusion, AR1 (or even higher order) schemes do not appear to be commonly used in oil spill modeling. In addition to the slight increase in mathematical and numerical complexity, a practical problem in using an AR1 scheme is that one can no longer use the eddy diffusivity directly but must instead obtain estimates of the parameters τ and η in Eq. (4.56). The interested reader is referred to Lynch et al. (2014) and references therein, as well as Duran (2016), Gillespie (1996), and the literature on atmospheric dispersion (see, e.g., Thomson, 1987; Wilson & Flesch, 1993).

4.9.3 Reconstructing a concentration field from particles

As discussed in Section 4.6, our random walk scheme is in some sense equivalent to the advection–diffusion equation. The link is that each particle, at time t_n, represents a *sample* from the distribution at that time, where the distribution develops according to the advection–diffusion equation. If we want to (approximately) reconstruct the distribution from the particles, there are several different approaches, and which is most suitable may depend on the application. We discuss these in one dimension in this section, but generalization to several dimensions is natural.

The simplest approach is the so-called box count or bin count, which consists of dividing the region of interest into discrete bins and counting the number of particles in each bin. The concentration in each bin is then proportional to that number, weighted by the particle mass if each particle represents a different mass. This is exactly the same as a weighted histogram of particle positions.

We let our cells have constant size Δz, and define cell i by $(i-1)\Delta z \leq z < i\Delta z$. Furthermore, let particle j have position z_j and represent a mass m_j. Then the concentration C_i in cell i, is given by

$$C_i = \frac{1}{\Delta z}\sum m_j \text{ for all } j \text{ such that } (i-1)\Delta z \leq z_j < i\Delta z, \tag{4.58}$$

where N_p is the total number of particles. A natural question to ask is then how the error in the concentration scales with the number of particles and the cell size.

We recall that the particle positions are essentially random samples, and two simulations will in general give somewhat different concentration fields due to this randomness. The difference between the *true* distribution (which is usually unknown) and the reconstructed distribution based on N_p samples is called the sampling error. One may see from the Central Limit Theorem (Billingsley, 1979, p. 308) that the sampling error scales as $1/\sqrt{N_p}$ where N_p is the number of *independent* samples. This means that increasing the number of particles by a factor of 10 will only reduce the error by a factor of $\sqrt{10}$.

Regarding cell size, there is a choice to be made between resolution and sampling error. In the case where the entire domain is covered by just one cell, then all the particles will be inside that cell, which is of course correct but also a useless result. On the other hand, if there are so many cells that most cells have either 0 of 1 particles, then the result is completely dominated by random sampling noise. The challenge is to use enough cells to resolve those changes in concentration that are of interest, and enough particles to give a reasonably smooth result.

Box counting often leads to very noisy concentration fields. In particular, if one is interested in the most dilute concentrations, the results are guaranteed to be noisy, because the most dilute concentrations are by definition represented by only a small number of particles. However, even the higher concentrations may be noisy. A common way to tackle this problem is to use

Booij, N., Holthuijsen, L., & Ris, R. (1997). The "SWAN" wave model for shallow water. In *Proceedings of the 25th International Conference on Coastal Engineering*, Orlando, FL, 1996, pp. 668–676.

Boufadel, M, Liu, R, Zhao, L, Özgökmen, T, Nedwed, T, & Lee, K (2020). Transport of oil droplets in the upper ocean: Impact of the eddy diffusivity. *Journal of Geophysical Research: Oceans*, *125*, e2019JC015727.

Bowden, K. (1965). Horizontal mixing in the sea due to a shearing current. *Journal of Fluid Mechanics*, *21*(1), 83–95.

Bowles, P., Burns, R., Hudswell, F., & Whipple, R.. (1958). Sea disposal of low activity effluent. In *Proceedings of the Second United Nations International Conference on the Peaceful Uses of Atomic Energy*, Geneva, Switzerland, volume 18, pp. 376–389.

Bozzano, G., & Dente, M. (2001). Shape and terminal velocity of single bubble motion: A novel approach. *Computers & Chemical Engineering*, *25*(4), 571–576.

Brandvik, P. J., Johansen, Ø., Leirvik, F., Farooq, U., & Daling, P. S. (2013). Droplet breakup in subsurface oil releases—Part 1: Experimental study of droplet breakup and effectiveness of dispersant injection. *Marine Pollution Bulletin*, *73*(1), 319–326.

Broström, G., Drivdal, M., Carrasco, A., Christensen, K., & Mattsson, J. (2014). The golden trader oil spill; evaluation of operational oil spill models. In *EGU General Assembly Conference Abstracts*, volume 16.

Brzeźniak, Z., & Zastawniak, T. (1999). *Basic stochastic processes*. London: Springer-Verlag.

Carslaw, H. S., & Jaeger, J. C. (1959). *Conduction of heat in solids*. Oxford, UK: Oxford University Press (Clarendon).

Carter, D. (1982). Prediction of wave height and period for a constant wind velocity using the JONSWAP results. *Ocean Engineering*, *9*, 17–33.

Clift, R., Grace, J. R., & Weber, M. E. (1978). *Bubbles, drops, and particles*. New York: Academic Press.

Craig, P. D., & Banner, M. L. (1994). Modeling wave-enhanced turbulence in the ocean surface layer. *Journal of Physical Oceanography*, *24*(12), 2546–2559.

Csanady, G. (1973). *Turbulent diffusion in the environment*. Dordrecht, Holland: D. Reidel Publishing Company.

Cui, F., Boufadel, M. C., Geng, X., Gao, F., Zhao, L., King, T., & Lee, K. (2018). Oil droplets transport under a deep-water plunging breaker: Impact of droplet inertia. *Journal of Geophysical Research: Oceans*, *123*(12), 9082–9100.

Cui, F., Zhao, L., Cosan, D., King, T., Lee, K., Katz, J., & Boufadel, M. C. (2020). Modeling oil dispersion under breaking waves. Part II: Coupling Lagrangian particle tracking with population balance model. *Environmental Fluid Mechanics*, *20*, 1553–1578.

Davidson, P. A. (2015). *Turbulence: An introduction for scientists and engineers* (2nd ed). Oxford, UK: Oxford University Press.

De Dominicis, M., Bruciaferri, D., Gerin, R., Pinardi, N., Poulain, P., Garreau, P., ... Manganiello, C. (2016). A multi-model assessment of the impact of currents, waves and wind in modelling surface drifters and oil spill. *Deep Sea Research Part II: Topical Studies in Oceanography*, *133*, 21–38.

Delnoij, E., Lammers, F., Kuipers, J., & van Swaaij, W. P. M. (1997). Dynamic simulation of dispersed gas-liquid two-phase flow using a discrete bubble model. *Chemical Engineering Science*, *52*(9), 1429–1458.

Delvigne, G., & Sweeney, C. (1988). Natural dispersion of oil. *Oil and Chemical Pollution*, *4*(4), 281–310.

Duran, R. (2016). *Sub-grid parameterizations for oceanic oil-spill simulations*. NETL-TRS-0-2016; EPAct Technical Report Series. Albany, OR: US Department of Energy, National Energy Technology Laboratory.

Einstein, A. (1905). Über die von der molekularkinetischen theorie der wärme geforderte bewegung von in ruhenden flüssigkeiten suspendierten teilchen. *Annalen der Physik*, *322*(8), 549–560.

Elliott, A. J. (1986). Shear diffusion and the spread of oil in the surface layers of the North Sea. *Deutsche Hydrografische Zeitschrift*, *39*(3), 113–137.

Fernandez, D. M., Vesecky, J. F., & Teague, C. C. (1996). Measurements of upper ocean surface current shear with high-frequency radar. *Journal of Geophysical Research C: Oceans*, *101*(C12), 28615–28625.

Gihman, I., & Skorohod, A. (1972). *Stochastic differential equations*. Berlin: Springer.

Gillespie, Daniel T. (1996). Exact numerical simulation of the Ornstein-Uhlenbeck process and its integral. *Physical Review E*, *54*(2), 2084–2091.

Gräwe, U. (2011). Implementation of high-order particle-tracking schemes in a water column model. *Ocean Modelling*, *36*(1–2), 80–89.

Gräwe, U., Deleersnijder, E., Shah, S. H. A. M., & Heemink, A. W. (2012). Why the euler scheme in particle tracking is not enough: the shallow-sea pycnocline test case. *Ocean Dynamics*, *62*(4), 501–514.

Haidvogel, D. B., & Beckmann, A. (1999). *Numerical ocean circulation modeling*. London, UK: Imperial College Press.

Hansen, B. H., Sørensen, L., Carvalho, P. A., Meier, S., Booth, A. M., Altin, D., ... Nordtug, T. (2018). Adhesion of mechanically and chemically dispersed crude oil droplets to eggs of Atlantic cod (*Gadus morhua*) and haddock (*Melanogrammus aeglefinus*). *Science of the Total Environment*, *640*, 138–143.

Holloway, G. (1994). Comment: On modelling vertical trajectories of phytoplankton in a mixed layer. *Deep Sea Research Part I: Oceanographic Research Papers*, *41*(5), 957–959.

Hundsdorfer, W. (2003). *Numerical solution of time-dependent advection-diffusion-reaction equations, volume 33 of Springer series in computational mathematics* (1st ed.). Berlin: Springer.

Hunter, J., Craig, P., & Phillips, H. (1993). On the use of random walk models with spatially variable diffusivity. *Journal of Computational Physics*, *106*(2), 366–376.

Ichiye, T. (1967). Upper ocean boundary-layer flow determined by dye diffusion. *The Physics of Fluids*, *10*(9), S270–S277.

ITOPF 2011. http://www.itopf.org/in-action/case-studies/case-study/golden-trader-denmark-2011/. Accessed 10.12.2019.

Johansen, Ø. (1982). *Drift of submerged oil at sea*. Technical Report P 319/1. Continental Shelf Institute, Trondheim, Norway. http://hdl.handle.net/11250/2601119.

Johansen, Ø. (2000). Deepblow–A Lagrangian plume model for deep water blowouts. *Spill Science & Technology Bulletin*, *6*(2), 103–111.

Johansen, Ø., Brandvik, P. J., & Farooq, U. (2013). Droplet breakup in subsea oil releases—Part 2: Predictions of droplet size distributions with and without injection of chemical dispersants. *Marine Pollution Bulletin*, *73*(1), 327–335.

Johansen, Ø., Reed, M., & Bodsberg, N. R. (2015). Natural dispersion revisited. *Marine Pollution Bulletin*, *93*(1), 20–26.

King, T. L., Robinson, B., Boufadel, M., & Lee, K. (2014). Flume tank studies to elucidate the fate and behavior of diluted bitumen spilled at sea. *Marine Pollution Bulletin*, *83*(1), 32–37.

Kloeden, P. E., & Platen, E. (1992). *Numerical solution of stochastic differential equations*. Berlin: Springer-Verlag.

Langevin, P. (1908). Sur la théorie du mouvement brownien. *Comptes Rendus de l'Académie des Sciences (Paris), 146*, 530–533, Presented by M. Mascart.

Launder, B. E., & Spalding, D. B. (1983). *The numerical computation of turbulent flows. Numerical prediction of flow, heat transfer, turbulence and combustion* (pp. 96–116)). Elsevier.

Laxague, N. J., Özgökmen, T. M., Haus, B. K., Novelli, G., Shcherbina, A., Sutherland, P., ... Molemaker, J. (2018). Observations of near-surface current shear help describe oceanic oil and plastic transport. *Geophysical Research Letters, 45*(1), 245–249.

Lee S., Lee H., Lee I., & Tseng C., (2004). Ink diffusion in water. *European Journal of Physics, 25*(2), 331–336.

Leifer, I., & De Leeuw, G. (2006). Bubbles generated from wind-steepened breaking waves: 1. bubble plume bubbles. *Journal of Geophysical Research: Oceans, 111*(C6).

Lemons, D. S., & Gythiel, A. (1997). Paul Langevin's 1908 paper "on the theory of Brownian motion" ["Sur la théorie du mouvement brownien," CR Acad. Sci. (Paris) 146, 530–533 (1908)]. *American Journal of Physics, 65*(11), 1079–1081.

Li, C., Miller, J., Wang, J., Koley, S., & Katz, J. (2017). Size distribution and dispersion of droplets generated by impingement of breaking waves on oil slicks. *Journal of Geophysical Research: Oceans, 122*(10), 7938–7957.

Li, Z., Spaulding, M., French McCay, D., Crowley, D., & Payne, J. R. (2017). Development of a unified oil droplet size distribution model with application to surface breaking waves and subsea blowout releases considering dispersant effects. *Marine Pollution Bulletin, 114*(1), 247–257.

Li, Z., Spaulding, M. L., & French-McCay, D. (2017). An algorithm for modeling entrainment and naturally and chemically dispersed oil droplet size distribution under surface breaking wave conditions. *Marine Pollution Bulletin, 119*(1), 145–152.

Lynch, D. R., Greenberg, D. A., Bilgili, A., McGillicuddy, D. J., Jr, Manning, J. P., & Aretxabaleta, A. L. (2014). *Particles in the coastal ocean: Theory and applications*. Cambridge University Press.

Maruyama, G. (1955). Continuous Markov processes and stochastic equations. *Rendiconti del Circolo Matematico di Palermo, 4*(1), 48.

Mellor, G. L., & Yamada, T. (1982). Development of a turbulence closure model for geophysical fluid problems. *Reviews of Geophysics, 20*(4), 851–875.

Nissanka, I. D., & Yapa, P. D. (2016). Calculation of oil droplet size distribution in an underwater oil well blowout. *Journal of Hydraulic Research, 54*(3), 307–320.

Nissanka, I. D., & Yapa, P. D. (2017). Oil slicks on water surface: Breakup, coalescence, and droplet formation under breaking waves. *Marine Pollution Bulletin, 114*(1), 480–493.

Nordam, T., Kristiansen, R., Nepstad, R., & Röhrs, J. (2019). Numerical analysis of boundary conditions in a Lagrangian particle model for vertical mixing, transport and surfacing of buoyant particles in the water column. *Ocean Modelling, 136*, 107–119.

Nordam, T., Litzler, E., Rønningen, P., Aune, J., Hagelien, T. F., Beegle-Krause, C., & Brönner, U. (2018). Oil spill contingency and response modelling in ice-covered waters. In *Proceedings of the Forty-First AMOP Technical Seminar*, Victoria, BC, Canada.

Nordam, T., Nepstad, R., Litzler, E., & Röhrs, J. (2019). On the use of random walk schemes in oil spill modelling. *Marine Pollution Bulletin, 146*, 631–638.

Okubo, A. (1968). Some remarks on the importance of the "shear effect" on horizontal diffusion. *Journal of the Oceanographical Society of Japan, 24*(2), 60–69.

Pletcher, R. H. (2013). *Computational fluid mechanics and heat transfer* (3rd ed.). Boca Raton: CRC Press.

Proctor, R., Elliot, A., & Flather, R. (1994). Forecast and hindcast simulations of the Braer oil spill. *Marine Pollution Bulletin, 28*(4), 219–229.

Reed, M., Johansen, Ø., Brandvik, P. J., Daling, P., Lewis, A., Fiocco, R., . . . Prentki, R. (1999). Oil spill modeling towards the close of the 20th century: Overview of the state of the art. *Spill Science & Technology Bulletin, 5*(1), 3–16.

Reed, M., Leirvik, F., Johansen, O., & Brørs, B. (2009). *Numerical algorithm to compute the effects of breaking waves on surface oil spilled at sea.* Final Report Submitted to the Coastal Response Research Center.

Röhrs, J., Christensen, K. H., Vikebø, F., Sundby, S., Saetra, Ø., & Broström, G. (2014). Wave-induced transport and vertical mixing of pelagic eggs and larvae. *Limnology and Oceanography, 59*(4), 1213–1227.

Silverman, B. W. (1986). *Density estimation for statistics and data analysis*. Chapman & Hall.

Skognes, K., & Johansen, Ø. (2004). Statmap—A 3-dimensional model for oil spill risk assessment. *Environmental Modelling & Software, 19*(7–8), 727–737.

Spaulding, M., Kolluru, V., Anderson, E., & Howlett, E. (1994). Application of three-dimensional oil spill model (wosm/oilmap) to hindcast the Braer spill. *Spill Science & Technology Bulletin, 1*(1), 23–35.

Spaulding, M. L. (2017). State of the art review and future directions in oil spill modeling. *Marine Pollution Bulletin, 115*(1), 7–19.

Spivakovskaya, D., Heemink, A. W., & Deleersnijder, E. (2007a). The backward Îto method for the Lagrangian simulation of transport processes with large space variations of the diffusivity. *Ocean Science, 3*(4), 525–535.

Spivakovskaya, D., Heemink, A. W., & Deleersnijder, E. (2007b). Lagrangian modelling of multi-dimensional advection-diffusion with space-varying diffusivities: Theory and idealized test cases. *Ocean Dynamics, 57*(3), 189–203.

Stokes, G. G. (1856). On the effect of the internal friction of fluids on the motion of pendulums. *Transactions of the Cambridge Philosophical Society, 9*, 8–106.

Sundby, S., & Kristiansen, T. (2015). The principles of buoyancy in marine fish eggs and their vertical distributions across the world oceans. *PLoS One, 10*(10), e0138821.

Tennekes, H., & Lumley, J. L. (1972). *A first course in turbulence*. MIT Press.

Thomson, D. (1987). Criteria for the selection of stochastic models of particle trajectories in turbulent flows. *Journal of Fluid Mechanics, 180*, 529–556.

Thorpe, S. A. (2005). *The turbulent ocean.* Cambridge, UK: Cambridge University Press.

Thorpe, S. A. (2007). *An introduction to ocean turbulence.* Cambridge, UK: Cambridge University Press.

Tkalich, P., & Chan, E. S. (2002). Vertical mixing of oil droplets by breaking waves. *Marine Pollution Bulletin, 44*(11), 1219–1229.

Transport Malta. (2012). *Joint safety investigation report into the collision between the Maltese bulk carrier Golden Trader and the Belgian fishing vessel Vidar.* Technical Report. Marine Safety Investigation Rport No. 12/2012.

Turrell, W. (1994). Modelling the Braer oil spill—A retrospective view. *Marine Pollution Bulletin, 28*(4), 211–218.

Turrell, W. (1995). Modelling the Braer oil spill—A comment on Procter et al. (1994). *Marine Pollution Bulletin, 1*(30), 92–93.

Umlauf, L., Burchard, H., & Bolding, K. (2005). *GOTM—Scientific Documentation: version 3.2*, Marine Science Reports. Warnemuende, Germany: Leibniz-Institute for Baltic Sea Research. Please see up-to-date version on http://www.gotm.net.

Versteeg, H. (2007). *An introduction to computational fluid dynamics: The finite volume method* (2nd ed.). Harlow: Pearson/Prentice Hall.

Visser, A. W. (1997). Using random walk models to simulate the vertical distribution of particles in a turbulent water column. *Marine Ecology Progress Series, 158*, 275–281.

Warner, J. C., Perlin, N., & Skyllingstad, E. D. (2008). Using the model coupling toolkit to couple earth system models. *Environmental Modelling & Software, 23*(10), 1240–1249.

Wilcox, D. C. (2008). Formulation of the *k-w* turbulence model revisited. *AIAA journal, 46*(11), 2823–2838.

Wilson, J. D., & Flesch, T. K. (1993). Flow boundaries in random-flight dispersion models: Enforcing the well-mixed condition. *Journal of Applied Meteorology, 32*(11), 1695–1707.

Wilson, J. D., & Yee, E. (2007). A critical examination of the random displacement model of turbulent dispersion. *Boundary-Layer Meteorology, 125*(3), 399–416.

Zeinstra-Helfrich, M., Koops, W., Dijkstra, K., & Murk, A. J. (2015). Quantification of the effect of oil layer thickness on entrainment of surface oil. *Marine Pollution Bulletin, 96*(1–2), 401–409.

Zeinstra-Helfrich, M., Koops, W., & Murk, A. J. (2016). How oil properties and layer thickness determine the entrainment of spilled surface oil. *Marine Pollution Bulletin, 110*(1), 184–193.

Zhao, L., Torlapati, J., Boufadel, M. C., King, T., Robinson, B., & Lee, K. (2014). VDROP: A comprehensive model for droplet formation of oils and gases in liquids—Incorporation of the interfacial tension and droplet viscosity. *Chemical Engineering Journal, 253*, 93–106.

Appendix A: Equivalence between Eulerian and Lagrangian pictures

The development of a concentration field under transport and mixing may be described by the Partial Differential Equation (PDE) known as the advection–diffusion equation. This is called an Eulerian approach and is characterized by an equation that describes how the concentration at fixed locations changes in time. The same process may also be described by an ensemble of "particles," which experience directed motion due to advection, and random motion due to diffusion. This approach is called Lagrangian and is characterized by an equation that describes how the position of a particle changes with time.

The link between the Eulerian and the Lagrangian picture is that the concentration field described in the Eulerian picture, if normalized, describes a probability distribution for where the Lagrangian particles will be found. Conversely, calculating the position of a Lagrangian particle is the same as drawing a sample from the probability distribution, and with a large number of samples, the distribution can be reconstructed approximately. We will here demonstrate how to obtain a random walk which is equivalent to the advection diffusion equation.

Consider a diffusion process described by the general Stochastic Differential Equation (SDE)

$$\mathrm{d}z = a(z,t)\,\mathrm{d}t + b(z,t)\,\mathrm{d}W(t), \qquad (4.\mathrm{A}1)$$

where $a(z, t)$ and $b(z, t)$ are "moderately smooth functions" (Kloeden & Platen, 1992, p. 37), and $dW(t)$ are the increments of a standard Wiener process (Kloeden & Platen, 1992, p. 40). Further conditions also apply, though these may be less important in practice. For details, see Gihman and Skorohod (1972, pp. 96–102).

For this diffusion process, the Fokker–Planck equation (also known as the Kolmogorov Forward equation) for evolution of the transition probability density, $p(z_0, t_0, z, t)$, from an initial position z_0 at time t_0, to a position z at a later time t, is (Kloeden & Platen, 1992, p. 37):

$$\frac{\partial p(z_0, t_0, z, t)}{\partial t} = \frac{1}{2}\frac{\partial^2}{\partial z^2}(b^2(z, t)p(z_0, t_0, z, t)) - \frac{\partial}{\partial z}(a(z, t)p(z_0, t_0, z, t)). \quad (4.A2)$$

We observe that the Fokker–Planck equation is a PDE, and like the advection–diffusion equation, it describes the time development of a distribution. For a particle initially at position z_0 at t_0, undergoing the random motion described by Eq. (4.A1), the probability density function for the position, z, at a later time, t, may be obtained by the Fokker–Planck equation. If we consider instead a large ensemble of particles, all starting out at z_0 at t_0, then at a later time t, they will be distributed according to $p(z_0, t_0, z, t)$, with many particles in areas of high probability and few particles in areas of low probability.

This is equivalent to the evolution of a concentration field from an instantaneous point source, as described by the advection–diffusion equation. Hence, our goal is to obtain the SDE that has the advection–diffusion equation as its Fokker–Planck equation. Then we know that the distribution of an ensemble of particles will develop according to the advection–diffusion equation, and thus we can use the distribution of particles to approximately reconstruct the concentration field.

Going back to Eq. (4.A2), we drop the arguments to a, b, and p for brevity, rewrite the equation a bit, and we get

$$\frac{\partial p}{\partial t} = \frac{1}{2}\frac{\partial}{\partial z}\left(b^2\frac{\partial p}{\partial z}\right) - \frac{\partial}{\partial z}\left[\left(a - \frac{1}{2}\frac{\partial b^2}{\partial z}\right)p\right]. \quad (4.A3)$$

We then compare Eq. (4.A3) to the advection–diffusion equation, with advection $w(z, t)$ and diffusion $K(z, t)$:

$$\frac{\partial C}{\partial t} = \frac{\partial}{\partial z}\left(K\frac{\partial C}{\partial z}\right) - \frac{\partial}{\partial z}(wC). \quad (4.A4)$$

By demanding that C should be proportional to p at all times, we find that each term in Eq. (4.A3) must be equal to the corresponding term in Eq. (4.A4). We thus obtain

$$K = \frac{b^2}{2} \Rightarrow b = \sqrt{2K} \quad (4.A5a)$$

$$w = a - \frac{1}{2}\frac{\partial b^2}{\partial z} \Rightarrow a = w + \partial_z K \tag{4.A5b}$$

Hence, the SDE whose probability density is described by the advection–diffusion equation is

$$\mathrm{d}z = (w + K'(z))\,\mathrm{d}t + \sqrt{2K(z)}\,\mathrm{d}W, \tag{4.A6}$$

where $K'(z) = \partial_z K$.

Note that since both $a(z, t)$ and $b(z, t)$ in Eq. (4.A1) must be continuous, we find that both $K(z)$ and $\partial_z K(z)$ must be continuous for the conditions mentioned above to be satisfied. Hence, the equivalence with the advection–diffusion does not hold for, for example, step-function diffusivity, or piecewise linear diffusivity profiles with discontinuous first derivatives, such as a linearly interpolated profile.

Chapter 5

Operational oil spill modelling assessments

George Zodiatis[1,2], Robin Lardner[2,3], Katerina Spanoudaki[1], Sarantis Sofianos[4], Hari Radhakrishnan[2], Giovanni Coppini[5], Svitlana Liubartseva[5], Nikos Kampanis[1], George Krokos[6], Ibrahim Hoteit[6], Joaquín Tintoré[7], Tatiana Eremina[8] and Aldo Drago[9]
[1]Coastal & Marine Research Lab, Institute of Applied and Computational Mathematics, Foundation for Research and Technology-Hellas, Heraklion, Greece, [2]MEDSLIK (Mediterranean Oil Spill and Trajectory Prediction Model), Nicosia, Cyprus, [3]Simon Fraser University, Burnaby, BC, Canada, [4]Ocean Physics and Modelling Group, University of Athens, Athens, Greece, [5]CMCC (Centro Euro Mediterraneo sui Cambiamenti Climatici), Bologna, Italy, [6]King Abdullah University of Science and Technology (KAUST), Thuwal, Saudi Arabia, [7]SOCIB and IMEDEA (CSIC-UIB), Palma, Spain, [8]Russian State Hydrometeorological University, Saint Petersburg, Russia, [9]Department of Geosciences, University of Malta, Msida, Malta

5.1 Introduction

Oil spills at sea is a matter of concern due to the damaging effect that can have on coastal resources and the marine environment, as clearly documented following the catastrophic accidents of the oil tankers Exxon Valdez in March 1989 in the Alaskan coastal waters, Haven in April 1991 in the Liguria Sea, Prestige in November 2002 in the North Eastern Atlantic, the Lebanon oil pollution crisis in July 2006, as well as the explosion of the BP Deepwater Horizon offshore drilling platform in April 2010 in the Gulf of Mexico. The risks associated with oil pollution from the maritime traffic and from the coastal and offshore installations have led the coastal countries to adopt, implement and strengthen regional and international protocols against oil pollution, as for example the Barcelona Convention for the Mediterranean Sea and the Convention on the Protection of the Black Sea Against Pollution. The protocols encourage the parties to develop Impact Damage Assessments taking into account all the elements that can affect the marine and coastal environments, due to oil/gas industry and maritime traffic activities.

The increasing risks associated with the exploration and exploitation of the continental shelf and the seabed include early detection and control of

Marine Hydrocarbon Spill Assessments. DOI: https://doi.org/10.1016/B978-0-12-819354-9.00010-7

spillage, reorganization and redistribution of resources available to combat spillages in the early stages, possible consequences in environmental economic and social terms. Each risk management plan should assess the risk of an oil spill incident and propose mechanisms to minimize the overall response time of the involved organizations. In order to assess the consequences of oil spill leakages from existing and planned maritime activities, the responsible authorities are requesting studies on Impact Damage Assessment. Such studies should be based on the results of seasonal and interannual oil spill simulations using well-established oil spill models and metocean data, which could be obtained from international data distribution centers, such as CMEMS (Copernicus Marine Environmental Monitoring Service), NOAA (National Oceanic and Atmospheric Administration), or other relevant databases (Tintoré et al., 2019).

To mitigate spill consequences, it is common to deploy such equipment as booms, skimmers, and apply dispersants from boats or planes. To use such measures optimally, numerical models are applied to predict possible impacts, that is, where the slick/s will most likely move to, how soon it will get there, which resources may be threatened/affected, and what will be its impact on sensitive infrastructure such as recreational facilities, desalination plants, etc. Oil spill models require as input data the type of oil and its characteristics, and the metocean conditions, that is, winds, sea currents, sea surface temperature and wave conditions. Such spill models may predict the expected state of the oil, that is, how much will have evaporated the degree of emulsification, the viscosity change in the spilled oil, the remaining oil on the water surface, and how much dispersed as fine droplets throughout the water column.

With the availability of metocean forecasts in near real time at present (e.g., from the CMEMS or other relevant subregional and coastal forecasting systems), oil spill models can be implemented operationally to predict the transport and weathering of the oil. The majority of the oil spill models used for operational applications are the Lagrangian ones, as were proven to be more efficient than the Eulerian ones (Ahlstrom, 1975; De Dominicis, Pinardi, Zodiatis, & Lardner, 2013; Hunter, 1987), especially during emergencies predicting the displacement of each particle amount released, representing the spilled oil.

Some of the most known Lagrangian oil spill models are: OSCAR (oil spill contingency and response model) (Reed, Aamo, & Daling, 1995), GNOME (Zelenke, O'Connor, Barker, Beegle-Krause, & Eclipse, 2012), SEATRACK WEB (Ambjörn, 2007), PISCES (potential incident simulation control and evaluation system) (Delgato, Kumzerova, & Martynov, 2006), OILMAP (Spaulding, Kolluru, Anderson, & Howlett, 1994; ASA, 1997), GULFSPILL (Al-Rabeh, Lardner, & Gunay, 2000), MOTHY (Modèle Océanique de Transport d'Hydrocarbures) (Daniel, Marty, Josse, Skandrani, & Benshila, 2003), MOHID (Modelo Hidrodinâmico) (Carracedo et al., 2006), OD3D (oil drift 3-dimensional) (Hackett, Breivik, & Wettre, 2006), SPILLMOD (Ovsienko, Zatsepa, & Ivchenko, 2005), MEDSLIK (oil spill trajectory and pollutant transport prediction model) (Lardner & Zodiatis, 2016, 2017; Zodiatis et al., 2018; Zodiatis, Lardner, Solovyov, Panayidou, & De Dominicis, 2012; Lardner et al., 2006; Lardner et al., 1998; Zodiatis et al., 2008), OILTRANS (oil spill modelling software application) (Berry et al., 2012), OSERIT (oil spill evaluation and response integrated tool) (Legrand & Dulière, 2012), and MEDSLIK-II (De Dominicis, Pinardi, Zodiatis, & Archetti, 2013; De Dominicis, Pinardi, Zodiatis, & Lardner, 2013). These models more or less incorporate the same or similar parameterization, shown schematically in Fig. 5.1, but only few of them examine the effect of biodegradation and oil plume parameterization.

Usually the oil spill models include the processes of advection and diffusion, beaching and sedimentation, together with a standard set of parameterizations for fate/weathering processes, such as evaporation, emulsification (water in oil), spreading, and dispersion of oil in the water column (Fig. 5.1). Weathering processes are determined by the physical and chemical properties of the spilled oil type under the influence of the metocean conditions and are described by similar semiempirical relationships derived from laboratory and field experiments (e.g., Reed et al., 1999). Most oil spill models use modified versions of Mackay's fate algorithms for evaporation, emulsification and dispersion (Mackay, Paterson, & Nadeau, 1980; Mackay, Paterson, & Trudel, 1980) to predict the contribution of each of the above fate weathering process.

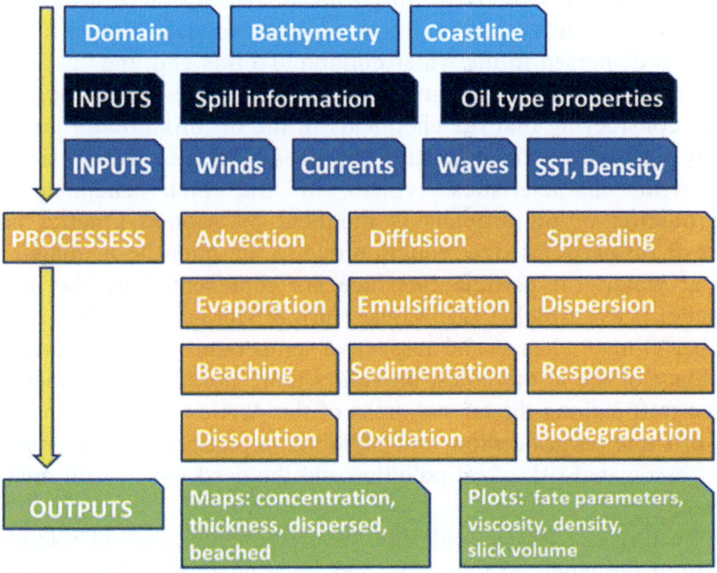

FIGURE 5.1 Schematic of a Lagrangian oil spill model's components.

Deep-sea oil releases from accidents during offshore exploratory drilling or production activities are of particular concern, as the potential for such accidents increases with the expansion of the offshore activities. Important knowledge gaps in terms of the spreading of dispersants at subsurface and of the long lasting biodegradation processes remain.

In order to predict the fate of the oil plume from subsea releases reaching the water surface, Lagrangian deep-water oil spill modules have been recently developed and implemented within the well-established Mediterranean community models: MEDSLIK (Lardner & Zodiatis, 2017) and MEDSLIK-II (Spanoudaki, 2016). The Lagrangian plume model is represented there by elements that trace the plume trajectory, where each Lagrangian element represents a mixture of water, oil, and gas, and where the gas might be present in different states, for example, as free gas in gas bubbles, gas dissolved in water, or gas in hydrates. Changes in the mass and composition of the element are accounted for by the turbulent entrainment of ambient water, by leakage of gas bubbles and oil droplets from the plume, dissolution of gas in seawater, and formation or disintegration of gas hydrates. The motion of the element is computed from the conservation equations for mass, momentum and buoyancy.

Although oil biodegradation by native bacteria at sea is an important natural processes that can attenuate the environmental impacts in medium and long terms (months to years), only few models include biodegradation kinetics of spilled oil, mostly represented as a first order decay process neglecting the effects of oil composition and oil droplets-water interface (Reed and Hetland, 2002; French McCay, 2003, 2004). The open source oil spill model MEDSLIK-II (http://medslik-ii.org) has been modified recently by Spanoudaki (2016) to incorporate biodegradation kinetics of oil droplets dispersed in the water column. To this end, the "pseudo-component" (PC) approach for simulating the oil weathering processes was adopted, considering that the chemicals in the oil mixture are grouped by their physical–chemical properties and behave as they were a single substance with characteristics typical of the their chemical group. Biodegradation of oil droplets is modeled by Monod kinetics and the kinetics of oil particles size reduction due to the microbe-mediated degradation at water–oil particle interface are represented by the shrinking core model (Levenspiel, 1999; Vilcáez, Li, & Hubbard, 2013).

Global climate change and respective temperature increase have caused the melding of ice at the poles, providing more possibilities to increase the maritime traffic along the coastal areas of the Arctic Ocean and the installation of offshore oil/gas drilling and exploitation platforms (Bazilchuk, 2018). There is associated risk of oil pollution and the need for oil spill predictions at present.

The accuracy of the oil spill advection predictions depend solely on the reliability of the winds, sea currents, and wave data, while the predictions of the horizontal surface spreading, vertical dispersion, and weathering processes depend on the characteristics of the spilled oil (Zodiatis et al., 2017). The metocean conditions drive the oil spill to drift at sea surface, with approximately 3% of the wind speed (Fig. 5.2). Below the sea surface, the spill moves by subsurface sea currents. Oil spills at the sea surface spread during the initial stages of the spillage, while the increase in density and viscosity, as a result of evaporation, cause the remaining nonevaporated oil to spread (De Dominicis, Pinardi, Zodiatis, & Lardner, 2013; Zodiatis et al., 2017).

This chapter provides an overview of the physical and chemical processes causing oil spill transport and transformation at sea, both offshore and coastal, regardless of its source. The following three major questions are addressed:

1. What are the processes affecting and controlling the transport and weathering of oil spills after their spillage?
2. What basic equations are used in oil spill models?
3. What is considered good practice for operational implementation of oil spill models to support the response mitigation authorities?

Advanced oil spill models are in general quite complex and a complete understanding of the modeled processes are presented in the chapters preceding this chapter. Instead, the basic equations included in most of the used well-established Lagrangian oil spill models will be described. Section 5.2 is

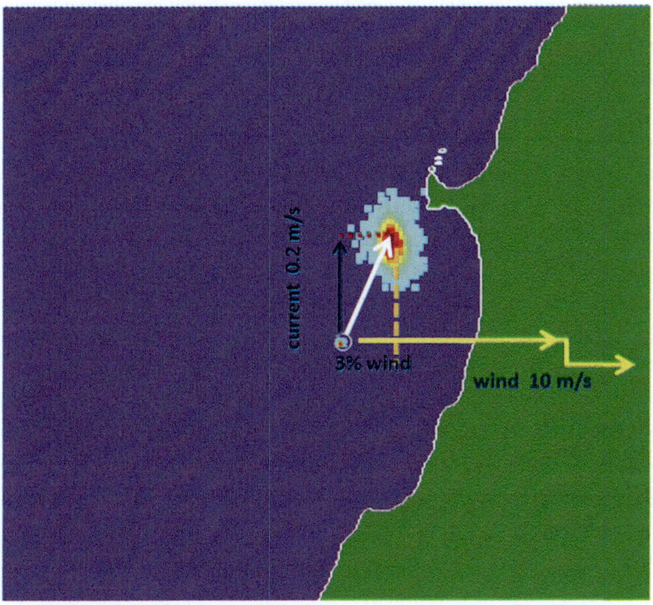

FIGURE 5.2 Schematic illustration showing how wind (yellow arrow) and sea surface current (black arrow) affect the drift of the oil spill (white arrow) at the sea surface.

presenting the convection, diffusion and beaching of oil slicks modeling. Section 5.3 is focusing on the weathering processes. Section 5.4 is explaining the biodegradation processes. Section 5.5 is describing the modeling approach for deep sea oil releases. Section 5.6 is outlining the approach of oil spill prediction in areas with ice. Section 5.7 is providing examples of the nowadays operational implementation of oil spill models to support the response agencies.

5.2 Convection, diffusion, and beaching

The horizontal transport/drift of oil on the water surface, caused by the combined action of drag exerted by winds, currents, and Stokes drift, determines the trajectory of the slick (Spaulding, 1988; Reed et al., 1993). The oil surface drift velocity is usually modeled as the vector sum of velocities induced by currents, winds, and waves. Based on empirical data, it is common to take the wind component to be around 3% of the wind speed at 10 m above the surface and the current component to be 100% of the speed of the surface current (ITOPF, 2002). Fig. 5.2 illustrates how the wind and sea surface current can influence the drift of an oil spill. Below the surface the spill moves by subsurface currents.

Sea currents act directly on the oil spill particles. The wind also acts on the oil spills that are on the sea surface, causing them to move relative to the sea water body.

Sea currents vary relatively slow, and the ocean forecasts provide hourly to daily mean currents, while winds on the other hand vary over much faster time scales, of the order of hours, as often made available by weather forecasting services. In order to account for the effect of wind on the thin skin of the oil slick floating on the sea surface, a simple "wind factor" approach is commonly used.

It is assumed that the surface oil spill is transported at a speed that is a certain fraction α of the wind speed and at a certain angle β to the right (in the northern hemisphere) of the wind direction. Different values for these parameters were suggested, as for example, the values $\alpha = 0.031$ and $\beta = 26$ degrees that were determined by Al Rabeh (1994) on the basis of a least squares analysis of the motion of a group of floating buoys in the Arabian Gulf (Henaidi, 1984), or the values $\alpha = 0.0$ and $\beta = 30$ degrees (Lardner, 2017).

Surface currents driving the oil spill can be parameterized as a function of wind intensity, and of the angle between winds and currents through the following equations:

$$\begin{aligned} U_w &= \alpha\left(W_x \cos\beta + W_y \sin\beta\right) \\ V_w &= \alpha\left(-W_x \sin\beta + W_y \cos\beta\right) \end{aligned} \tag{5.1}$$

where W_x, W_y are the wind velocity components, U_w, V_w, is the surface current on x, y axis induced by the wind, α is the percentage of the wind to be added to the sea-current velocity, and β is the deviation angle between sea currents and wind.

Most Lagrangian oil spill models follow a Monte Carlo approach, where the pollutant is divided into a large number of Lagrangian parcels, from few thousands to hundreds of thousands. At each time step of the numerical integration, each parcel is assigned a convective and a diffusive displacement as described below (De Dominicis, Pinardi, Zodiatis, & Lardner, 2013; Zodiatis et al., 2017).

Let (X_i, Y_i, Z_i) be the position of the ith parcel at the beginning of a particular step, Z_i being its depth measured vertically downward from the surface. Then at the end of a time step τ, the parcel is displaced to the point (X_i', Y_i', Z_i') through the following equations:

$$
\begin{aligned}
X_i' &= X_i + \left\{u(X_i, Y_i, Z_i) + \alpha(W_x\cos\beta + W_y\sin\beta)\right\}\tau + \Delta X_i^{(d)} \\
Y_i' &= Y_i + \left\{v(X_i, Y_i, Z_i) + \alpha(-W_x\sin\beta + W_y\cos\beta)\right\}\tau + \Delta Y_i^{(d)} \\
Z_i' &= Z_i + \Delta Z_i^{(d)}
\end{aligned}
\tag{5.2}
$$

where $u(x,y,z)$ and $v(x,y,z)$ are the sea water velocity components in the x and y directions, and ΔX_i^d, $\Delta \Upsilon_i^d$, ΔZ_i^d are the diffusive displacements in the three directions. The vertical sea water velocity w is not included in the model since it is generally very small compared to the x, y axis components of the sea currents. The diffusive displacements are given by a random walk model:

$$
\begin{aligned}
\Delta X_i^d &= \left[2rand\ (0,1) - 1\right]\sqrt{6K_h\tau} \\
\Delta \Upsilon_i^d &= \left[2rand\ (0,1) - 1\right]\sqrt{6K_h\tau} \\
\Delta Z_i^d &= \left[2rand\ (0,1) - 1\right]\sqrt{6K_v\tau}
\end{aligned}
\tag{5.3}
$$

where K_h and K_v are the horizontal and vertical diffusivities and *rand* (0,1) is a random number between 0 and 1.

The vertical displacements are applied only to the parcels dispersed in the water column (see Subsection on Dispersion) and the wind transport terms are excluded for these parcels. In the event that the vertical displacement of such a parcel takes it either above the water surface or below the bottom level, the parcel is reflected back into the water column. It can easily be seen that the root mean square (rms) values of the diffusive displacements are:

$$
\mathrm{rms}\left\{\Delta X^{(d)}, \Delta Y^{(d)}, \Delta Z^{(d)}\right\} = \left\{\sqrt{2K_h\tau}, \sqrt{2K_h\tau}, \sqrt{2K_v\tau}\right\} \tag{5.4}
$$

It can be shown that a cloud of such particles undergoing random walks with these rms displacements satisfies the convection-diffusion equation with K_h and K_v the horizontal and vertical diffusion coefficients (Csanady, 1973; Hunter, 1987).

One of the most important issues of interest for the decision makers and the response agencies in oil spill predictions is to provide information regarding the beaching of oil slick, particularly the first impact on the coast, that is, the minimum time the slick needs to reach any coastline. Advanced oil spill models can predict the amount of oil that permanently deposits on the coast and the percentage that is potentially free to return back to the sea. Depending on the type of oil, the simulated level of oil depositions on the coastline and the type of the shoreline (sandy or rocky), the response agencies can plan and organize the clean-up operations to remove the beached oil and to mitigate impacts.

Lagrangian oil spill models approximate the coastline by boundary segments of a rectangular grid. For instance, the MEDLSIK model uses 250 m for the whole Eastern Mediterranean Levantine Basin, 150 m for the NE Levantine basin and 50 m in smaller domains. Intersect of the displacement with coastal segments is examined at each time step of the model integration. In case of an intersection, the parcel is moved to the nearest coastal segment, where it is considered "beached."

Each beached parcel is assigned a probability of washing back into the sea (Shen, Yapa, & Petroski, 1987; Torgrimson, 1980),

$$Pr = 1 - 0.5^{\tau/T_W} \tag{5.5}$$

where Pr is the probability of release, τ is the time step of numerical integration; T_w is the half-life for oil to remain on the beach before washing off again. For each beached parcel, the random number generator is called and the parcel is released back into the water if *rand* (0,1) $<Pr$ and provides the parcel's new position on the same side of the coastal segment as that from which it arrived. The parcel is then returned back to the coastal segment side where it originated from. The value of T_w is assigned depending on the type of the coast (e.g., sandy beach, rocky coastline, etc.) according to Shen et al. (1987).

A probability is also assigned for permanent beaching. A certain fraction of the oil may permanently beach, for example, by seeping into the sand or being adsorbed on rock or pebbles. On each time step, it is assumed that

$$Frs = 1 - 0.5^{\tau/T_s} \tag{5.6}$$

where Frs is the fraction of seeping, T_s is a half-life for seepage or other mode of permanent attachment. The amount of oil remaining in any parcel that is on the beach is then reduced by this fraction, while the volume of oil lost is counted as remaining permanently on the coastal segment. Again the half-life T_s can be assigned to each coastal segment depending on coast type.

Predicting the *sedimentation* of the dispersed oil at the sea bottom, especially coastal shallow areas, is also important for the decision makers and the response agencies. Interaction of the dispersed oil with suspended particles may also lead to sedimentation. The process is further enhanced for emulsified oil or parcels washed back from the shore, due to increased density and adhesivity.

Sedimentation may also occur in the case of blowout at the sea bottom, either due to the absorption of dispersed oil to suspended sediment/particles that eventually settle on the sea bottom, or due to direct contact of oil with the sea bottom. In this case, the sedimentation of oil droplets at the bottom occurs in the vicinity of the blowout release. It is assumed that the subsea dispersant treatment will reduce the potential for such sedimentation, due to lower adsorption/adhesiveness to sediment particles.

5.3 Weathering processes

In addition to the convective and diffusive displacements and the interaction with the coastline and sediment, oil undergoes physical and chemical changes (Figs. 5.3 and 5.4), which modify the spilled oil characteristics (ITOPF, 2002). During the first several hours, the spill *spreads* mechanically over the water surface under the action of gravitational forces. In addition, the lighter fractions of the spilled oil evaporated through *evaporation* and the remaining fractions begin to absorb water, or *emulsify*. In most of Lagrangian oil spill models, these changes are reflected in changes of the properties linked to each parcel. Finally, some of the oil is driven below the water surface, that is, undergoes *dispersion* in the sea water column by wind/wave action. This dispersion of oil is treated as a random process that may

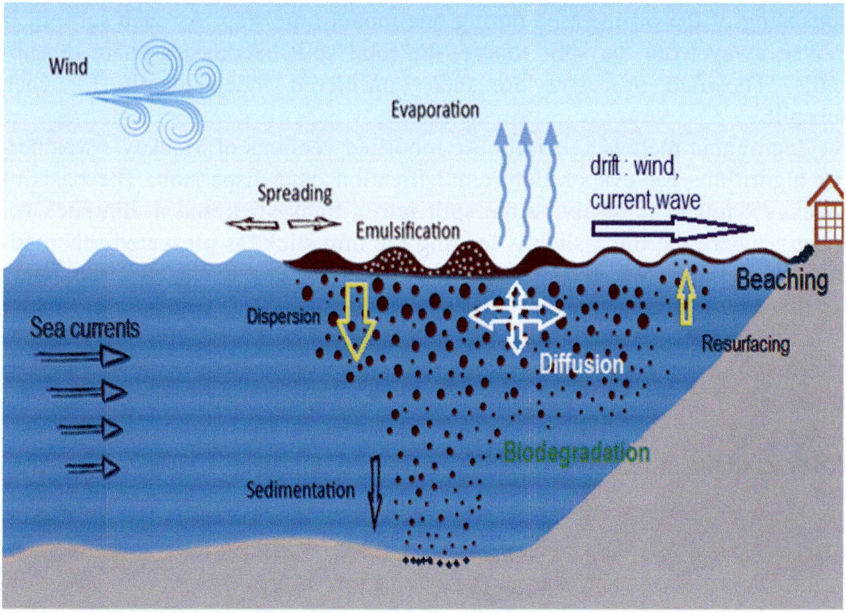

FIGURE 5.3 Schematic representation of the processes acting after oil spillage at sea surface. *Modified after Zodiatis et al. (2017) and Zodiatis et al. (2019).*

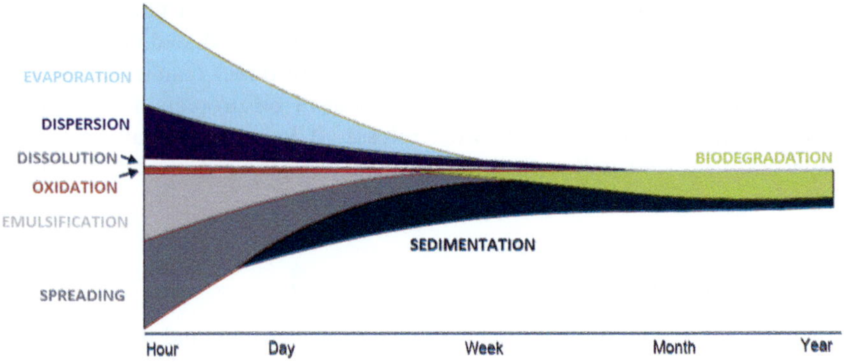

FIGURE 5.4 Schematic representation of the oil spill fate weathering processes for a typical crude oil in time and their importance indicated by the width on *Y*-axis of each process. *Modified from ITOPF, 2002.*

drive any parcel into the water column. Once there, the subsequent movement of the parcel is driven by the action of three-dimensional velocity field of the water.

In the case of an oil spill for which leakage may last for several hours or even days, the initial volumes of spilled oil will have been transported away from the site of the spill by wind and sea currents. In order to model the weathering of oil, the total spill is divided into a number of subspills each consisting of the oil released during one model time step. As each subspill is moved away from the spill source, the total spill becomes a chain of subspills. The fate processes are then considered independently for each subspill.

Lagrangian models generally use modified versions of Mackay's weathering algorithms for evaporation, emulsification, and dispersion. The basis of Mackay's model is to divide the spill into a thick slick and a thin slick (or sheen), where the thick slick is feeding the thin slick, as presented schematically in the Fig. 5.5. Evaporation and dispersion are considered separately for these two parts of the slick.

For any subspill at a given time step, denoted by V_{tk} and V_{tn}, the volumes of remaining oil in the thick and the thin slicks, A_{tk} and A_{tn} their two surface areas, and T_{tk} and T_{tn} their thicknesses, respectively. It is assumed that the thickness T_{tn} of the thin slick is considered to be 10 μm (Fingas & Brown, 2018; Zodiatis et al., 2012). At any time step, the two volumes are updated as

$$\begin{aligned} V'_{tk} &= V_{tk} - \Delta V_{tk}^{(e)} - \Delta V_{tk}^{(d)} - \Delta V_{tn}^{(s)} \\ V'_{tn} &= V_{tn} - \Delta V_{tn}^{(e)} - \Delta V_{tn}^{(d)} + \Delta V_{tn}^{(s)} \end{aligned} \tag{5.7}$$

where $\Delta V_{tk}^{(e)}$and $\Delta V_{tn}^{(e)}$ are the volumes lost by evaporation, and $\Delta V_{tk}^{(d)}$and$\Delta V_{tn}^{(d)}$ the volumes lost by dispersion. $\Delta V_{tn}^{(s)}$ is the amount of oil

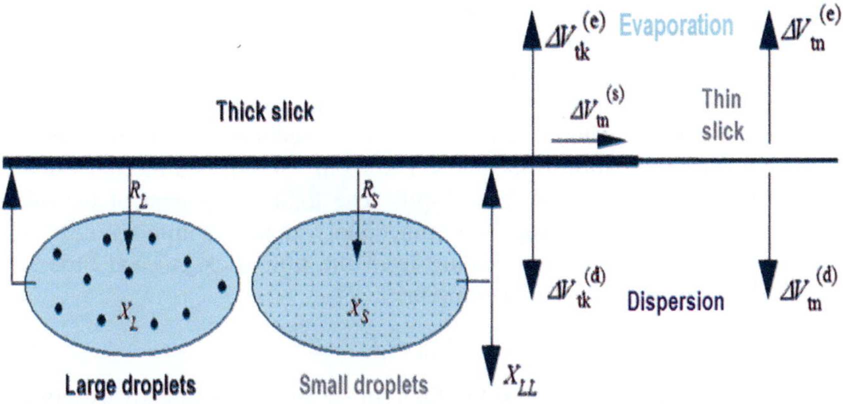

FIGURE 5.5 Right: Schematic diagram showing expected volume transfers between thin and thick slicks based on Mackay's concepts of weathering processes. The changes in surface-oil volumes of the thin ΔV_{tn} and thick ΔV_{tk} parts of a slick result from evaporation (*e*), dispersion (*d*), and spreading (*s*). Left: Schematic representation of dispersion of large and small oil droplets below the water surface. The model of dispersion of oil into the water column is based on the work of Buist (1979) and Mackay, Buist, Mascarenhas, and Paterson (1979), where R_L and R_S are the downward volume fluxes of oil per unit area of the slick entering the water as large (X_L) and small (X_S) droplets, respectively.

flowing from the thick to the thin parts of the slick. These transfers of oil are illustrated in Fig. 5.5.

After updating the volumes of the two parts of the slick, their respective areas are updated at each step using semiempirical spreading formulas. The new thickness of the thick slick is then computed as:

$$T_{tk} = V_{tk}A_{tk} \tag{5.8}$$

Evaporation is one of the most important weathering processes contributing to the removal of the light oil components from the sea surface. The amount of the spilled oil removed by evaporation depends primarily on the type of oil, particularly on its density and second on the wind speed and sea surface temperature conditions. The evaporation of the light components of oil starts immediately after the oil was spilled on the sea surface. In cases of light density oil products, such as gasoline and diesel, the spillage at sea surface may completely evaporate after few hours/days. Condensates may lose up to 60% of their original volume during the first 2–3 days after the spillage. Evaporation causes the loss of volatile and semivolatile oil components, resulting in the changes of the chemical and physical properties of the remaining oil, particularly an increase of its density and viscosity. The evaporation process is one or two orders of magnitude (approximately 10–100 times) faster than the dispersion of oil droplets in the water column (Mackay, Paterson, & Nadeau, 1980; Mackay, Paterson, & Trudel, 1980; Mackay et al., 1979).

Each parcel of spilled oil in numerical oil spill modeling is viewed as consisting of light evaporative and a heavy, nonevaporative, components. The initial fraction of evaporative component is set according to the type of the spilled oil. On each time step of numerical integration, the fraction of the light component remaining in each subspill is reduced based on Mackay's algorithm for evaporation (Mackay & Paterson, 1980a), and the reduction is applied to all the parcels of the subspill. The light component of the oil in the thin slick assumed to evaporate immediately, that is, within the next time step, so that the volume evaporating from the thin slick at each time step equals the total content of light component:

$$\Delta V_{tn}^{(e)} = V_{tn}(f_{\max} - f_{tn})/(1 - f_{tn}) \tag{5.9}$$

where f_{tn} is the fraction of the oil in the thin slick that has already evaporated at the beginning of the numerical integration and $f_{\max}$ is the initial fraction of light component, which represents the maximum value that f_{tn} can attain.

For the thick slick, the increment in the fraction f_{tk} of the oil that has evaporated is expressed as a product of the vapor pressure P_{oil} and the change in an evaporative exposure ΔE_{tk}

$$\Delta f_{tk} = P_{\mathrm{oil}} \Delta E_{tk} \tag{5.10}$$

The vapor pressure is expressed in exponential form

$$P_{\mathrm{oil}} = P_0 \exp(-c f_{tk}) \tag{5.11}$$

where P_0 is the initial vapor pressure and c is a constant that measures the rate of decrease of vapor pressure. The increment in exposure is expressed as a product of a mass transfer coefficient k_m, the time step of the numerical integration τ, the slick area A_{tk}, and the molar volume of the oil V_{mol}, divided by the gas constant R, the temperature T in °K and the initial volume of the subspill, $V_{(0)}$:

$$\Delta E_{tk} = \frac{K_m V_{\mathrm{mol}} A_{tk} dt}{\mathrm{RTV}^{(0)}} = \frac{K_m V_{\mathrm{mol}} A_{tk}(1 - f_{tk}) dt}{\mathrm{RTV}_{tk}} \tag{5.12}$$

where V_{tk} is the current volume of oil in the thick slick, equal to $V(0)(1-f_{\mathrm{tk}})$, the molar volume of oil is $V_{\mathrm{mol}} = 0.0002$, $R = 0.000082$ and

$$K_m = C^{(e)} \left(W_{kph}\right)^{\gamma} \tag{5.13}$$

where W_{kph} is the wind speed in km per hour and $C^{(e)}$ and γ are constants.

Finally, the volume loss by evaporation per time step of the numerical integration is equal to the increment in the fraction evaporated multiplied by the original volume:

$$\Delta V_{tk}^{(e)} = \Delta f_{tk} V^{(0)} = \Delta f_{tk} V_{tk}/(1 - f_{tk}) \tag{5.14}$$

Although the evaporative component in the thin slick has been assumed to disappear immediately, the thin slick is fed by oil from the thick slick as long as it has not been fully evaporated. Thus the fraction f_{tn} of oil in the thin slick that has evaporated must be reduced from the maximum value $f_{\max}$. Equating the oil content of the thin slick before and after its fed from the thick slick, we have

$$V'_{tn}(1-f_{tn}) = \left(V'_{tn} - \Delta V_{tn}^{(s)}\right)(1-f_{\max}) + \Delta V_{tn}^{(s)}(1-f_{tk}) \quad (5.15)$$

where V'_{tn} is the updated volume. This leads to

$$f_{tn} = f_{\max} - \Delta V_{tn}^{(s)}(f_{\max} - f_{tk})/V'_{tn} \quad (5.16)$$

Having updated the evaporated volumes from the thick and thin slicks, the total fraction of the oil that has been lost by evaporation can be computed. This lost fraction is assumed to apply to all the parcels in the particular subspill and their fraction of light component is adjusted accordingly. Evaporation is stopped when the fraction evaporated reaches the fraction $f_{\max}$ of light component in the original spilled oil.

Evaporation further leads to an increase in the viscosity of the oil, which is often expressed as:

$$\eta_{\text{oil}} = \eta_0 \exp\left(K^{(e)} f_{tk}\right) \quad (5.17)$$

where η_0 is the initial viscosity and $K_{(e)}$ is a constant that determines the increase of viscosity of oil with evaporation (η_{oil}).

As suggested by Zodiatis et al. (2017, 2019) in the case of a very-light-grade oil type with API 42 (American Petroleum Institute unique identifier number assigned to each gas/oil type), the evaporation usually is as high as 50% of the spilled oil, while in the case of a moderately heavy-grade oil type with API 26, the evaporation is as low as 30% of the spilled oil. Therefore, in the case of heavy and moderated grade oil type spillage, the expected spill concentration on the sea surface and spill deposition (beaching) on the coast will be higher, compared to light- and very-light-grade oil type.

Most of Lagrangian oil models simulate the beached oil considering two subcategories, one that is permanently beached and another potentially free to reenter the sea from the coast. For example for an oil spillage with API 43.2, from a coastal pipe rupture located 1 km from the coast (Fig. 5.6), after a period of 30 h: 0.55% of the slick remains at the sea surface, while 54.99% was evaporated and a total of 43.79% impacted the coast, of which 31.42% is potentially free to reenter the sea.

After 49 h from the pipe rupture spillage: 0.11% of the oil slick remained at the sea surface, while 59.35% evaporated and a total of 39.88% impacted the coast, of which 16.22% is potentially free to reenter the sea (Fig. 5.7).

Evaporation continues up to 49 h after the spill from the pipe rupture, after which it nearly stops. The evaporation is predicted to continue within the time interval between the 30th and 49th hours, even after a small trace of

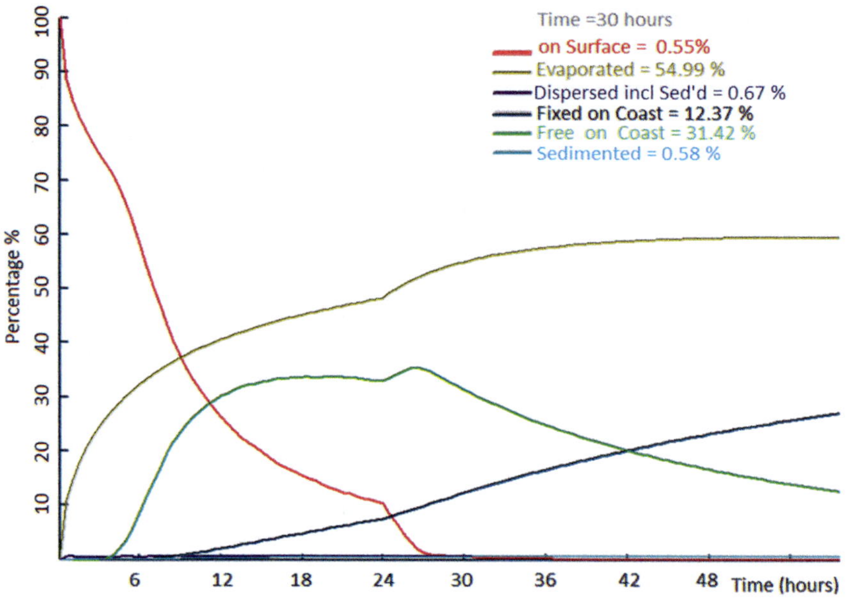

FIGURE 5.6 Simulation of weathering processes of API 43.2 hydrocarbon after 30 h from the pipe rupture located 1 km from the coast, showing the predicted percentage of oil slick on sea surface, evaporated, permanently fixed on coast, on coast free to reenter the sea, dispersed and sedimented at 30th hour after the spill event. *Adapted from Zodiatis, G., Liubartseva, S., Loizides, L., Pellegatta, M., Coppini, G., Lardner, R., … Brillant, A. (2020). Evaluation of the Leviathan offshore platform environmental studies in the Eastern Mediterranean Sea. EGU-European Geosciences Union General Assembly 2020, Vienna, 3–8 May.*

slick is predicted at the sea surface (0.55%) and >40% has beached on the coast. The evaporation increases by almost 5% between the 30th and 49th hours (after the spillage), as a result of evaporation of 31.42% of the slick that potentially was available to reenter the sea from the shore, leading to an increase in the oil slick that permanently stuck on the coast from 12.37% to 23.66%.

5.3.1 Emulsification

Wave action mixes oil with water and forms water-in-oil or oil-in-water emulsion types often referred to as "chocolate mousse." The process oil viscosity and density may increase up to four times (Azevedo, Oliveira, Fortunato, Zhang, & Baptista, 2014; Fingas, 2011). Formation of emulsion reduces the rate of other weathering processes (such as evaporation, biodegradation, and dissolution) by orders of magnitude. The rate of emulsification is determined by the sea state condition, oil viscosity, and by the presence of a minimum percentage of concentration of asphaltenes or resins compounds

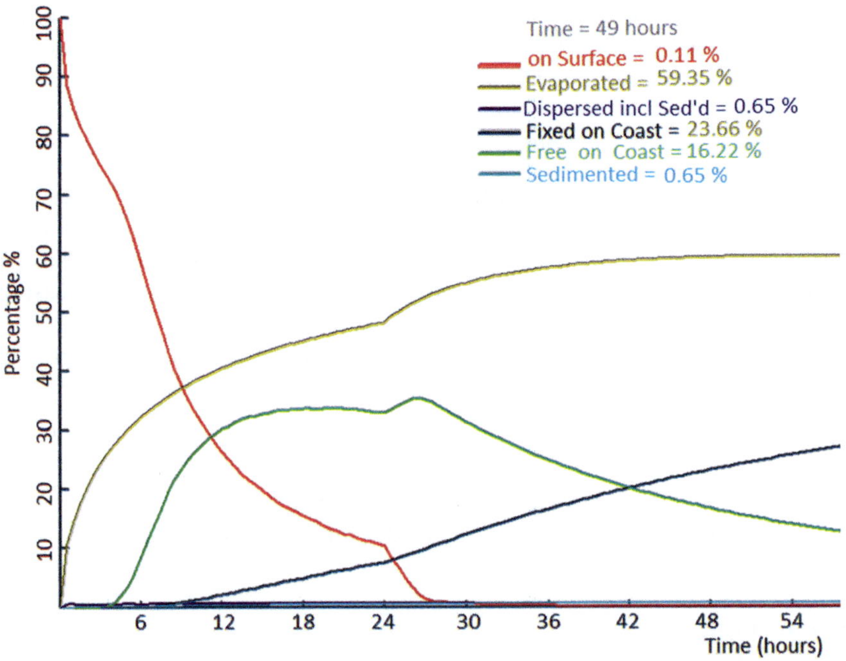

FIGURE 5.7 Simulation of weathering processes of API 43.2 hydrocarbon after 49 h from the pipe rupture located 1 km from the coast, showing the predicted percentage of oil slick on sea surface, evaporated, permanently fixed on coast, on coast free to reenter the sea, dispersed, and sedimented at 49th hour after the spill event. *Adapted from Zodiatis, G., Liubartseva, S., Loizides, L., Pellegatta, M., Coppini, G., Lardner, R., ... Brillant, A. (2020). Evaluation of the Leviathan offshore platform environmental studies in the Eastern Mediterranean Sea. EGU-European Geosciences Union General Assembly 2020, Vienna, 3–8 May.*

(ASCE Task Committee on Modelling of Oil Spills of the Water Resources Engineering Division, 1996) in the spilled oil (Mackay, Paterson, & Nadeau, 1980; Mackay, Paterson, & Trudel, 1980). The emulsification process is slower for high viscosity oils compared to less viscous oil types (Mackay, Paterson, & Nadeau, 1980; Mackay, Paterson, & Trudel, 1980; Mackay et al., 1979). The emulsion may expand under certain conditions in a semi-solid state consisting from up to 70%–80% water-in-oil and then remain emulsified for a long time. In contrast, under calm sea conditions and under sunlight or when beached, the emulsions may separate into oil and water again.

Emulsification is modeled using the Mackay, Paterson, and Nadeau (1980), Mackay, Paterson, and Trudel (1980) algorithms, which expresses the emulsification rate as a function of wind speed, the emulsion's water content, and the oil type (ASCE Task Committee on Modelling of Oil Spills of the Water Resources Engineering Division, 1996; Reed et al., 1999).

Let f_w be the fraction of water in the oil–water mousse. Then Mackay's model (Mackay et al., 1979) for the change in this fraction of water in oil–water mousse per time step of numerical integration is:

$$\Delta f_w = C_2^{(m)}\left(1 - C_3^{(m)} f_w\right) dt \tag{5.18}$$

The wind wave action significantly increase the rate of emulsification; therefore, the Δf_w is multiplied by the factor $(W + 1)2$ where W is the wind speed (m/s) and $C_2^{(m)}$ and $C_3^{(m)}$ are constants. The water-in-oil fraction has an upper limit of $C_3^{(m)}$ (default value taken as 75% for light oils but decreasing for heavy oils).

The principal effect of emulsification is creation of a mousse with greatly increased viscosity. The viscosity of the mousse η_{em} is modeled as:

$$\eta_{\mathrm{em}} = \eta_{\mathrm{oil}} \exp\left\{2.5 f_{\mathrm{w}} / (1 - C_1^{(m)} f_{\mathrm{w}})\right\} \tag{5.19}$$

where $C_1^{(m)}$ is a constant that determines the increase of viscosity with emulsification. The value of η_{oil} already includes the effect of evaporation.

5.3.2 Dispersion

The action of the waves results in the dispersion of oil in the water column, breaking the slick into small and large droplets, with diameters between 1 μm and 1 mm. The small oil droplets are mixed into the water column, while the larger ones resurface, either combined with other droplets to reform a slick, or spread out in a very thin film, often referred to as "sheen." This thin oil film rapidly disperses again by breaking waves as smaller droplets into the water column and is subject to rapid biodegradation, dissolution, and sedimentation. The dispersion rate depends highly on the oil type and can contribute to the main processes that determine the decrease of oil concentration and the lifetime of the slick on the sea surface. The dispersion process gradually decreases due to the fact that the evaporation of the lighter compounds increases the viscosity of the remaining oil. Within the water column, the dispersed particles move solely under the action of sea currents and buoyancy.

The model of dispersion of oil into the water column used in MEDSLIK is based on the work of Buist (1979) and Mackay et al. (1979). The process is illustrated in Fig. 5.5. Wave action drives oil from sea surface into the water column, forming a cloud of droplets beneath the water surface. The droplets are classified as either large droplets that rapidly rise and coalesce again with the surface oil spill, or small droplets that rise more slowly, and may be immersed long enough to diffuse into the lower layers of the water column. In the latter case, they are lost from the surface oil spill and are considered to be permanently dispersed. The criterion that distinguishes the small droplets is that their rising velocity under buoyancy forces is comparable to their diffusive velocity, while for large droplets, the rising velocity is much larger.

Consider first the thick slick at a given instant and let R_L and R_S be the downward volume fluxes of oil per unit area of the slick entering the water column as large and small droplets, respectively. Let the corresponding concentrations be c_L and c_S. If the rising velocities of the large and small droplets are v_L and v_S, then for a quasi-steady state, one can equate the downward and upward fluxes, that is,

$$v_L c_L = R_L \text{ and } \left(v_S + C_1^{(d)}\right) c_S = R_S \tag{5.20}$$

where $C_1^{(d)}$ is the (upward) diffusive velocity of the small droplets. That is,

$$c_L = R_L / v_L \text{ and } c_S = R_S / \left(v_S + C_1^{(d)}\right). \tag{5.21}$$

The total volumes of oil beneath the thick slick in the form of large and small droplets are then given by

$$X_L = c_L u_m A_{tk} = u_m A_{tk} R_L / v_L \text{ and } X_S = c_S u_m A_{tk} = u_m A_{tk} R_S / \left(v_S + C_1^{(d)}\right) \tag{5.22}$$

where u_m is the vertical thickness of the droplet cloud and A_{tk} is the area of thick oil. At each time step of numerical integration, a fraction of the small droplets is assumed to be lost by diffusion to the lower layers of the water column. The total volume lost is taken as:

$$\Delta X_{LL} = C_1^{(\mathrm{d})} c_S A_{tk} \tau \tag{5.23}$$

where again $C_1^{(d)}$ is the diffusive velocity of the small droplets. This leads to a new value of X_S,

$$X_S' = X_S - \Delta X_{LL}$$

The total volume of dispersed oil beneath the thick slick is then incremented according to

$$\Delta V_{\mathrm{tk}}^{(d)} = \Delta X_{LL} + (X_L' - X_L) + (X_S' - X_S) \tag{5.24}$$

where the last two terms represent the changes in oil content of the droplet clouds.

To complete this part of the model, expressions are required for the downward fluxes, R_L and R_S. For this, the fraction of the oil in either the thick or thin slick that is dispersed at each time step of numerical integration is taken as:

$$\Delta f_d = C_3^{(d)} (W_{\mathrm{m/s}} + 1)^2 \tau \tag{5.25}$$

where $W_{\mathrm{m/s}}$ is wind speed in m/s.

For the thick slick, the fraction of small droplets is taken as

$$f_s = \left\{1 + C_4^{(d)}(\eta_{\text{em}}/10)^{1/2}(T_{\text{tk}}/0.001)(\sigma/24)\right\}^{-1} \tag{5.26}$$

where σ is the interfacial surface tension between oil and water. The downward fluxes per unit area of the slick per time step of the numerical integration are then given by

$$R_S = f_s(\Delta f_d/\tau) \quad \text{and} \quad R_L = (1 - f_s)(\Delta f_d/\tau) \tag{5.27}$$

For the thin slick, the following simpler expression is used for small droplets:

$$f_s = \left\{1 + C_5^{(d)}(\sigma/24)\right\}^{-1} \tag{5.28}$$

and it is assumed that all the small droplets below the thin slick are permanently dispersed. Thus, the volume loss is given by

$$\Delta V_{tn}^{(d)} = f_s \Delta f_d V_{tn}^{(d)} \tag{5.29}$$

The total volume of oil dispersed from both thick and thin slicks, $V^{(d)}$, is then incremented according to

$$\Delta V^{(d)} = \Delta V_{tk}^{(d)} + \Delta V_{tn}^{(d)} \tag{5.30}$$

This gives a probability of any particular Lagrangian parcel being dispersed into the water column at the given time step of the numerical integration equal to

$$p^{(d)} = \Delta V^{(d)} / V^{(0)} \tag{5.31}$$

For each oil parcel, the random number generator is called, and the oil parcel is dispersed if

$$rand(0, 1) < p^{(d)} \tag{5.32}$$

Dispersion is assumed to stop when the viscosity η_{em} of the mousse reaches a value η_{max}.

5.3.3 Spreading

Spreading is taking place during the first hour of an oil spill and affects the horizontal extent of the oil spill and its surface thickness. The process is driven by gravity and the surface tension between the interface between oil and water, and dragging opposed by the oil's inertia and viscosity (ASCE Task Committee on Modelling of Oil Spills of the Water Resources Engineering Division, 1996; Fingas, 2011; James, 2002). The spreading continues until a certain thickness limit is reached, after which the slick breaks into patches. As the slick spreads and its surface area increases, the weathering processes such as evaporation, dissolution, dispersion, and oxidation tend to speed up (Spaulding, 1988).

To complete the fate algorithms, changes in the thick and thin slicks and the rate of flow between the two are finally computed (Mackay & Leinonen, 1977; Mackay, Paterson, & Nadeau, 1980; Mackay, Paterson, & Trudel, 1980).

For the thick slick, spreading consists of two parts: a loss of area due to oil flowing from the thick to the thin slick, and a change corresponding to Fay's gravity-viscous phase of the spreading (Fay, 1969, 1971). Thus, the change of area of the thick slick per time step of the numerical integration is

$$\Delta A_{tk}^{s} = \frac{\Delta V_{tn}^{s}}{T_{tk}} + C_2^{(s)} A_{tk}^{1/3} T_{tk}^{4/3} \tau \tag{5.33}$$

where $C_2^{(s)}$ is a constant and $\Delta V_{tn}^{(s)}$ is the volume increment flowing from thick to thin slick. This volume is related to the increment in area of the thin slick:

$$\Delta V_{tn}^{(s)} = \Delta A_{tn}^{(s)} T_{tn} \tag{5.34}$$

Once a value of $\Delta A_{tn}^{(s)}$ is available, one can update the area of the thick slick $A_{tk}^{(s)}$.

Mackay approximates the increment in area of the thin slick similar to the Fay formula: proportional to the cube root of the area, the time from the release, and an exponential function of the thickness of the thick slick that reflects the tendency of the slicks to stop spreading when they become very thin:

$$\Delta A_{tn}^{(s)} = C_1^{(s)} A_{tn}^{1/3} dt \ \exp(-\, C_3^{(s)} / (T_{tk} + 0.00001)) \tag{5.35}$$

Spreading is considered to occur for an initial period of 48 h after the release of each subspill. In case the thickness of the thick slick becomes equal to that of the thin slick, spreading terminates, and the remaining oil parcel is transferred to the thick slick and the droplet clouds beneath it to the thin slick. From that point on, evaporation and dispersion are not further considered for the thick slick.

5.3.4 Dissolution

Oil contains very small amounts of soluble compounds (<1 mg/L), which may dissolve in the water. Oil can dissolve in the water column from the surface slick or from dispersed oil droplets. Dissolution and evaporation are competitive processes. The lower molecular weight hydrocarbon components that are most soluble in sea water are also those that are relatively volatile, such as the light hydrocarbons benzene and toluene, which can dissolve within a few hours (ASCE Task Committee on Modelling of Oil Spills of the Water Resources Engineering Division, 1996; ITOPF, 2002). Since there is a strong correlation between volatility and evaporation, volatile products evaporate very quickly and a higher proportion of their mass is lost through evaporation than dissolution (ASCE Task Committee on Modelling of Oil Spills of the Water Resources Engineering Division, 1996; ITOPF, 2002). For surface slicks, since the partial pressures tend to exceed the solubility of these lower molecular weight compounds, evaporation accounts for a larger portion of the mass than dissolution, except perhaps under ice (McAuliffe, 1987). Dissolution can be significant from dispersed oil droplets due to the lack of atmospheric exposure and the higher available oil surface area per unit of volume (French, Schuttenberg, & Isaji, 1999).

The rate and extent to which oil dissolves depends on the oil type, the sea conditions (sea surface temperature and waves), the spreading, and dispersion. The solubility of oil decreases with the increase of the sea surface temperature (ASCE Task Committee on Modelling of Oil Spills of the Water Resources Engineering Division, 1996), while the rate of dissolution will increase with the increase of the dispersion. Dissolution rate increases with the enlargement of the surface area of the oil relative to its volume, that is, the smaller the droplet (French et al., 1999; ITOPF, 2002). Oil spill models typically do not include a realistic description of dissolution and degradation, although these weathering processes are significant for estimating the impact of an oil spill on to marine ecosystems and for risk assessment for the mid and long term. The algorithm developed by Mackay and Leinonen (1977) is usually applied in oil spill modeling for estimating dissolution from the surface slick, which treats dissolution as a mass flux related to solubility and temperature. For subsurface (dispersed) oil, dissolution is usually treated as a mass flux across the surface area of a droplet (treated as a sphere) in a calculation analogous to the Mackay and Leinonen (1977) algorithm (French et al., 1999; Mackay & Leinonen, 1977). Dissolution rate increases with the decrease of the oil droplets' size (French et al., 1999; French McCay, 2003).

Heavy components of crude oil are insoluble in the water, while lighter oil compounds such as benzene can dissolve in few hours (ASCE Task Committee on Modelling of Oil Spills of the Water Resources Engineering Division, 1996; ITOPF, 2002). Benzene is most volatile and can evaporate 10–1000 times faster than dissolve. Concentrations of dissolved oil do not make significant contribution to the removal of the oil slick from the sea surface. The amount of dissolved oil is usually much less than 1% of the spilled mass (ASCE Task Committee on Modelling of Oil Spills of the Water Resources Engineering Division, 1996; Fingas, 2011), and its effect on oil advection and weathering is negligible.

Importantly, the lower molecular weight aromatic and aliphatic hydrocarbons are both the more volatile and more soluble than those of higher molecular weight. These lower molecular weight aromatic compounds [monoaromatic (MAHs) and polycyclic aromatic hydrocarbons (PAHs)] are the most toxic components of oil to aquatic organisms. Therefore, it is important to model their fate in the water column and sediments, in order to predict the impact of oil spills to marine ecosystems. The fact that oil consists from a large amount of different components, each with its own specific characteristics, makes the prediction of dissolution difficult, especially of oil droplets dispersed in the water column (ASCE Task Committee on Modelling of Oil Spills of the Water Resources Engineering Division, 1996). The dissolution of oil components into the water column does not contribute to removing the oil from the sea surface. However, the water-soluble fraction has a high bioavailability and therefore has the potential to cause acute toxic effects on marine organisms.

5.3.5 Photooxidation

Photooxidation can change the composition of an oil (Fingas, 2011). It occurs when solar radiation and the presence of oxygen cause oxidation of oil spilled at sea, which results in the production of a variety of oxygenated hydrocarbons and sulfur compounds, including aliphatic and aromatic ketones, aldehydes, carboxylic acids, fatty acids, esters, epoxides, sulfoxides, sulfones, phenols, anhydrides, quinones, and aliphatic and aromatic alcohols (Lee, 2003; Payne & Phillips, 1985; Plata, Sharpless, & Reddy, 2008; Tarr et al., 2016). Payne and Phillips (1985) published one of the first extensive reviews on the photochemistry of petroleum in water that identified a variety of oxidation products.

As described in detail in the report of 2003 of the National Research Council (NRC) Committee on Oil in the Sea and in Lee (2003), the mechanism of photooxidation of petroleum hydrocarbons includes both direct photolysis, where petroleum hydrocarbons absorb light energy, to form less-stable reactive intermediate products, like hydroxyl radicals, and indirect photoreactions, where other chemical species in solution absorb light energy. Both produce reactive intermediates (e.g., solvated electrons, hydroxy radicals) that attack the hydrocarbon molecule or transfer energy directly to the reactant hydrocarbon. The necessary factors for photooxidation are solar radiation and light-absorbing molecules (chromophores). Since few petroleum hydrocarbons absorb sunlight efficiently, most photooxidation occurs via indirect photoreactions. The degree and extent of photooxidation of petroleum hydrocarbons at sea depend on: (1) the spectrum and intensity of incident light, (2) the optical properties of the surface water as modified by the petroleum hydrocarbons and other dissolved and particulate constituents, and (3) the optical properties of the hydrocarbons themselves (National Research Council United States Committee on Oil in the Sea: Inputs, Fate, & Effects, 2003). Modeling the photooxidation of petroleum hydrocarbons is complex because the surface slick alters the intensity and spectrum of the incident sunlight. As weathering proceeds, individual components of the petroleum hydrocarbon mixture degrade by photooxidation at different rates and to different products, further altering the spectral environment (National Research Council United States Committee on Oil in the Sea: Inputs, Fate, & Effects, 2003). Besides the effects on surface slicks, photooxidation can also act on petroleum components in the water, including dissolved hydrocarbons (the aromatic and unsaturated fractions of dissolved petroleum hydrocarbons undergo both direct and indirect photolysis in seawater), dispersed oil droplets, and water-in-oil emulsions (Lee, 2003).

Photooxidation may not be very important from a mass balance perspective; however, products of photooxidation may be more toxic than those in the parent material and contribute to the marine biota toxicity observed after an oil spill (Lee, 2003). Many of the photooxidized compounds of oil are also much more rapidly degraded by native bacteria than the parent compounds. Photooxidation also plays an important role in the weathering of dissolved petroleum hydrocarbons (National Research Council United States Committee on Oil in the Sea: Inputs, Fate, & Effects, 2003). Aliphatic and aromatic fractions of petroleum are oxidized to more polar ketones, aldehydes, carboxylic acids, and esters. Since these products are more soluble in seawater, photooxidation enhances the overall dissolution of intact petroleum. These dissolved products can undergo further oxidation by either direct or indirect photolysis (National Research Council NRC, 1985; National Research Council United States Committee on Oil in the Sea: Inputs, Fate, & Effects, 2003; Schwarzenbach, Gschwend, & Imaboden, 1993; Zitka & Cooper, 1987).

Polycyclic aromatic hydrocarbons (PAHs) are chromophores that undergo direct photolysis and likely dominate oil photooxidation at sea. Higher-molecular-weight PAHs typically absorb at longer wavelengths, so PAHs with three or more rings have a greater overlap with sunlight and are more likely to be involved in photooxidation processes (Tarr et al., 2016). In Lee (2003) the rate of direct photolysis of PAHs is described by the equation developed by Zepp and Schlotzheuer (1979):

$$\frac{d\mathrm{PAH}_d}{dt} = \phi \cdot k_\alpha [\mathrm{PAH}_d]$$

where ϕ is a molar yield coefficient, k_α the sum of the k_α values for all wavelengths (λ) of sunlight absorbed by the PAH, PAH_d is the concentration of dissolved PAHs (μmoles/L). Reported yield coefficients ranged between 0.001 and 0.01 for PAHs (Zepp & Schlotzheuer, 1979). Using information on the attenuation of solar radiation in natural waters, a depth-specific rate of photolysis, k_α, of PAHs can be calculated (Bartell, Landrum, Giesy, & Leversee, 1981).

In addition to water-soluble photooxidation products, photooxidation may also result in higher-molecular-weight insoluble polymers produced by photo-initiated free radical reactions acting on an oil slick (Thominette & Verdu, 1984). Photo-produced insoluble polymers contribute to the formation and stability of water-in-oil emulsions (Cormack, 1999) and after a long period of weathering at sea, tar balls, mainly consisting of asphaltenes, may be formed. Stable and persistent emulsions with their increased density and viscosity can make for difficult cleanup operations, both at sea and on beaches (Lee, 2003).

5.4 Biodegradation

Seawater contains an abundance of microorganisms that can biodegrade almost all types of oil components. The various microorganisms prefer specific oil components as their energy source. Bacteria can only degrade oil in contact with water and biodegradation depends on the water–oil interface area. The interface area increases as the oil is spread over the sea surface in a thin layer or by chemical or natural dispersion of oil in the water column.

Important factors influencing the biodegradation rate are temperature, nutrients, oxygen supply, oil type, and the degree of weathering. Low-molecular compounds are degraded more rapidly than the heavier compounds in the oil, thus giving the following order for biodegradation: straight-chain n-alkanes > branched isoalkanes > cyclic alkanes > cyclic naphthenes > aromatics > resins > asphaltenes (Perry et al. 1984). PAHs dissolved in water can be degraded within a few days (Brakstad and Faksness, 2000). At sea, the formation of oil droplets by natural or chemical enhanced dispersion increases the biodegradation rate in the water mass by 10 to >100 times compared to surface oil due to increased emulsification interfacial area, and it has been found that n-alkanes biodegrade within 2–4 weeks at North Sea conditions (Brakstad & Lødeng, 2005). Other higher molecular weight oil compounds are biodegraded more slowly and some very high molecular weight compounds (equivalent to the heavy residues in crude oil that are used to make bitumen) may not biodegrade to any significant degree (e.g., Atlas, 1981; Atlas & Hazen, 2011; Overton et al., 2016).

Oil spill models do not typically include a realistic description of biological degradation of hydrocarbons, although biodegradation is significant for estimating the impact of an oil spill. An attempt to include these processes has recently been carried out using a modified version of MEDSLIK-II model (Spanoudaki, 2016).

Oil biodegradation by native bacteria is one of the most important natural processes that can attenuate the environmental impacts of marine oil spills. During the Exxon Valdez and BP Deepwater Horizon oil spills, for example, and despite using numerous physical means to remove or disperse the oil, natural and enhanced biodegradation greatly reduced oil concentrations and played a major role in mitigating environmental impacts of these two worst oil spills in the US history (Atlas & Hazen, 2011). Furthermore, in models that account for biodegradation (SIMAP, French McCay, 2003, 2004), it is mostly represented as a first-order decay process neglecting the effect of several important parameters that control the biodegradation rate, such as oil composition, microbial population, dispersed oil droplets–water interface, and availability of dissolved oxygen and nutrients. There is generally a need for a more realistic description of biodegradation kinetics in oil spill models, to enable for a more accurate prediction, evaluation of possible bioremediation strategies, and risk assessment in the mid and long term.

5.4.1 The pseudo-component approach

The fate algorithms of Mackay, Paterson, and Nadeau (1980) and Mackay, Paterson, and Trudel (1980), which are typically used in oil spill models, consider the oil as a uniform substance whose properties change as the slick weathers, an approach that can lead to reduced accuracy, especially in the estimation of oil evaporation and biodegradation. Oil is a complicated mixture of a large number of different types of chemical compounds; therefore, as the slick weathers the more volatile oil compounds will partition to the air causing the slick to become rich in higher molecular weight compounds. In addition, the easily biodegradable chemicals (e.g., alkanes, small aromatic molecules—with one or two rings) will be depleted first leaving the recalcitrant components in the water column. Therefore, it is important to differentiate between different chemical groups in order to accurately estimate evaporation and evaluate biodegradation kinetics (Spanoudaki, 2016).

Based on the above considerations, it is important to differentiate between different chemical groups in oil spill modeling based on their physical, chemical, and toxicological characteristics and track their fates separately. To this end, the PC approach (Jones, 1997; Lehr et al., 2000) has been adopted in several oil spill models for simulating certain weathering processes such as evaporation, dissolution, and degradation (e.g., French McCay, 2004; Spanoudaki et al., 2019). Under this approach, the chemical compounds that constitute the oil are grouped into a relatively small number of discrete noninteracting components (PCs), based on physical–chemical properties (volatility, solubility, biodegradability). The resulting PCs behave as if they were single substances with characteristics typical of the chemical group. The fate of each component is then tracked separately. Distillation cuts can be used for generating the PCs, based on the volatility and biodegradability of different chemical groups. An example is depicted in Table 5.1.

In the modified version of MESLIK-II, evaporation is simulated following the PC evaporation model of Jones (1997), while biodegradation of the different PCs in oil droplets and dissolved oil is modeled by Monod kinetics. The kinetics of oil particles size reduction due to the microbe-mediated degradation at water–oil particle interface is represented by the shrinking core model (Levenspiel, 1999; Vilcáez et al., 2013).

TABLE 5.1 Example of pseudo-components used in the modified MEDSLIK-II model to track the fate of different chemical groups based on volatility and biodegradability.

Distillation cut	1	2	3	4
Boiling point	<180°C	180–265°C	265–380°C	>380°C
Molecular weight	50–125	125–168	152–215	>215
Aliphatic compounds	Volatile aliphatics: C4–C10	Semivolatile aliphatics: C10–C15	Low-volatility aliphatics: C15–C20	Nonvolatile aliphatics: >C20
Aromatic compounds	MAHS:BTEX and substituted benzenes	2 ring PAHs (napthalenes)	3 ring PAHs	≥C4 ring aromatics

BTEX, benzene, toluene, ethylbenzene, xylene; *MAHs*, monoaromatic hydrocarbons; *PAHs*, polynuclear aromatic hydrocarbons.

5.4.2 Pseudo-component evaporation model

In the Jones (1997) PC evaporation model, each PC is treated as a single substance with an associated vapor pressure and relative mole fraction. The total evaporation rate of the (thick and thin) slick is the sum of the individual rates. However, the individual rate for a particular component is coupled to the other PCs by the relative mole fraction. The volumetric evaporation rate for a single PC can be written as a function of the volume of the oil, and the mole fraction and molar volume of the component:

$$\left(\frac{dV}{dt}\right)_i = \frac{K_\alpha}{RT_a}\frac{U^{\frac{7}{9}}V(t)P_i\overline{V}_i f_i(t)}{T} \tag{5.36}$$

where i specifies the particular PC, P_i is the vapor pressure of the PC, $\overline{V}_i$ is the molar volume of the PC, $f_i(t)$ is the time-varying molar fraction of the PC, $V(t)$ is the oil volume, U is the wind speed, T is the (thick and thin) slick thickness, K_α is the reference mass transfer coefficient at 1 m/s, R is the gas constant, and T_a is the ambient water temperature. The vapor pressure of each PC is based on Antoine's equation as discussed in Lyman et al. (1990):

$$\ln\frac{P_i}{P_{\mathrm{atm}}} = \frac{\Delta S_i(BP_i - C_{Fi})^2}{RBP_i}\left(\frac{1}{BP_i - C_{Fi}} - \frac{1}{T_a - C_{Fi}}\right) \tag{5.37}$$

where ΔS_i is the change in entropy for vaporization of the *PC*, BP_i is the boiling point, and C_{Fi} is a parameter correlating viscosity with fraction evaporated:

$$C_{Fi} = 0.19BP_i - 18 \tag{5.38}$$

$$\Delta S_i = 8.75 + 1.987 \log BP_i \tag{5.39}$$

Jones (1997) also related molar volume and molecular weight of the PCs to the boiling point of each component. The equations describing the biodegradation kinetics for dissolved oil and oil droplets dispersed in the water column are described in Appendix A.

5.4.3 Biodegradation test case

The predictive ability of the modified version of MEDSLIK-II has been tested for the Lebanon coastal spill, which occurred in mid-July 2006, the characteristics and spreading of which are known. In addition to successfully predicting the spreading of the oil slick, the total fate of the oil spill has been simulated both with and without biodegradation kinetics, for comparison. The oil spill occurred as a result of two bombing raids by Israeli missiles, on fuel storage tanks at the Jiyeh power plant 30 km south of Beirut on the mornings of the 13 and 15 of July 2006, causing the leakage of an estimated 12,000–15,000 tons of fuel oil into the Mediterranean Sea (UNEP, 2007). Unfortunately, clean-up operations were delayed for 5 weeks due to the conflict, during which time the oil contamination spread over 150 km of the Lebanese coastline, reaching as far as the Syrian coastline in the north (UNEP, 2007). According to UNEP (United Nation Environmental Pollution), the type of oil spilled was medium/heavy oil that can be compared to IFO 150 (Intermediate Fuel Oil, with a viscosity of 150 cSt at 50°C). CEDRE (Centre de Documentation, de Recherche et d'Expérimentation sur les pollutions accidentelles des eaux) analyzed several oil samples and found that the spilled Jiyeh oil had high saturate levels (around 50%), low aromatics (around 28%), and around 22% resins and asphaltenes (CEDRE, 2006). The spill was modeled as a continuous leakage of oil over a period of 6 days, starting July 13, 2006, at a rate of 130 tons/h, for a total mass of 18,770 tons of oil (Coppini et al., 2011). Water currents (hourly mean fields) were obtained from the CMEMS Med MFC (Copernicus Marine Environmental Monitoring Service of the Mediterranean Monitoring and Forecasting Centre) and wind forcing from ECMWF (European Centre for Medium-Range Weather Forecasts).

Chemicals in the oil mixture have been grouped into eight PCs (four for aliphatic compounds and four for aromatic compounds), defined based on four distillation cuts (Table 5.2). Kinetic parameters used for calculations are outlined in Table 5.3. Values are based on averaged literature data (Choi, Katsutoshi, Yasunori, & Unno, 1999; Walter et al., 1991; Desai, Autenrieth, Dimitriou-Christidis, & McDonald, 2008; Dimitriou-Christidis & Autenrieth,

TABLE 5.2 PCs used in the modified MEDSLIK-II for the Lebanon oil pollution test case.

Distillation cut	1 (High volatility and solubility)	2 (Semivolatile and soluble)	3 (Low volatility and solubility)	4 (Nonvolatile and insoluble)
Boiling point	<180°C	180–265°C	265–380°C	>380°C
Molecular weight	50–125	125–168	152–215	>215
Aliphatic compounds	C4–C10 (PC 1)	C10–C15 (PC 2)	C15–C20 (PC 3)	>C20 (PC 4)
PC % in oil mixture	4%	5%	18%	23%
Aromatic compounds	MAHS:BTEX and substituted benzenes (-to C3- benzenes) (PC 5)	C4 benzenes, 2 ring PAHs (- to C2- naphthalenes) (PC 6)	C3-, C4- naphthalenes, 3-ring PAHs (PC 7)	≥C4 ring aromatics (PC 8)
PC % in oil mixture	3%	3%	12%	10%

BTEX, benzene, toluene, ethylbenzene, xylene; *MAHs*, monoaromatic hydrocarbons; *PAHs*, polynuclear aromatic hydrocarbons; *PCs*, pseudo-components.

TABLE 5.3 Kinetic parameters used for Lebanon case study.

Pseudo-components	1	2	3	4	5	6	7	8
μmax (h^{-1})	0.6	0.6	0.55	0.05	0.32	0.23	0.1	0.053
K_s (mg/L)	27.28	55.57	86	28.65	32.3	23.75	28.65	28.65
Y_S	0.119	0.119	0.119	0.485	0.4	0.4	0.485	0.485

2007; Lin & Cheng, 2007). Initial concentrations for biomass, dissolved O_2, N, and P of 2.73×10^4 cells/mL, 6 mg/L, 2 mg/L and 1 mg/L, respectively, have been assumed (typical for the summer period of July–August in Southeastern Mediterranean).

In Fig. 5.8, the oil slick concentration and oil adhered to the shoreline (bold black line) are shown after 50, 150, 216, 250, 300, and 400 hours from

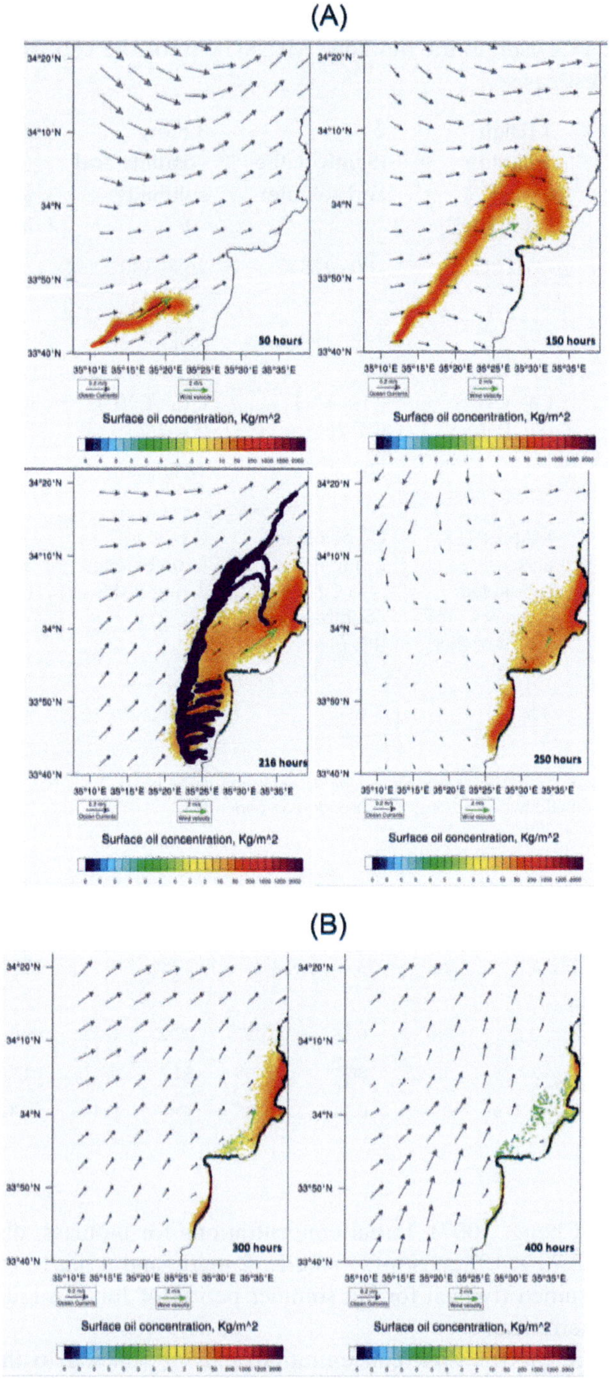

FIGURE 5.8 Oil-slick concentration, coastal oil impact (bold black coast line), current fields (black arrows), wind in the gravity center of each slick (green arrows), and satellite detected observations on July 22, 2006 at 07:55 (blue slicks).

the beginning of the spill. Comparison with satellite detected observations for the surface slick on July 22, 2006 is also shown (blue slick). As shown in the figure, after 400 hours almost all oil has been adhered to coast or dispersed in the water column.

The total fate of the spill is shown in Fig. 5.9. Around 14% (2781 tons) of the total oil spill mass evaporated in the first 6 days; evaporated oil is composed of the lighter components of the spilled oil (mainly from PCs 1, 2, 5, & 6). The remaining oil moved on the sea surface due to spreading, advective, and diffusive displacements; dispersed in the water column; and adhered to the coastline. After 6 days, when the total volume of the spill was released, around 15,000 tons of oil was in the oil slick on the sea surface. As time progressed, due to the direction of sea surface currents, most of the sea surface oil adhered to the coastline, polluting over 150 km of the Lebanese coastline (a total of 16,000 tons, from

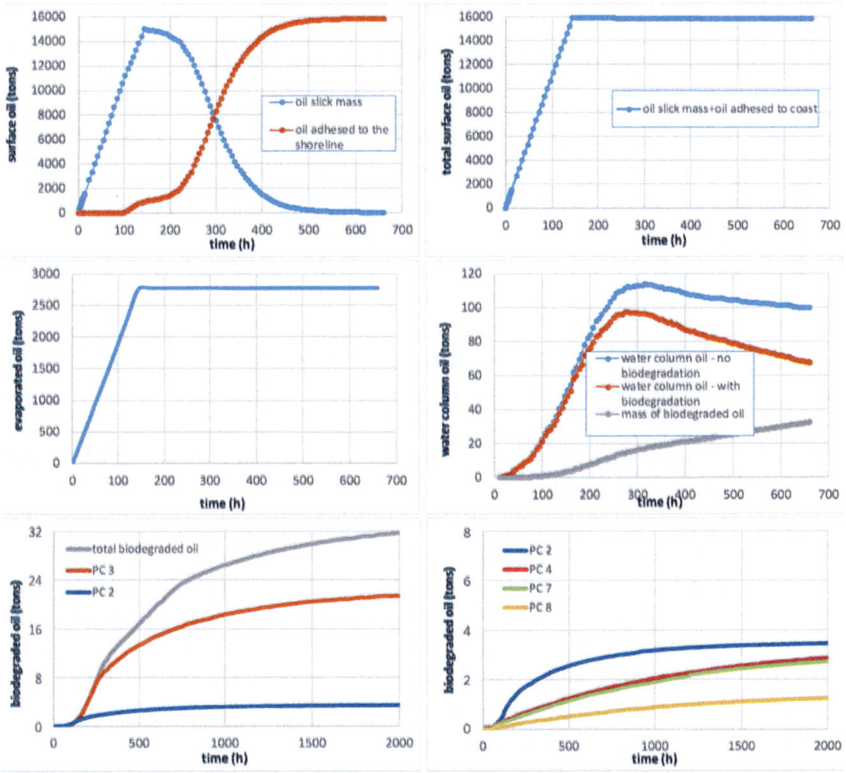

FIGURE 5.9 Total fate of the oil spill: sea surface oil (oil slick mass), oil adhered to the coastline, total sea surface oil (oil slick mass + oil adhered to coast), evaporated oil, oil in the water column, biodegraded oil mass.

the 18,770 tons of spilled oil). Approximately 110 tons of spilled oil was dispersed in the water column (in oil droplets and dissolved form). Fig. 5.9 shows the mass of oil in the water column with and without biodegradation. From the total amount of oil in the water column, approximately 32 tons are biodegraded after 1500 hours (62 days). The biodegradation rate differs significantly between the different PCs, as shown in Fig. 5.8. The total mass of PC 2 remaining in the water column is biodegraded after 1,000 hours (41 days), while 85% of the mass of PC 3 remaining in the water column is biodegraded after 1500 hours (62 days). For the more recalcitrant PCs 4, 7, and 8, biodegradation reaches only 8.5%, 15%, and 8%, respectively, of the total PC mass in the water column after 2000 hours (82 days). This is an indication that bioremediation approaches would be needed to enhance the biodegradation process. The model estimated that approximately 32 tons of oil from PC 4, 15.5 tons of oil from PC 7, and 13.8 tons of oil from PC 8 remained in the water column 82 days after the spill.

5.5 Modeling of oil spills below the sea surface

Although deep-sea oil well blowouts are not as frequent as oil spills from maritime traffic and tanker accidents, oil accidentally maybe released during exploratory drilling or production activities at deep water depth is of particular concern, especially with the expansion of the offshore oil/gas industry to deeper environments. The large drilling platform, known as the Deepwater Horizon, experienced a gas blowout on April 20, 2010 and started releasing oil into the Gulf of Mexico at a rate of 1000 barrels per day (US Coast Guard, 2011; Klemas, 2010; Liu et al., 2011a,b,c). This event caused long-term damage to the marine and coastal environment of the Gulf. Despite many studies to develop effective response strategies following the Deepwater Horizon accident, there still are significant knowledge gaps.

As an example of recent developments to address the oil spill modeling issues from a seabed blowout, the MEDSLIK module for predicting the oil plume until reaching the sea surface is described below. The MEDSLIK plume module incorporates a modified model originally proposed by Yapa and Zheng (1997), Zheng and Yapa (1998), and Malačič (2001) for entrainment of sea water into the oil plume and a revised Yapa and Zheng model of detrainment of oil into the water body (Lardner & Zodiatis, 2017). The mass and momentum equations in the MEDSLIK plume module differ from those provided by Yapa and Zheng (1997), and no gas–oil mixture is considered.

The rising of an oil plume generates a region of high turbulence around its lateral boundaries, causing loss of momentum and leading to the permanent entrainment of sea water into the oil plume and detrainment of oil into the water body.

Therefore, the oil plume is assumed to consist of bubbles of oil interspersed with entrained sea water (Fig. 5.10). The whole plume admixture rising under the driving force of the buoyancy of the oil component, depending on the density (ρ) of the plume:

$$\rho = c\rho_o + (1 - c)\rho_a \tag{5.40}$$

where ρ_o is the oil density, ρ_α is the sea water density, and c is the mean oil concentration.

Let r be the radius and A the cross-sectional area of the plume at the point s.

If w is the mean velocity of the plume matter (oil + water) at this cross section, then the volume flux Φ is:

$$\Phi = Aw = \pi r^2 w \tag{5.41}$$

The mass flux is equal to $\Psi = \rho\ \Phi$ and therefore the mass flux of oil is equal to $\Psi o = c\ \rho_o\ \Phi$.

The horizontal and vertical equations of motion of oil admixture in the plume with mass M due to the buoyancy (B') and drag forces (F') from the turbulence of the surrounding fluid are given by:

$$\frac{d}{dt}(Mw\cos\theta) = -F'\cos\theta \qquad \frac{d}{dt}(Mw\sin\theta) = -F'\sin\theta + B'$$
$$N = \rho A w^2, F' = C_D(2\pi r)w^3\delta t_r \qquad B' = g\Delta\rho A w\delta t_r, M = \rho A w \delta t_r \tag{5.42}$$
$$\frac{dN}{ds} = -C_D\rho_a 2\pi r w^2 + gA(\Delta\rho)\sin\theta \qquad N\frac{d\theta}{ds} = gA(\Delta\rho)\cos\theta$$

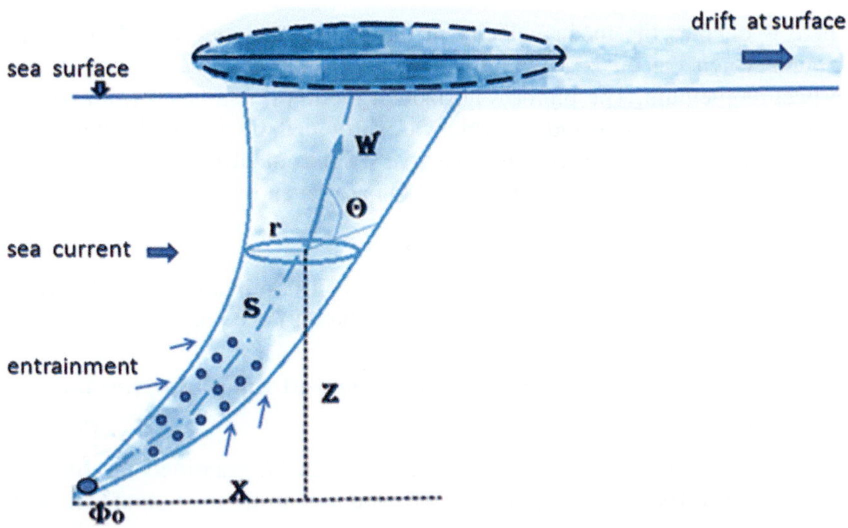

FIGURE 5.10 Schematic of the oil plume problem description with water entrainment.

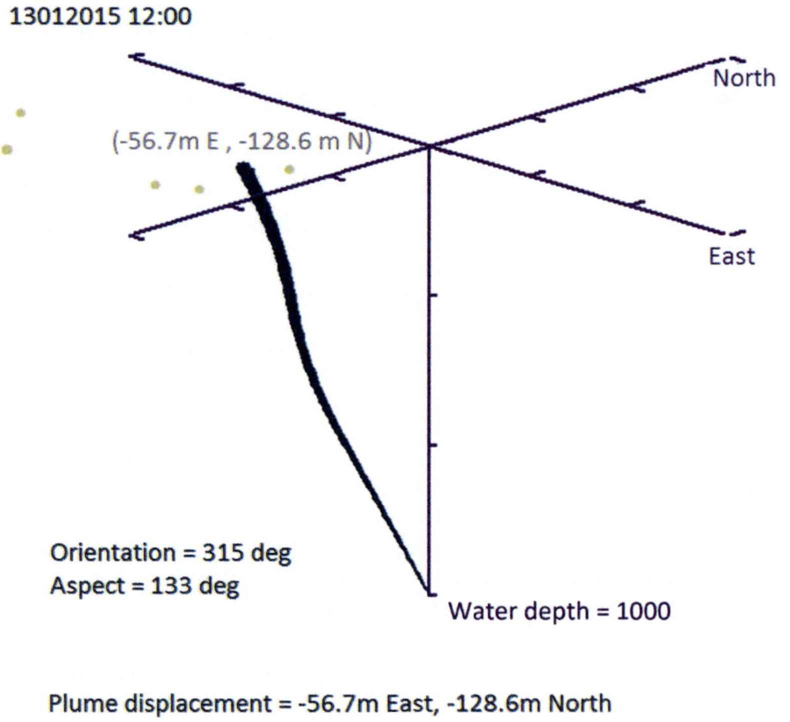

FIGURE 5.11 A typical example of the *X, Y, Z* displacements in an oil plume modeling. The displacement from the vertical axis becomes more and more rapid as the plume approaches the sea surface, being especially pronounced in the upper 50 m, due to higher current speeds, compared to the weaker subsurface currents.

where w denotes the mean velocity at the across-sectional area A of plume through which the oil is flowing/leaking, C_D is the drag coefficient, Mw is vertical momentum. The plume's horizontal motion is estimated from the sea currents components u and v, which are provided by the hydrodynamical forecasting data. The horizontal advective displacement of the plume in the easterly and northerly directions (Fig. 5.11) are given by,

$$dx/ds = u/w, \quad dy/ds = v/w \tag{5.43}$$

where $w = \Phi_0/A_0$ is the mean velocity of the oil admixture at the A_0 cross-sectional area of the opening through which the oil is leaking.

5.5.1 Experiments to access the sensitivity of the plume model parameters

For a testing experiment, Lardner & Zodiatis, 2017 examined the oil spill blowout with fixed C_D- drag of 0.08 and α-entrainment coefficient of

0.00005, using realistic sea density and currents. The MEDSLIK plume model was run using a variety of oil plume parameters such as: oil density, oil leakage depth, and leakage rate. The resulting oil thickness at the sea surface with oil being released from a deep water source was always smaller compared to the case when the release source was located at sea surface.

Table 5.4 presents the results for different oil densities, from light to very heavy oil. Under the same set of environmental conditions, the heavier the oil, the slower it rises, and the greater the horizontal displacement, but there are relatively small changes in the final cross section of the plume and the final oil fraction.

Table 5.5 outlines the corresponding results for five different water depths of the source location, between 500 and 2500 m. The rise time increases with depth and the final oil fraction decreases. For the larger depths, the final oil fraction becomes very low.

Table 5.6 presents the results of experiments considering leakage rates of 10, 100, 1000, and 8000 m^3/h through an opening with 0.01 m^2 diameter. The last case is probably unrealistic because it corresponds to an initial jet speed of over 200 m/s; the interesting point is that the jet becomes almost vertical (with inclination over 89°) within less than 10 m from the source of discharge. The first case, corresponding to a very low initial jet speed, buoyancy causes the velocity initially to increase to a value of 0.62 m/s, before it decreases with the increasing water fraction.

TABLE 5.4 Comparison of plume structure for different densities of oil. After Lardner and Zodiatis (2017).

API	Density (kg/m^3)	RiseTime (minutes)	Velocity (m/s)	X-section (m^2)	OilFrac	PlumeVol (m^3)	Displ (m)
38.0	834.8	17.7	0.86	0.077	0.42	29.5	0.87
33.5	857.6	18.7	0.82	0.083	0.41	31.1	0.93
28.0	887.1	20.2	0.76	0.093	0.40	33.7	1.02
18.0	946.5	25.4	0.60	0.128	0.36	42.3	1.28

Rise time: The time in minutes for the oil to reach the surface.
Velocity: The final vertical velocity of the plume at the surface in m/s.
X-section: The cross-sectional area of the plume at the surface in m^2.
OilFrac: The final oil concentration when the plume reaches the surface.
PlumeVol: The total oil content of the plume in m^3.
Displ: The horizontal displacement of the plume at the surface in m.

TABLE 5.5 Comparison of plume structure at different depths of spill. After Lardner and Zodiatis (2017).

Depth (m)	RiseTime (minutes)	Velocity (m/s)	X-section (m^2)	OilFrac	PlumeVol (m^3)	Displ (m)
500	8.9	0.89	0.051	0.61	14.9	0.93
1000	18.7	0.82	0.083	0.41	31.1	0.93
1500	29.2	0.76	0.13	0.29	48.6	0.93
2000	40.4	0.72	0.18	0.21	67.3	0.93
2500	52.2	0.68	0.25	0.16	87.1	0.93

Rise time: The time in minutes for the oil to reach the surface.
Velocity: The final vertical velocity of the plume at the surface in m/s.
X-section: The cross-sectional area of the plume at the surface in m^2.
OilFrac: The final oil concentration when the plume reaches the surface.
PlumeVol: The total oil content of the plume in m^3.
Displ: The horizontal displacement of the plume at the surface in m.

TABLE 5.6 Comparison of plume structure for different leakage rates. After Lardner and Zodiatis (2017).

Discharge rate (m^3/h)	InitialVel (m/s)	RiseTime (minutes)	FinalVel (m/s)	X-section (m^2)	OilFrac	PlumeVol (m^3)	Displ (m)
10	0.28	38.5	0.34	0.054	0.09	6.4	0.13
100	2.78	18.7	0.82	0.083	0.41	31.1	0.93
1000	27.8	11.0	1.48	0.242	0.77	183.	3.21
8000	222.2	7.1	2.34	1.02	0.92	951.	8.13

Rise time: The time in minutes for the oil to reach the surface.
Velocity: The final vertical velocity of the plume at the surface in m/s.
X-section: The cross-sectional area of the plume at the surface in m^2.
OilFrac: The final oil concentration when the plume reaches the surface.
PlumeVol: The total oil content of the plume in m^3.
Displ: The horizontal displacement of the plume at the surface in m.

5.6 Oil spill prediction in areas with ice

Oil spill models may include the effect of ice on the movement of an oil slick. The presence of ice on the sea surface affects the transport of an oil slick, that is, the advection, the Stoke's drift, the spreading and dispersion, as well as the fate/weathering processes of evaporation and emulsification. For oil spill predictions, the ice may be considered in two main forms: drift ice that consists of separate floes that gradually move over the water surface and restrict the movement of the oil and one or more ice shelves that consist of packed ice that moves only slowly through melting or additional freezing and behaves like a coastal boundary.

In the case of high concentration of surface drifting ice (>0.7 for example), the velocity of the oil will be equal to the velocity of the drifting ice, while with low ice concentrations (<0.25), the advection of the oil is little affected by the ice, but will be the usual combination of the sea currents, wind drift and Stokes' drift. For intermediate concentrations of the ice (>0.25 and <0.7) the oil advection will vary from the one to the other.

The total advection velocity of the oil parcels ui, vi as from the Eqs. (5.1) and (5.2) are the sum of advection velocities ui_{adv}, vi_{adv} and wind drift velocities W_{drftx}, W_{drfty} and in addition plus the stokes drift velocities S_x, S_y:

$$\begin{aligned} ui &= ui_{\text{adv}} + W_{\text{drftx}} + S_x \\ vi &= vi_{\text{adv}} + W_{\text{drfty}} + S_y \end{aligned} \tag{5.44}$$

then the parcel total advection velocities are to be a combination of this and the ice drift velocities U_{icedrift}, V_{icedrift}:

$$\begin{aligned} ui &= ui(1 - fi) + fi\, U_{\text{icedrift}} \\ vi &= vi(1 - fi) + fi\, V_{\text{icedrift}} \end{aligned} \tag{5.45}$$

where fi is the concentration of ice.

The Stoke' drift in oil spill modeling usually is calculated using the JONSWAP (Joint North Sea Wave Project spectrum) wind parameterization (Hasselmann et al., 1973), while the availability of wave forecast from CMEMS, NOAA, and regional wave forecasting systems made possible to calculate the Stoke's drift velocity (S) directly from the significant wave height (H), the wave zero crossing period (T), and the water depth (z) according to the equation,

$$\left(S_x, S_y\right) = \frac{1}{8} H k_2 \omega d \tag{5.46}$$

where k is the wave number, ω is the waves' angular frequency, and d is the vector wave direction.

Similarly the fate/weathering processes of evaporation and emulsification are also reduced according to the ice concentration, toward zero as the concentration of drift ice increases from a low ice concentration value (0.25) to a high ice concentration value (0.7). Similarly, the expressions for the spreading of the thick and thin slicks, in the presence of ice, are modified as below

$$\Delta A_{tk}^{(s)} = -\frac{\Delta V_{tn}^{(s)}}{T_{tk}} + C_2^{(s)} A_{tk}^{1/3} T_{tk}^{4/3} \tau \tag{5.47}$$

$$\Delta A_{tn}^{(s)} = C_1^{(s)} A_{tn}^{1/3} \tau \exp\left(- C_3^{(s)} / (T_{tk} + 0.00001) \right) \tag{5.48}$$

the two constants $C^{(s)}{}_1$ and $C^{(s)}{}_2$ are modified by multiplying them by the factor $(1-f)$ where f is the concentration of the drift ice. This reduction of the effective areas has the effect of reducing the rate of evaporation (but no effect on the rate of emulsification or the change in oil viscosity).

To obtain the physical boundaries and the ice concentration for the drift and/or shelf ice coverage at sea, the corresponding data are provided from remote observations (by airplanes and satellites), or from numerical models. The CMEMS, ESA, and NOAA portals report in near real-time data on ice coverage at regional polar sea areas (Fig. 5.12). To be useful for the models, the drift/floe ice data should consist of the geographical

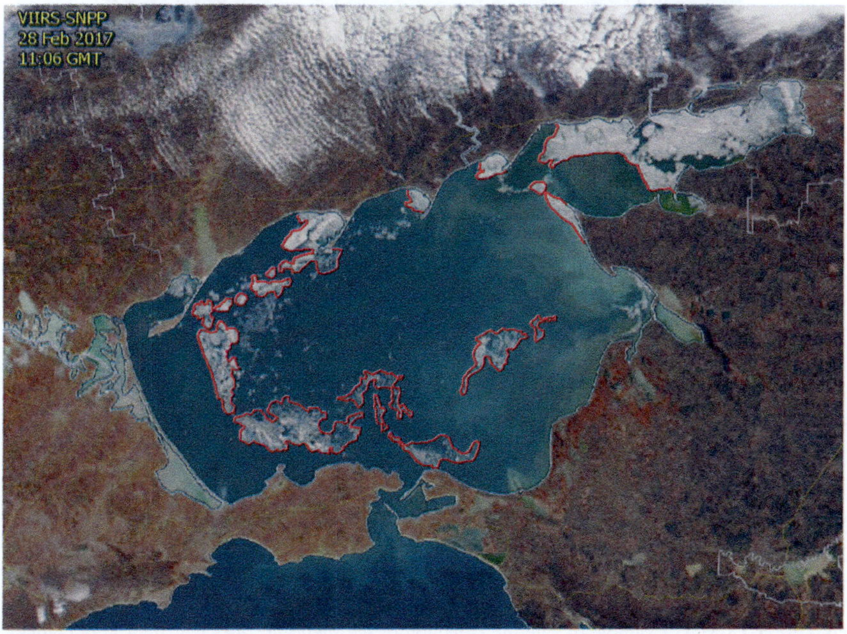

FIGURE 5.12 Example of ice shelf and ice floes in the Azov Sea, as observed by VIIRS (visible infrared imaging radiometer suite) satellite image on the 28 February 2017 at 11:06 GMT. *Source: image processed by Dmitry Soloviev, MHI-RAS.*

coordinates at given data points, the fraction of ice cover and the east and north components of the ice drift velocity at that given data point (Fig. 5.13). The data points should be chosen so as to surround the anticipated location of the oil slick. Similarly, the ice shelf data may consist of

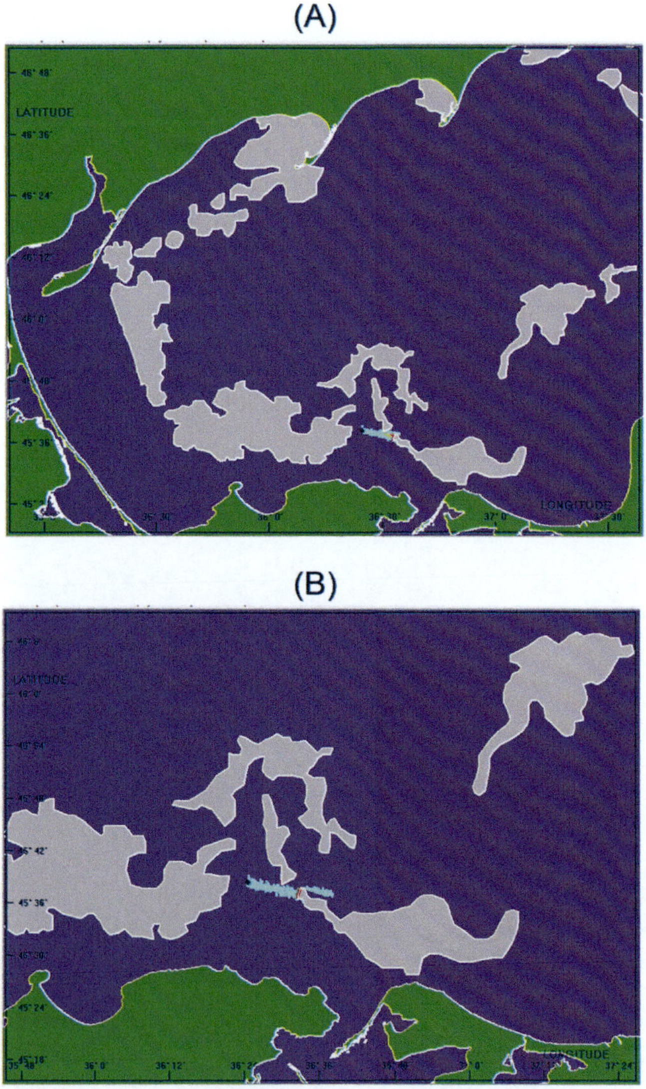

FIGURE 5.13 (A) Example of satellite detection of ice shelf and floes ice (light gray color) in the Azov Sea on the February 28, 2017 at 11:06 GMT, integrated in MEDLSIK oil spill modeling prediction, after 18 h of simulations. (B) Enlarged domain north from Kerch strait with satellite detected floes ice (light gray color) on the February 28, 2017 at 11:06 GMT and the integrated MEDLSIK oil spill prediction, after 22 h of simulations.

several separate sectors and the physical boundary of each must be provided separately for the oil spill simulations. Each record in the data consists of the geographical coordinates of a point on the boundary of the ice shelf.

Parcels of oil that hit the boundary of an ice shelf are assumed to become stuck there, just as on a usual coastal boundary, but may be released back into the water area under the influence of wind and currents later in the numerical modeling simulations. During a long simulation, there may also be changes in the position of an ice shelf due to melting or further accretion of ice. In such a case, parcels of oil that are stuck on the boundary are assumed to move with the ice to the nearest position on the new boundary.

5.7 Good practice for operational implementation of oil spill models

The European Commission has established the European Marine Observation and Data Network (http://www.emodnet.eu) for providing a convenient, single entry point for accessing and retrieving marine data derived from the EMODNET thematic portals, the CMEMS, and other initiatives for the European seas. The EMODNET and CMEMS portals aim to provide readily available marine data sets for operational tasks and scenarios such as oil spill predictions to determine the likely trajectory of an oil slick and the statistical likelihood of affecting sensitive coastal habitats, species and tourist beaches. For instance, an "oil platform leak" scenario concerns the ability to produce oil spill predictions in the Mediterranean, Black Sea, Baltic Sea, and other European seas, where the EC (European Commission) generates the oil leak alert online. In the framework of this EU(European Union) activity, Lagrangian oil spill models such as MEDSLIK, MEDSLIK II, and SEATRACK were implemented using the environmental data from CMEMS and or other downscaled metocean forecasting systems.

The Synthetic Aperture Radar (SAR) satellite observations (ENVISAT, Sentinel, RADARSAT, etc.) have become a useful means for detecting oil slicks on the sea surface [Klemas, 2010; Liu et al., 2011a,b,c; De Dominicis et al., 2013; Zodiatis et al., 2012; European Maritime Safety Agency portal of the CleanSeaNet (EMSA-CSN)], due to its broad spatial coverage, independence of the lighting and weather conditions. Detection of oil spills by SAR systems is based on the dampening effect that oil has on sea surface waves. An oil slick at sea "smoothes" the water surface and thus reduces the radar backscatter. This creates a darker signature in the image, which can be interpreted as a possible oil slick. A condition for detecting oil slicks on the sea surface is that the wind is strong enough (between 2.5 and 13 m/s) for the generation of waves, above which oil films become invisible to SAR.

EMSA-CSN (http://cleanseanet.emsa.europa.eu) provides oil slick detection to support the European Union Member States response agencies with routine monitoring of the European seas, using SAR satellite data provided by the European Space Agency (ENVISAT, Sentinel) and other satellites (RADARSAT) for illegal oil discharges (Fig. 5.14).

The downscaled CMEMS metocean forecasts are used along with SAR satellite data for short forward and backward predictions (Fig. 5.15). Such predictions can also be superimposed on the Automatic Identification of Ships (AIS) traffic information to assist the response agencies to indentify the ship responsible for the detected oil slicks, contributing in this way to the implementation of the EU Directive 2005/35.

The simplest approach to use SAR data in oil spill models is to predict the future and past movements of the boundary points of the slick outlined on the satellite image. The advantage of this approach is that it is fast and requires no extra programming. The disadvantage is that it does not allow including physical processes other than advection.

A more sophisticated approach (Zodiatis et al., 2012) is to fill randomly the slick outline with a large number of parcels of oil and to follow the movements of the parcels. This strategy allows including the diffusion in the estimation of future slick movement. It also enables the prediction of the future and past states of the oil. This approach requires however significantly

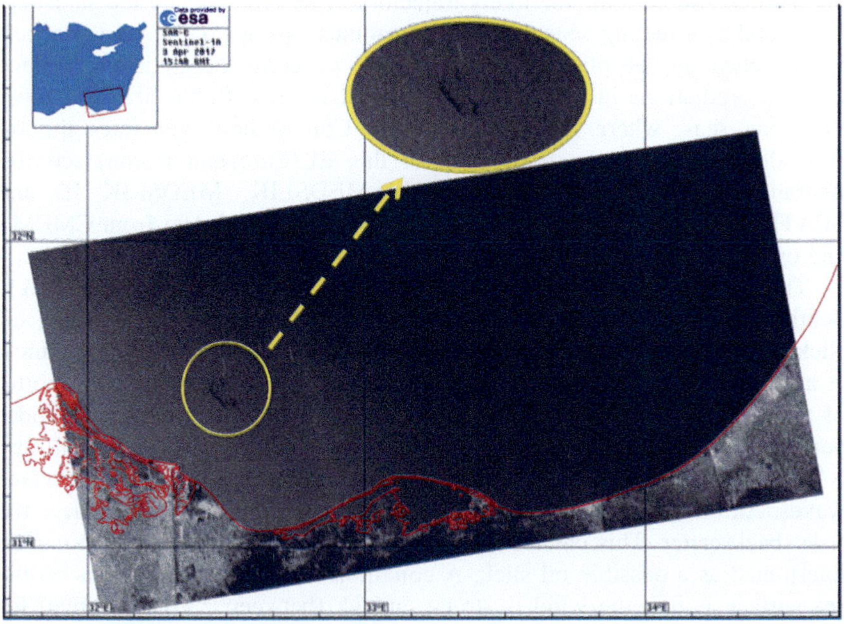

FIGURE 5.14 Example of satellites remote sensing monitoring for oil slicks using ESA SAR-C data from Sentinel 1 A, detected on the April 3, 2017 at 15:48 GMT, offshore Suez Canal in the Mediterranean Sea. *Source: image processed by Dmitry Soloviev, MHI-RAS.*

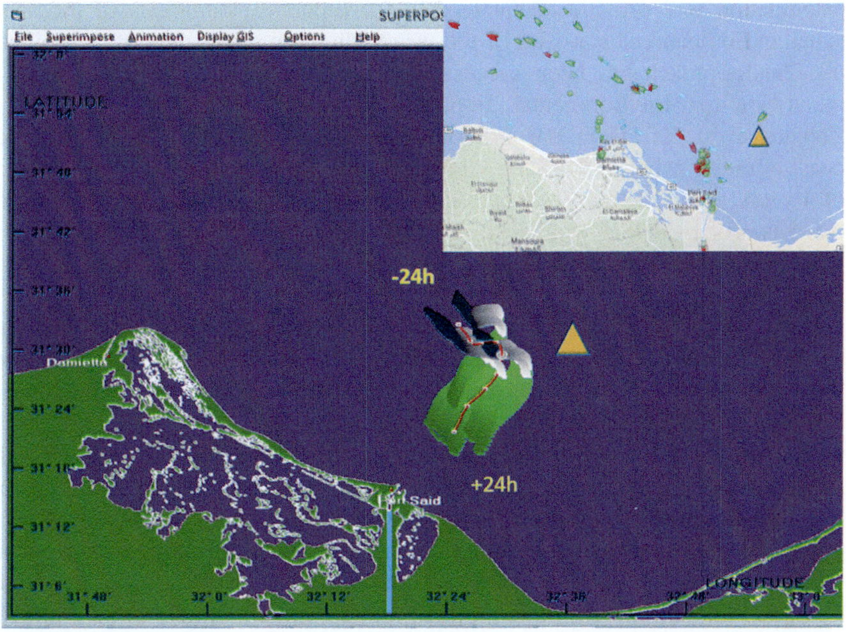

FIGURE 5.15 Example of MEDSLIK hourly oil slicks predictions for 24 h forward (green color) and backward (gray to black color) using ESA SAR-C data from Sentinel 1A, detected (white color) on April 3, 2017 at 15:48 GMT, offshore Suez Canal in the Mediterranean Sea. The orange color triangle indicates the location of an offshore oil/gas platform in the exclusive economical zone (EEZ) of Egypt, while the superimposed map the Automatic Identification of Ships shipping traffic in the area with the ships (colored dots south-west from the offshore platform) waiting to enter the Suez Canal.

more computing resources, though is certainly preferable in that diffusion is included.

Moreover, oil slick forecasts are useful in determining if any valuable resources may be threatened by the oil and in suggesting the most appropriate methods for combating it. Slick hind casts may help find the source of the oil, assisting in this way the response agencies to implement the EU Directive 2005/35. By plotting data on ship movements obtained from local or regional AIS, advanced oil spill models can assess whether any ship was in the same place and at the same time at any hind-cast position of the oil slick (De Dominicis, Pinardi, Zodiatis, & Archetti, 2013; Zodiatis et al., 2012).

With respect to the preparation of Impact Damage Assessment from offshore oil/gas industry, stochastic oil spill simulations are usually carried out using climatology and hind cast metocean data from CMEMS, ECMWF, NOAA, or other systems (Alves et al., 2016). It is suggested that in the case of long-run oil spill simulations for the needs of Impact Damage Assessment, a large number of oil spill predictions, over the entire period under examination will provide adequate data for accurate estimations of the seasonal and interannual statistics of the weathering fate parameters of the spilled oil. This approach has been successfully implemented for the Evaluation of the Leviathan Environmental Impact Assessment (Zodiatis et al., 2020; Zodiatis et al., 2021) in the easternmost part of the Mediterranean, off the coast of Israel, where 5844 oil spill simulations were carried out over a period of 4 years, using the daily sea currents and SST from CMEMS Med MFC and the 6-hourly ECMWF wind data. The stochastic oil spill simulations in this particular study were initiated every 6 hours for 10 days predictions (Fig. 5.16).

From Fig. 5.16 is demonstrated that the 12 simulations scenarios carried out in the frame of the Leviathan Environmental Impact Assessment (Brenner, 2019) do not provide the adequate time coverage for seasonal and interannual statistics of the fate parameters, compared to the 5844 scenarios covering the entire period of the 4 years carried out in the Evaluation of the Leviathan Environmental Impact Assessment (Zodiatis et al., 2020; Zodiatis et al., 2021). The statistics of the data obtained from the 5844 oil spill simulations were inter compared with those resulting from 12 in total oil spill simulations scenarios (Brenner, 2019) over a period of 4 years during the Evaluation of the Leviathan Environmental Impact Assessment (Zodiatis et al., 2020; Zodiatis et al., 2021).

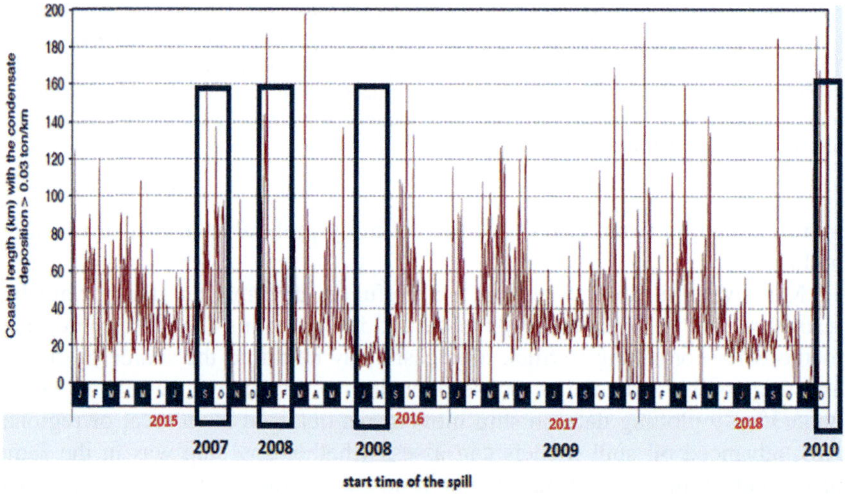

FIGURE 5.16 Time series of the impacted coastal length from the condensate spillage during the period 2015–18 carried out in the frame of the Evaluation of the Leviathan Environmental Impact Assessment, superimposed with the time windows of the 12 simulations for the period 2007–10 carried out previously in the frame of the Leviathan Environmental Impact Assessment. *After Zodiatis, G., Liubartseva, S., Loizides, L., Pellegatta, M., Coppini, G., Lardner, R., ... Brillant, A. (2020). Evaluation of the Leviathan offshore platform environmental studies in the Eastern Mediterranean Sea. EGU2020-5386-European Geosciences Union General Assembly 2020, Vienna, 3–8 May.*

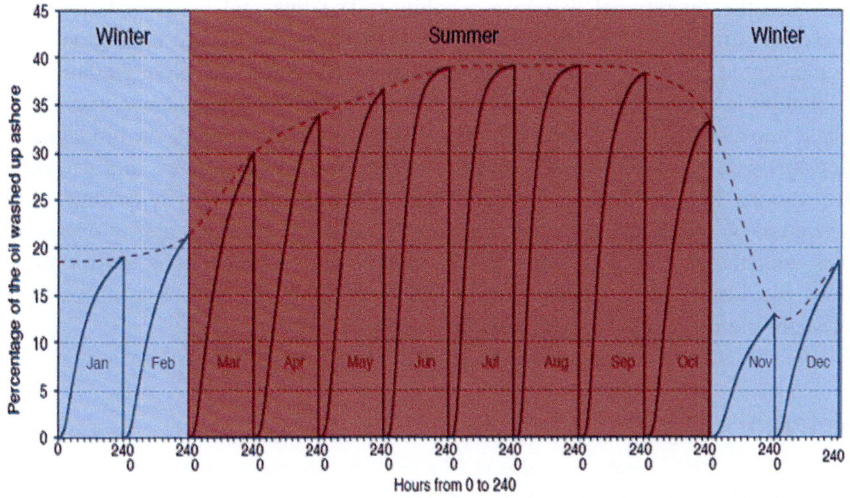

FIGURE 5.17 Monthly and seasonal variations of condensate onshore carried out in the framework of the Evaluation of the Leviathan Environmental Impact Assessment. *After Zodiatis, G., Liubartseva, S., Loizides, L., Pellegatta, M., Coppini, G., Lardner, R., ... Brillant, A. (2020). Evaluation of the Leviathan offshore platform environmental studies in the Eastern Mediterranean Sea. EGU2020-5386-European Geosciences Union General Assembly 2020, Vienna, 3–8 May.*

The obtained results demonstrate the importance of the implemented approach for the identification of the monthly and seasonal variation of the weathering fate processes, for example, for the assessment of the monthly variability of the impacted shoreline (Fig. 5.17).

5.8 Conclusions

Generally, the movement of oil spill on the sea surface, with and without ice and in the water column is a complex process, as the chemical and biological changes in the oil affect its physical properties, which, in turn, affect the fate of spilt oil in the marine environment. Several well established oil spill models were developed and implemented during the last two decades, all of which have incorporated the advection diffusion and the fate processes of evaporation, emulsification, dispersion, beaching and sedimentation, but only few of them include the parameterization of biodegradation, the oil plume and the effect of ice. During the same period, the development of the operational metocean forecasting systems (CMEMS, NOAA, etc.) and the satellite technology for detecting oil slicks (SAR) and monitoring the shipping routes (AIS) made possible the near real-time use of quality controlled forecasting for winds, sea currents, waves, and SST data. Combined with the metadata information of a spill, these allowed improving the accuracy of the oil spill advection-diffusion predictions and thus the operational response capability. The fate and biodegradation processes described in this chapter provide the basic equations that should be implemented in oil spill modeling; further research of the biodegradation of spilled oil is required in the laboratory and in the field to assess toxic impacts of oil to ecosystems, to describe the effect of different parameters on biodegradation kinetics, to assess the performance of microbial communities, and to evaluate various bioremediation strategies. Nevertheless, the oil spill models were proven to be a valuable tool for assisting the response agencies during emergencies with the provision of short predictions, while ensemble long spill simulations provide valuable information related to the possible impact damages from offshore platforms at monthly, seasonal, and annual scales.

References

Ahlstrom, S. (1975). *A mathematical model for predicting the transport of oil slicks in marine waters*. Richland Energy Service: Rept. Richland, WA: Batelle Laboratories.

Al Rabeh, A. H. (1994). Estimating surface oil spill transport due to wind in the Arabian Gulf. *Ocean Engineering, 21*, 461–465.

Al-Rabeh, A. H., Lardner, R. W., & Gunay, N. (2000). Gulfspill Version 2.0: A software package for oil spills in the Arabian Gulf. *Environmental Modelling & Software, 15*, 425–442.

Alves, M. T., Kokinou, E., Zodiatis, G., Radhakrishnan, H., Panagiotakis, C., & Lardner, R. (2016). Multidisciplinary oil spill modeling to protect coastal communities and the environment of the Eastern Mediterranean Sea. *Scientific Reports (Nature SR), 6*, 36882. Available from https://doi.org/10.1038/srep36882.

Ambjörn, C. (2007). Seatrack web, forecasts of oil spills, a new version. *Environmental Research, Engineering and Management, 3*, 60–66.

ASA. (1997). *OILMAP for Windows (technical manual)*. Narrangansett, RI: ASA Inc.

ASCE Task Committee on Modelling of Oil Spills of the Water Resources Engineering Division. (1996). State-of-the-art review of modeling transport and fate of oil spills. *Journal of Hydraulic Engineering, 122*(11), 594–610.

Atlas, R. M. (1981). Microbial degradation of petroleum hydrocarbons: An environmental perspective. *Microbiological Reviews, 45*(1), 180–209.

Atlas, R. M., & Hazen, T. C. (2011). Oil biodegradation and bioremediation: A tale of the two worst spills in United States history. *Environmental Science & Technology, 45*, 6709–6715.

Azevedo, A., Oliveira, A., Fortunato, A. B., Zhang, J., & Baptista, A. M. (2014). A cross-scale numerical modeling system for management support of oil spill accidents. *Marine Pollution Bulletin, 80*(1–2), 132–147.

Bartell, S. M., Landrum, P. F., Giesy, J. P., & Leversee, G. J. (1981). Simulated transport of polycyclic aromatic hydrocarbons in artificial streams. In W. J. Mitsch, R. W. Bosserman, & J. M. Klopatek (Eds.), *Energy and ecological modelling* (pp. 133–143). New York: Elsevier.

Bazilchuk, N. (2018). *Predicting the fate of oil spills in a frozen world*. Published on 15 February 2018 by the Norwegian SciTechNews. Available from https://norwegianscitechnews.com/2018/02/predicting-fate-oil-spills-frozen-world/.

Berry, A., Dabrowski, T., & Lyons, K. (2012). The oil spill model OILTRANS and its application to the Celtic Sea. *Marine Pollution Bulletin*, 64. Available from https://doi.org/10.1016/j.marpolbul.2012.07.036.

Brakstad, O., & Faksness, L.-G. (2002). Biodegradation of water-accommodated fractions and dispersed oil in the seawater column. In *SPE international conference on health, safety and environment in oil and gas exploration and production*. Available from https://doi.org/10.2523/61466-MS.

Brakstad, O., & Lødeng, A. G. G. (2005). Microbial diversity during biodegradation of crude oil in seawater from the North Sea. *Microbial Ecology*, *49*, 94–103. Available from https://doi.org/10.1007/s00248-003-0225-6.

Brenner, S. (2019). The risk of potential cross border transport of oil spills in the semi-enclosed Eastern Mediterranean Sea. Available from https://doi.org/10.5772/intechopen.86205.

Buist, I. (1979). *An experimental study of the dispersion of oil slicks into the water column* (MASc thesis). Department of Chemical Engineering, University of Toronto.

Carracedo, P., Torres-Lopez, S., Barreiro, M., Montero, P., Balseiro, C. F., Penabad, E., et al. (2006). Improvement of pollutant drift forecast system applied to the Prestige oil spills in Galicia Coast (NW of Spain): Development of an operational system. *Marine Pollution Bulletin*, *53*, 350–360. Available from https://doi.org/10.1016/j.marpolbul.2005.11.014.

CEDRE (Centre de Documentation, de Recherche et d'Expérimentation sur les pollutions accidentelles des eaux). (2006). *Report of analyses, oil pollution of the Lebanese shoreline*. Report GC.06–21. Brest: CEDRE.

Choi, D. H., Katsutoshi, H., Yasunori, T., & Unno, H. (1999). Microbial degradation kinetics of solid alkane dissolved in nondegradable oil phase. *Biochemical Engineering Journal*, *3*, 71–78.

Coppini, G., De Dominicis, M., Zodiatis, G., Lardner, R., Pinardi, N., Santoleri, R., et al. (2011). Hindcast of oil-spill pollution during the Lebanon crisis in the Eastern Mediterranean, July–August 2006. *Marine Pollution Bulletin*, *62*, 140–153. Available from https://doi.org/10.1016/j.marpolbul.2010.08.021.

Cormack, D. (1999). *Response to marine oil pollution—Review and assessment* (p. 385) Boston: Kluwer.

Csanady, G. T. (1973). *Turbulent diffusion in the environment*. Dordrecht: D. Reidel Publ Co.

Daniel, P., Marty, F., Josse, P., Skandrani, C., & Benshila, R. (2003). Improvement of drift calculation in MOTHY operational oil spill prediction system. In: *International oil spill conference* (vol. 6). Vancouver: Canadian Coast Guard and Environment Canada.

De Dominicis, M., Pinardi, N., Zodiatis, G., & Lardner, R. (2013a). MEDSLIK-II, a Lagrangian marine surface oil spill model for short-term forecasting—Part 1: Theory. *Geoscientific Model Development*, *6*, 1851–1869. Available from. Available from https://doi.org/10.5194/gmd-6-1851-2013.

De Dominicis, M., Pinardi, N., Zodiatis, G., & Archetti, R. (2013b). MEDSLIK-II, a Lagrangian marine surface oil spill model for short-term forecasting—Part 2: Numerical simulations and validations. *Geoscientific Model Development*, *6*, 1871–1888. Available from. Available from https://doi.org/10.5194/gmd-6-1871-2013.

Delgato, L., Kumzerova, E., & Martynov, M. (2006). Simulation of oil spill behavior and response operations in PISCES. *WIT Transactions on Ecology and the Environment*, *88*, 279–292. Available from. Available from https://doi.org/10.2495/CENV060271.

Delvigne, G. A. L., & Sweeney, C. E. (1988). Natural dispersion of oil. *Oil and Chemical Pollution*, *4*, 281–310.

Desai, A. M., Autenrieth, R. L., Dimitriou-Christidis, P., & McDonald, T. J. (2008). Biodegradation kinetics of select polycyclic aromatic hydrocarbon (PAH) mixtures by Sphingomonaspaucimobilis EPA505. *Biodegradation*, *19*, 223–233.

Dimitriou-Christidis, P., & Autenrieth, R. L. (2007). Kinetics of biodegradation of binary and ternary mixtures of PAHs. *Biotechnology & Bioengineering*, *97*, 788–800.

Fay, J. A. (1969). The spread of oil slicks on a calm sea. In Hoult, D. (Ed.), *Oil on the sea* (pp. 53–64). New York: Plenum Press.

Fay, J. A. (1971). Physical processes in the spread of oil on a water surface. In *Proceedings of joint conference on prevention and control of oil spills* (pp. 463–467). Washington, DC: American Petroleum Institute.

Fingas, M. F. (2011). *Oil spill science and technology [electronic resource]: Prevention, response, and clean up*. Burlington, MA: Gulf Professional Pub./Elsevier.

Fingas, M. F., & Brown, C. E. (2018). A Review of oil spill remote sensing. *Sensors (Basel, Switzerland), 18*, 91. Available from https://doi.org/10.3390/s18010091.

French McCay, D. P. (2003). Development and application of damage assessment modeling: Example assessment for the North Cape oil spill. *Marine Pollution Bulletin, 47*, 341–359.

French McCay, D. P. (2004). Oil spill impact modeling: Development and validation. *Environmental Toxicology and Chemistry, 23*, 2441–2456.

French, D., Schuttenberg, H., & Isaji, T. (1999). Probabilities of oil exceeding thresholds of concern: Examples from an evaluation for Florida power and light. In *Proceedings of the 22nd Arctic and Marine Oil Spill Program (AMOP)* (vol. 1. pp. 243– 270), 2–4 June. Technical Seminar. Calgary: Environment Canada.

Hackett, B., Breivik, Ø., & Wettre, C. (2006). Forecasting the drift of objects and substances in the ocean. In *Ocean weather forecasting* (pp. 507–523). The Netherlands: Springer.

Hasselmann, K., Barnett, T., Bouws, E., Carlson, H., Cartwright, D., Enke, K., et al. (1973). Measurements of wind-wave growth and swell decay during the Joint North Sea Wave Project (JONSWAP). *Ergänzungsheft zur Deutschen Hydrographischen Zeitschrift Reihe*, A8–12.

Henaidi, A. K. (1984). *Preliminary report on drifting buoy movements*. MEPA, Gulf Area Oil Companies Mutual Aid Organization Doc. No. GO-86/87-07.

Hunter, J. R. (1987). The application of Lagrangian particle tracking technique to modelling of dispersant in the sea. In J. Noy (Ed.), *Numerical modelling: Application to marine systems*. North Holland.

ITOPF (2002). *Technical Information Paper: Fate of marine oil spills*. No. 2. London: ITOPF.

James, I. D. (2002). Modelling pollution dispersion, the ecosystem and water quality in coastal waters: A review. *Environmental Modelling & Software, 17*(4), 363–385.

Jones, R. K. (1997). A simplified pseudo-component of oil evaporation model. In *Proceedings of the 20th Arctic and Marine Oil Spill Program (AMOP)* (pp. 43–61). Technical Seminar. Vancouver, BC: Environment Canada.

Klemas, V. (2010). Tracking oil slicks and predicting their trajectories using remote sensors and models: Case studies of the Sea Princess and deepwater Horizon oil spills. *Journal of Coastal Research, 26*, 789–797.

Lardner, R. (2017). MEDSLIK User Manual. Nicosia.

Lardner, R., Zodiatis, G., Hayes, D., & Pinardi, N. (2006). *Application of the MEDSLIK oil spill model to the Lebanese Spill of July 2006*. European Group of Experts on Satellite Monitoring of Sea Based Oil Pollution, European Communities. ISSN: 1018-5593.

Lardner, R., Zodiatis, G., Loizides, L., & Demetropoulos, A. (1998). An operational oil spill model for the Levantine Basin (Eastern Mediterranean Sea). In *International symposium on marine pollution*. Malta: IAEA-SM-354/166P.

Lardner, R., & Zodiatis, G. (2016). *MEDSLIK oil spill model recent developments*. Geophysical Research Abstracts: 18. Vienna: EGU General Assembly. EGU2016-16240.

Lardner, R., & Zodiatis, G. (2017). Modelling oil plumes from subsurface spills. *Marine Pollution Bulletin*. Available from https://doi.org/10.1016/j.marpolbul.2017.07.018.

Layman, W. J., Reehl, W. F., & Rosenblatt, D. H. (1990). Handbook of chemical property estimation methods: Environmental behavior of organic compounds. United States: OSTI.GOV. Available from https://www.osti.gov/biblio/6902382.

Lee, R. F. (2003). Photo-oxidation and photo-toxicity of crude and refined oils. *Spill Science & Technology Bulletin*, *8*(2), 157–162.

Legrand, S., & Dulière, V. (2012). OSERIT: An oil spill evaluation and response integrated tool. In Troch, P. (Ed.), et al., Proceeding of the *Book of abstracts of the 4th International conference on the Application of Physical Modelling to Port and Coastal Protection*, Ghent, 17–20 September (pp. 275–276).

Lehr, W. J., Wesley, D., Simecek-Beatty, D., Jones, R., Kachook, G., & Lankford, J. (2000). Algorithm and interface modifications of the NOAA oil spill behavior model. In *Proceedings of the 23rd Arctic and Marine Oil Spill Program (AMOP)* (pp. 525–539). Technical Seminar. Vancouver, BC: Environmental Protection Service, Environment Canada.

Levenspiel, O. (1999). *Chemical reaction engineering—Industrial & engineering chemistry research*. Hoboken, NJ: John Wiley & Sons.

Li, Z., Lee, K., King, T., Boufadel, M. C., & Venosa, A. D. (2008). Assessment of chemical dispersant effectiveness in a wave tank under regular non-breaking and wave breaking wave conditions. *Marine Pollution Bulletin*, *56*, 903–912.

Lin, C. W., & Cheng, Y. W. (2007). Biodegradation kinetics of benzene, methyl tert-butyl ether, and toluene as a substrate under various substrate concentrations. *Journal of Chemical Technology and Biotechnology*, *82*, 51–57.

Liu, Y., MacFadyen, A., Ji, Z., & Weisberg, R. H. (2011). *Monitoring and modeling the deepwater horizon oil spill: A record-breaking enterprise*. Geophysical Monograph Series. Washington, DC: AGU/geopress, 195, 0–271. ISSN: 0065-8448, ISBN 978-0-87590- 485-6. 195, 0–271.

Liu, Y., Weisberg, R. H., Hu, C., & Zheng, L. (2011a). Tracking the Deepwater Horizon oil spill: A modeling perspective. *Eos, Transactions American Geophysical Union*, *92*, 45–46. Available from https://doi.org/10.1029/2011EO060001.

Liu, Y., Weisberg, R. H., Hu, C., & Zheng, L. (2011b). Trajectory forecast as a rapid response to the deepwater horizon oil spill. In *Monitoring and modeling the deepwater horizon oil spill: A record-breaking enterprise* (vol. 195, pp. 153–165). Geophysical Monograph Series.

Mackay, D., Buist, I., Mascarenhas, R., & Paterson, S. (1979). *Oil spill processes and models*. Research Report, Arctic Marine Oilspill Program, Environmental Protection Service. Ottawa: Fisheries and Environment Canada.

Mackay, D., & Leinonen, P.J. (1977). *Mathematical model of the behavior of oil spills on water with natural and chemical dispersion*. Prepared for Fisheries and Environment Canada. Economic and Technical Review Report EPS-3-EC-77–19, p. 39.

Mackay, D., & Paterson, S. (1980). Calculation of the evaporation rate of volatile liquids. In *Proceedings of the 1980 national conference on control of hazardous material spills*, Louiville, KY. Washington, DC: United States Environmental Protection Agency.

Mackay, D., Paterson, S., & Nadeau, S. (1980). Calculation of the evaporation rate of volatile liquids. In *Proceedings of the national conference on control of hazardous material spills* (pp. 361–368). Louisville, KY. Washington, DC: United States Environmental Protection Agency.

Mackay, D., Paterson, S., & Trudel, B. (1980). *A mathematical model of oil spill behaviour*. Report to Research and Development Division, Environment Emergency Branch, Environmental Impact Control Directorate, Environmental Protection Service. Ottawa: Environment Canada.

Malačič, V. (2001). Numerical modelling of the initial spread of sewage from diffusers in the Bay of Piran. *Ecological Modelling*, *138*, 173–191.

McAuliffe, C. D. (1987). Organism exposure to volatile/soluble hydrocarbons from crude oil spills—A field and laboratory comparison. In *Proceedings of the 1987 oil spill conference* (pp. 275–288). Washington, DC: American Petroleum Institute.

National Research Council (NRC). (1985). *Oil in the sea: Inputs, fates, and effects*. Washington, DC: National Academy Press.

National Research Council (United States) Committee on Oil in the Sea: Inputs, Fate and Effects. (2003). *Oil in the sea III: Inputs, fates, and effects*. Washington, DC: The National Academies Press. Available from http://doi.org/10.17226/10388.

Overton, E., Wade, T., Radović, J., Meyer, B., Miles, M., & Larter, S. (2016). Chemical composition of Macondo and other crude oils and compositional alterations during oil spills. *Oceanography*, *29*(3), 50–63.

Ovsienko, S.N., Zatsepa S.N., & Ivchenko A.A. (2005). Modeling of oil spills and environmental risk assessment. Trudy GOIN. In: *Proceedings of GOIN* (vol. 209, pp. 248–271). Moscow: Gidrometeoizdat.

Payne, J. R., & Phillips, C. R. (1985). Photochemistry of petroleum in water. *Environmental Science & Technology*, *19*, 569–579. Available from https://doi.org/10.1021/es00137a602.

Perry, R. H., Green, D. W., & Maloney, J. O. (1984). *Perry's chemical engineers' handbook*. McGraw-Hill. ISBN: 9780071422949.

Plata, D. L., Sharpless, C. M., & Reddy, C. M. (2008). Photochemical degradation of polycyclic aromatic hydrocarbons in oil films. *Environmental Science & Technology*, *42*, 2,432–2,438. Available from https://doi.org/10.1021/es702384f.

Reed, M., Aamo, O. M., & Daling, P. S. (1995). Quantitative analysis of alternate oil spill response strategies using OSCAR. *Spill Science & Technology Bulletin*, *2*(1), 67–74. Available from https://doi.org/10.1016/1353-2561(95)00020-5.

Reed, M., Daling, P. S., Brandvik, P. J., & Singsaas, I. (1993). Laboratory tests, experimental oil spills, models and reality: The Braer oil spill. In *Proceedings of the 16th Arctic and Marine Oil Spill Program* (pp. 203–209). Technical Seminar. Vancouver, BC: Environment Canada.

Reed, M., Johansen, Ø., Brandvik, P. J., Daling, P., Lewis, A., Fiocco, R., et al. (1999). Oil spill modeling towards the close of the 20th century: Overview of the state of the art. *Spill Science & Technology Bulletin*, *5*(1), 3–16. Available from https://doi.org/10.1016/S1353-2561(98)00029-2.

Reed, M., & Hetland, B. (2002). DREAM: A dose-related exposure assessment model technical description of physical-chemical fates components. In *Proceedings of SPE International Conference on Health, Safety and Environment in Oil and Gas Exploration and Production (Society of Petroleum Engineers)*. Available from https://doi.org/10.2118/73856-ms.

Schwarzenbach, R. P., Gschwend, P. M., & Imaboden, D. M. (1993). *Environmental organic chemistry* (pp. 436–484). New York: Wiley Interscience.

Shen, H. T., Yapa, P. D., & Petroski, M. E. (1987). A simulation model for oil slick transport in lakes. *Water Resources Resources*, *23*, 1949–1957.

Spanoudaki, K. (2016). *Mathematical modelling of oil spill fate and transport in the marine environment incorporating biodegradation kinetics of oil droplets*. Geophysical Research Abstracts, 18. EGU2016-13155. Vienna.

Spanoudaki, K., Kozyrakis, G., & Kampanis, N. (2019). *Long-term modelling of the Corsica oil spill fate and transport incorporating biodegradation kinetics*. Geophysical Research Abstracts, EGU2019-13632, 21. Vienna.

Spaulding, M. L. (1988). A state-of-the-art review of oil spill trajectory and fate modeling. *Oil and Chemical Pollution*, *4*(1), 39–55.

Spaulding, M., Kolluru, V., Anderson, E., & Howlett, E. (1994). Application of three-dimensional oil spill model (WOSM/OILMAP) to hindcast the Braer spill. *Spill Science & Technology Bulletin*, *1*, 23–35.

Tarr, M. A., Zito, P., Overton, E. B., Olson, G. M., Adhikari, P. L., & Reddy, C. M. (2016). Weathering of oil spilled in the marine environment. *Oceanography*, *29*(3), 126–135. Available from. Available from https://doi.org/10.5670/oceanog.2016.77.

Thominette, F., & Verdu, J. (1984). Photo-oxidative behaviour of crude oils relative to sea pollution. Part II. Photo-induced phase separation. *Marine Chemistry*, *15*, 105–115.

Tintoré, J., Pinardi, N., Álvarez-Fanjul, E., Aguiar, E., Álvarez-Berastegui, D., Bajo M., et al. (2019). Challenges for sustained observing and forecasting systems in the Mediterranean Sea. Frontiers in Marine Science, 6, 568. Available from http://doi.org/10.3389/fmars.2019.00568.

Torgrimson, G. M. (1980). *The on-scene spill model: A user's guide*. Technical Report, Hazardous Materials Response Branch. Seattle, WA: National Oceanic and Atmospheric Administration.

UNEP. (2007). *Lebanon post-conflict environmental assessment*. ISBN: 978-92-807-2794-4.

US Coast Guard. (2011). *On scene coordinator report on deepwater horizon oil spill*. Available from www.uscg.mil/foia/docs/dwh/fosc_dwh_report.pdf.

Vilcáez, J., Li, L., & Hubbard, S. S. (2013). A new model for the biodegradation kinetics of oil droplets: Application to the Deepwater Horizon oil spill in the Gulf of Mexico. *Geochemical Transactions*, *14*, 4.

Walter, M. F., Black, B. C., Afshar, G., Kermabon, A. Y., Wright, T. R., & Biessmann, H. (1991). Temporal and spatial expression of the yellow gene in correlation with cuticle formation and dopa decarboxylase activity in Drosophila development. *Developmental Biology*, *147*(1), 32–45.

Yapa, P., & Zheng, L. (1997). Simulation of oil spills from underwater accidents I: Model development. *Journal of Hydraulic Research*, *35*, 673–687.

Zelenke, B., O'Connor, C., Barker, C., Beegle-Krause, C.J., & Eclipse, L. (Eds.). (2012). *General NOAA Operational Modeling Environment (GNOME)*. Technical Documentation, United States Department of Commerce. NOAA Technical Memorandum NOS OR&R 40. Seattle, WA: Emergency Response Division, NOAA. 105 pp. Available from http://response.restoration.noaa.gov/gnome_manual.

Zepp, R. G., & Schlotzheuer, P. F. (1979). Photoreactivity of selected aromatic hydrocarbons in water. In Jones, P. W., & Leber, P. (Eds.), *Polynuclear aromatic hydrocarbons* (pp. 141–148). Ann Arbor, MI: Ann Arbor Sciences.

Zheng, Li, & Yapa, P. (1998). Simulation of oil spills from underwater accidents II: Model verification. *Journal of Hydraulic Research*, *36*, 117–134.

Zitka, R. G., & Cooper, W. J. (1987). *Photochemistry of environmental aquatic systems. ACS Symposium Series 327* (p. 288) Washington, DC: American Chemical Society.

Zodiatis, G., Coppini, G., Perivoliotis, L., Lardner, R., Alves, T., Pinardi, N., et al. (2018). Numerical modeling of oil pollution in the Eastern Mediterranean Sea. In Carpenter, A., & Kostianoy, A. G. (Eds.), *The handbook of environmental chemistry: Oil pollution in the Mediterranean Sea, Part I* (vol. 83, pp. 215–254). Springer. Available from http://doi.org/10.1007/698_2017_131.

Zodiatis, G., Lardner, R., Alves, T. M., Krestenitis, Y., Perivoliotis, L., Sofianos, S., et al. (2017). Oil spill forecasting (prediction). *Journal of Marine Research*, *75*, 923–953.

Zodiatis, G., Lardner, R., Alves, T. M., Krestenitis, Y., Perivoliotis, L., Sofianos, S., et al. (2019). Oil spill forecasting (prediction). *In THE SEA: The Science of Ocean Prediction* (vol. 17, pp. 923–953). Yale: Yale University.

Zodiatis, G., Lardner, R., Hayes, D., Georgiou, G., Pinardi, N., & De Dominicis, M. (2008). The mediterranean oil spill and trajectory prediction model in assisting the EU response agencie. In *Congreso Nacional de Salvamentoen la Mar* (pp. 535–547), Cadiz, 2–4 October, Libro deActas.

Zodiatis, G., Lardner, R., Solovyov, D., Panayidou, X., & De Dominicis, M. (2012). Predictions for oil slicks detected from satellite images using MyOcean forecasting data. *Ocean Science*, *8*, 1105–1115.

Zodiatis, G., Liubartseva, S., Loizides, L., Pellegatta, M., Coppini, G., Lardner, R., et al. (2020). Evaluation of the Leviathan offshore platform environmental studies in the Eastern Mediterranean Sea. EGU2020-5386- Vienna, 3–8 May.

Zodiatis, G., Liubartseva, S., Loizides, L., Pellegatta, M., Coppini, G., Lardner, R., et al. (2021). CMEMS and CYCOFOS assessing the pollution risk from the Leviathan offshore platform in the Eastern Mediterranean Sea. In *Proceedings of the 9th EuroGOOS: Advances in Operational Oceanography: Session: Expanding Europe's Ocean Observing and Forecasting Capacity: Extreme Events/Hazard Forecasting*, 3–5 May. (Submitted for publication in the Conference proceedings.)

Appendix A: Biodegradation for dissolved oil and oil droplets dispersed in the water column

The equations for the biodegradation kinetics for dissolved oil (multiple substrates can be simulated; no inhibition is assumed) and oil droplets dispersed in the water column are described by the following equations.

$$r_{S_i} = \frac{dS_i}{dt} = -\frac{\mu_{\max_i}}{Y_{S_i}}\left[\frac{S_i}{K_{S_i}+S_i}\right]\left[\frac{O_2}{K_{O_2}+O_2}\right]\left[\frac{N}{K_N+N}\right]\left[\frac{P}{K_P+P}\right]B, \quad (5.A1)$$

$$r_B = \frac{dB}{dt} = \sum_1^n \mu_{\max_i}\left[\frac{S_i}{K_{S_i}+S_i}\right]\left[\frac{O_2}{K_{O_2}+O_2}\right]\left[\frac{N}{K_N+N}\right]\left[\frac{P}{K_P+P}\right]B - k_dB, \quad (5.A2)$$

$$r_N = \frac{dN}{dt} = -\sum_1^n \frac{\mu_{\max_i}}{Y_{N_i}}\left[\frac{S_i}{K_{S_i}+S_i}\right]\left[\frac{S}{K_S+S}\right]\left[\frac{O_2}{K_{O_2}+O_2}\right]\left[\frac{N}{K_N+N}\right]\left[\frac{P}{K_P+P}\right]B + \psi_1 k_d B, \quad (5.A3)$$

$$r_P = \frac{dP}{dt} = -\sum_1^n \frac{\mu_{\max_i}}{Y_{P_i}}\left[\frac{S_i}{K_{S_i}+S_i}\right]\left[\frac{O_2}{K_{O_2}+O_2}\right]\left[\frac{N}{K_N+N}\right]\left[\frac{P}{K_P+P}\right]B + \psi_2 k_d B, \quad (5.A4)$$

$$r_{O_2} = -\sum_1^n \frac{\mu_{\max_i}}{Y_{O_{2_i}}}\left[\frac{S_i}{K_{S_i}+S_i}\right]\left[\frac{O_2}{K_{O_2}+O_2}\right]\left[\frac{N}{K_N+N}\right]\left[\frac{P}{K_P+P}\right]B, \quad (5.A5)$$

$$\frac{d_{O_2}}{dt} = K_{L\alpha}\left(O_2^s - O_2\right) + r_{O_2}, \quad (5.A6)$$

where S_i is the dissolved oil concentration of pseudo-component i [M/L^3]; r_{Si} is the rate of dissolved oil degradation for pseudo-component i [M/L^{3}T]; B is the bacteria concentration in the bulk fluid [M/L^3]; r_B is the bacterial growth rate [M/L^{3}T]; O_2 is the oxygen concentration [M/L^3]; r_{O2} is the rate of oxygen consumption [M/L^{3}T]; $O_2{}^s$ is the oxygen saturation concentration [M/L^3]; N is the nitrogen concentration [M/L^3]; r_N is the rate of nitrogen consumption [M/L^3-T]; P is the phosphorous concentration [M/L^3]; r_P is the rate of phosphorous consumption [M/L^{3}T]; μ_{maxi} is the maximum specific growth rate [T^{-1}]; Y_{Si} is the oil yield coefficient (microbial mass produced per mass of pseudo-component i consumed); Y_{O2i} is the oxygen yield coefficient (microbial biomass produced per mass of oxygen consumed for the biodegradation of hydrocarbons types in pseudo-component i); Y_{Ni} is the nitrogen yield coefficient (microbial biomass produced per mass of nitrogen consumed for the biodegradation of hydrocarbons types in pseudo-component i); Y_{Pi} is the phosphorous yield coefficient (microbial biomass produced per mass of phosphorous consumed for the biodegradation of hydrocarbons types in pseudo-component i; k_d is the biomass death rate [T^{-1}]; ψ_1 is the percentage content of N in biomass; ψ_2 is the percentage content of N in biomass; $K_{L\alpha}$ is the overall oxygen mass transfer coefficient [T^{-1}L^{-3}]; K_S is the oil half saturation constant [M/L^3]; K_{O2} oxygen half saturation constant [M/L^3]; K_N is the nitrogen half saturation constant [M/L^3]; K_N is the phosphorous half saturation constant [M/L^3].

Oil droplets biodegradation kinetics, where the rate of particle size reduction is controlled by the reaction rate at the water–oil particle surface, is described by the Eqs. (5.A7)–(5.A11), while the Eq. (5.A12) describes the rate of oil droplet diameter reduction.

$$\frac{1}{A}\frac{dN_{oil}}{dt} = \frac{\rho_{oil}}{2}\frac{dD}{dt}, \tag{5.A7}$$

$$r_{S_{di}} = \frac{1}{A}\frac{dN_i}{dt} = -\frac{\mu_{\max_i}}{Y_{S_i}}\left[\frac{S_{di}}{K_{S_i}+S_{di}}\right]\left[\frac{O_2}{K_{O_2}+O_2}\right]\left[\frac{N}{K_N+N}\right]\left[\frac{P}{K_P+P}\right]B_s, \tag{5.A8}$$

$$r_{B_d} = \frac{V}{A}\frac{dB_d}{dt} = \sum_1^n \mu_{\max_i}\left[\frac{S_{di}}{K_{S_i}+S_{di}}\right]\left[\frac{O_2}{K_{O_2}+O_2}\right]\left[\frac{N}{K_N+N}\right]\left[\frac{P}{K_P+P}\right]B_s, \tag{5.A9}$$

$$r_{N_d} = \frac{1}{A}\frac{dN_N}{dt} = -\sum_1^n \mu_{\max_i}\left[\frac{S_{di}}{K_{S_i}+S_{di}}\right]\left[\frac{O_2}{K_{O_2}+O_2}\right]\left[\frac{N}{K_N+N}\right]\left[\frac{P}{K_P+P}\right]B_s, \tag{5.A10}$$

$$r_{P_d} = \frac{1}{A}\frac{dN_P}{dt} = -\sum_{1}^{n}\mu_{\max_i}\left[\frac{S_{di}}{K_{S_i}+S_{di}}\right]\left[\frac{O_2}{K_{O_2}+O_2}\right]\left[\frac{N}{K_N+N}\right]\left[\frac{P}{K_P+P}\right]B_s, \tag{5.A11}$$

$$r_{O_{2d}} = \frac{1}{A}\frac{dN_{O_2}}{dt} = -\sum_{1}^{n}\mu_{\max_i}\left[\frac{S_{di}}{K_{S_i}+S_{di}}\right]\left[\frac{O_2}{K_{O_2}+O_2}\right]\left[\frac{N}{K_N+N}\right]\left[\frac{P}{K_P+P}\right]B_s, \tag{5.A12}$$

where N_{oil} is the mass of oil droplets [M]; A is the surface area of oil droplets [L^2]; D is the oil droplet diameter [L]; ρ_{oil} is the oil density [M/L^3]; r_{Sdi} is the rate of dispersed oil degradation for pseudo-component i [M/L^3T]; N_i is the mass of pseudo-component i in oil droplets [M]; S_{di} is the concentration of pseudo-component i in oil droplets [M/L^3]; B_s is the concentration of microbes at the oil droplet-water interface [M/L^3]; r_{Bd} is the bacterial growth rate due to oil droplets biodegradation [M/L^3T]; r_{Nd} is the nitrogen consumption rate for oil droplets degradation [M/L^3T]; N_N is the nitrogen mass [M]; r_{Pd} is the phosphorous consumption rate for oil droplets degradation [M/L^3T]; N_P is the phosphorous mass [M]; r_{O2d} is the oxygen consumption rate due to oil droplets degradation [M/L^3T]; N_{O2} is the oxygen mass [M]; V is the control volume [L^3].

Eq. (5.A13) describes the rate of oil droplets size shrinking due to oil biodegradation by native bacteria. Integration of Eqs. (5.A13) and (5.A14) gives the fraction of the biodegraded oil droplet volume as a function of time. The term X_1 quantifies the fraction of biodegraded oil for a single oil droplet:

$$1-(1-X_1)^{1/3} = \frac{k_{bD}}{D}t \tag{5.A13}$$

$$k_{bD} = \frac{2}{\rho_{oil}}\sum_{1}^{n}\frac{\mu_{\max_i}}{Y_{S_i}}\left[\frac{S_{di}}{K_{S_i}+S_{di}}\right]\left[\frac{O_2}{K_{O_2}+O_2}\right]\left[\frac{N}{K_N+N}\right]\left[\frac{P}{K_P+P}\right]B_s \tag{5.A14}$$

The fraction of biodegraded oil for a distribution of oil droplets, X, is a function of the oil droplet size distribution. Therefore, the overall conversion of oil droplets of various sizes is found by integrating Eq. (5.A15) with respect to the size of oil droplets (D):

$$X = 1 - \int_0^{D_{\max}}\left(\frac{1-k_{bD}}{D}t\right)P(D)dD, \tag{5.A15}$$

where $P(D)$ is the oil droplet size distribution function. At each time step, Eqs. (5.A1–5.A15) are solved to determine the biodegradation rate of each *PC* separately (for both dissolved oil and oil droplets), as well as biomass production and oxygen, nitrogen, and phosphorous consumption rates. The model produces as output the evolution of the concentration and the position of the surface slick, the dispersed oil and the oil adhered to coast.

A gamma distribution is often assumed for $P(D)$ since this droplet size distribution (for oil dispersed in the water column) has been reported in studies describing the effect of chemical dispersants on oil droplet size distribution (e.g., Li, Lee, King, Boufadel, & Venosa, 2008):

$$P(D) = \frac{1}{\beta^{\alpha}\Gamma(\alpha)} D^{\alpha-1} e^{-D/\beta}, \tag{5.A16}$$

where $\mu = \alpha\beta$ is the mean droplet diameter, $\sigma = \alpha^{0.5}\beta$ is the standard deviation, and $CV = \sigma/\mu$ is the coefficient of variation. Delvigne and Sweeney (1988) suggest, based on experimental work, that the number of droplets centered on diameter D per unit volume of water, can be expressed as:

$$P(D) = N_o \left(\frac{D_o}{D}\right)^{2/3}, \tag{5.A17}$$

where N_o, D_o are experimental values. The definition of droplet size distribution is a choice of the user.

Chapter 6

Assessment of oil toxicity in water

Trond Nordtug and Bjørn Henrik Hansen
SINTEF Ocean AS, Trondheim, Norway

6.1 Introduction

The need for establishing toxicity data for crude oil and related petroleum products has triggered the development of a variety of test procedures. Unfortunately, a large fraction of the reported data is of little value for risk assessments due to incomplete information on test conditions and limited verification of exposure concentrations. Reviews of literature data and test conditions have been issued with recommendations of future standardization of test regimes and reporting of results, and we recommend consulting review publications by Hodson, Adams, and Brown (2019), Adams et al., (2017), Echols, Smith, Gardinali, and Rand (2016), and Redman and Parkerton (2015). The current chapter discusses some of the background issues that make toxicity testing of oil challenging, including some examples on how experimental conditions and choice of exposure metrics can affect reported toxicity values. In this process, we have revisited some of our earlier work to illustrate the aspects discussed. Finally, we describe how, in our opinion, experiments can be conducted to supply data to risk and damage assessment models for oil spills.

6.2 Crude oil properties in water

Crude oil and related product such as bitumen and fuel oils show a huge variation in both physical properties and composition. Crude oil consists of thousands of compounds of which only a fraction is dissolved in water, and of those dissolved, only a tiny fraction is characterized with traditional chemical analysis. In case of an oil spill, the oil properties have a huge impact on the distribution and toxicity in the environment. Light crude oil will easily disperse into the water column, while heavier oils will stay on the water surface either as an oil slick, lumps of oil or mixed with water to form

Marine Hydrocarbon Spill Assessments. DOI: https://doi.org/10.1016/B978-0-12-819354-9.00003-X

emulsions (mousse). The main factors influencing the weathering degree of the spilled oil are the spill properties (e.g., surface release or blow-out), physical properties of the oil, chemical composition of the oil, environmental conditions (including weather, winds, sunlight, temperature, and currents), and the properties of the water (including salinity, temperature, density, oxygen, microbial communities, and presence of particles) (Daling et al., 2014; NRC, 2005). All these factors have a pronounced impact on the fate and spreading of the oil after a spill, which again affects the bioavailability of toxic components and consequently the impact on the local ecosystem.

After a surface oil spill into water, the oil immediately starts weathering, initially by losing a large portion of the lighter components such as the benzene, toluene, ethylbenzene, and xylenes (BTEX), lighter polycyclic aromatic hydrocarbons (PAHs), and small aliphatic substances fraction by evaporation. Further weathering includes biodegradation and photodegradation. Thus, fresh crude oil is only found the area in the immediate vicinity of a continuous oil spill and a short time window of an episodic spill, whereas the environment is mostly exposed to various degrees of weathered oil. Toxicity testing is therefore often done with artificially weathered oil.

For deepwater oil spills, dispersion, dissolution, spreading, and subsequent biodegradation will be the dominating weathering processes before the oil reaches the water surface. As for oil reaching the water surface and for the surface spills, the lighter components will be the first to leave the oil through evaporation.

In wind-exposed open waters and for high-velocity subsea blow outs creating turbulent conditions at the release point, such as the Deepwater Horizon accident, the oil will be dispersed into the water as oil droplets, and this will affect the dilution and bioavailability of toxic oil components.

To recreate a spill scenario in the laboratory is impossible and providing reliable risk assessment based on acute toxicity tests run in the laboratory should be done with caution. Toxicity threshold values from laboratory studies display large variations for oil, even for the same oil type. Much of the variation is suspected to be caused by the methodology used to prepare exposure solutions.

6.3 Approaches for characterizing oil toxicity

Characterizing the toxicity of oil in water is challenging due to the variety of physical properties of different crude oil. Different properties may significantly affect the concentration of bioavailable fraction of toxic oil components. It is therefore crucial how exposure solutions are made and how the expose concentrations are quantified. Standardized procedures such as those described by the Chemical Response to Oil Spills: Ecological Effects Research Forum (CROSERF) has been a necessary step toward being able to compare the toxicity of different oil qualities (Aurand & Coelho, 1996;

Singer et al., 2000). However, within these standard procedures there are also room for inconsistencies between tests.

There are two basic experimental approaches commonly used for determining oil toxicity:

(1) Mimicking the actual exposure conditions (e.g., spiked exposure)
(2) Standardized exposure conditions with constant exposure (e.g., 48 or 96 hours static, semistatic or flow-through toxicity tests)

These two approaches have different challenges. Mimicking environmental exposure conditions is a case study which at best is representative of a specific spill scenario and usually give very little general information that can be used for other spill situations. The second case is less environmentally realistic but may provide specific toxicity of specific oils (for specific species). However, there are numerous pitfalls in using this method that may significantly affect the estimated toxicity results. Furthermore, using these data in risk assessment require some kind of modeling to estimate the fate and spreading of the oil spill. Both approaches can to a certain extent be standardized as has been done by the CROSERF (Aurand & Coelho, 1996; Singer et al., 2000). Unfortunately, the standardization is in most cases related to the procedures such as stirring velocity and oil loading (oil to water ratio) and often very little information given about the actual exposure conditions.

In estimating the potential fate and effects of a large oil spill, there is no alternative to using numerical modeling, like the Oil Spill Contingency and Response (OSCAR) model (Reed et al., 2000). All models have in common that they need parameterized data input, and most of the existing toxicity data on oils do not provide data that support these models (environmental models cannot use stirring velocity or vortex depth). In the following, we discuss aspects related to toxicity characterization of oil for different organisms and how to provide parameterized data for use in oil spill risk and damage assessment.

6.4 Preparation of exposure solutions

6.4.1 Oil loading and impacts on partitioning of components between oil and water

The traditionally accepted exposure media preparation for oil toxicity testing adopted from CROSERF has involved generation of water accommodated fractions (WAFs) representing the dissolved components of the oil and chemically enhanced WAFs (CEWAFs) defined to represent mixtures containing entrained oil-added chemical dispersant (Singer et al., 2000). Preparation of WAF/CEWAF according to CROSERF is made by stirring of water with oil added on the surface. The oil to water ratio (OWR, oil loading) may be varied, and the toxicity is determined by testing of the water phase in a series of OWRs. The results of these tests when presented as specific toxicity (e.g., EC/LC_{50}) often show that both oil weathering and reduced oil loading increase the toxicity. This is related to solubility and partitioning of individual oil components between oil and water and does not necessarily reflect the environmental toxicity of the oil. Understanding dilution and dissolution of oil in water is therefore important for being able to evaluate reported toxicity results.

Most of the available literature indicate that toxicity of oil is predominantly related to the water-soluble fraction (WSF) of oil (Carls et al., 2008; Olsvik et al., 2011). Partitioning of oil components between oil and water is related to the ratio of oil and water, the composition of the oil, and the solubility of individual oil components. In addition, the contact time and area of the oil–water interface will determine the time needed to reach an equilibrium between oil and water. Gas charged oils during deepwater spills such as DWH, pressure, and temperature also affect the dissolution process (Jaggi et al., 2017).

The mass fraction of individual components associated with the oil phase in an oil dispersion increase with log K_{ow} of the components. Components with log $K_{ow} > \sim 6$ have very low seawater solubility, being almost exclusively associated with the oil droplets in an oil dispersion. Dilution of oil dispersions causes a redistribution of components between oil and water, increasing the mass fraction of less-soluble components in the water phase. This implies that even if the equilibrium between oil and water is maintained, there will be a net mass transfer of components from the oil to the water at each dilution step.

Although oil is not an ideal solution, the shift in mass distributions can be explained by Raoult's law (Guggenheim, 1937). A simplified expression to describe equilibrium partitioning between an organic phase and water is shown in Eq. (6.1) (Cline, Delfino, & Rao, 1991; Lee, Hagwall, Delfino, & Rao, 1992; Lee, Rao, & Okuda, 1992).

$$C_w = x_o S_l \tag{6.1}$$

where C_w is the chemical's concentration (mol/L) in the water in equilibrium with the oil, S_l is the aqueous solubility of the pure chemical (mol/L), and x_o is the mole fraction of the chemical in the oil. Aqueous solubility (S_l) for a given component is constant at similar physical conditions and thus the equilibrium water concentration depends only on its molar fraction (x_o) in the oil. More of the total mass of individual components will be transferred to the water during the dilution process as the oil:water ratio decreases. Consequently, the molar fraction in the oil matrix (x_o) will ultimately be reduced, causing a decline in the equilibrium water concentration. Since partition coefficients (e.g., measured as log K_{ow}) is inversely related to solubility, the depletion of the mass fraction contained in the oil first declines for the low log K_{ow} components (highest solubility). Further dilution causes successively more of the mass of high log K_{ow} and low solubility components to be transferred to the water phase. This results in a higher mass fraction of the less-soluble components (e.g., $K_{ow} > 5$) in the aqueous phase at lower oil:water ratios Fig. 6.1. Similar observations of a relative increase in dissolution of higher molecular weight oil components corresponding to an increase in their mole fraction have been reported in crude oil and coal–oil solubility studies (Eganhouse, Dorsey, Phinney, & Westcott, 1996; Picel, Stamoudis, & Simmons, 1988; Shiu, Maijanen, Ng, & Mackay, 1988), but to our knowledge few for crude oil droplet dispersions (Hansen, Olsen, et al., 2018). Thus, when dispersions are diluted, the composition and concentration of the WSF of crude oil dispersions will not follow a linear dilution.

If the dissolved and particulate oil components have different toxicological properties, the partitioning between the two phases will have an impact on the predicted toxicity of the dispersion. To illustrate this, water analyses from an experiment where Hansen, Parkerton, Nordtug, Størseth, and Redman (2019) compared the effects of dispersions to the corresponding (inline filtered) WSFs is shown in Fig. 6.2. The figure shows PAH distribution between oil and water in dispersions with different oil concentrations. The data are produced by simply subtracting the concentrations in the WSF from the corresponding concentrations in the dispersion to visualize the composition of the oil. The sum of the concentrations in the two compartments correspond to the

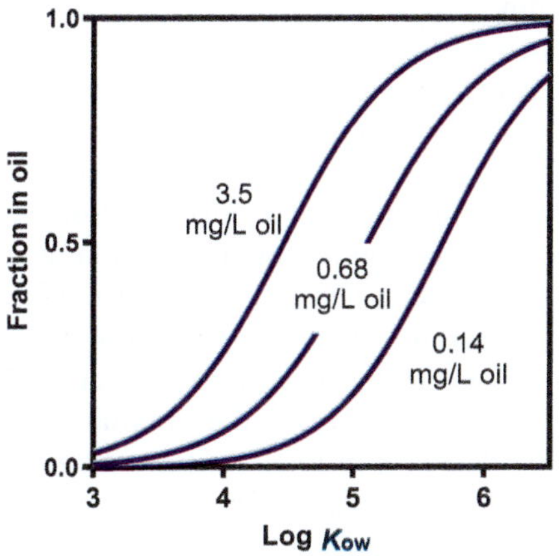

FIGURE 6.1 Distribution of polycyclic aromatic hydrocarbons between oil and water at different oil dispersion concentrations related to their log K_{ow} during dilution in clean seawater. Data based on comparing concentrations of 23 polycyclic aromatic hydrocarbons in dispersions and their corresponding inline-filtered water-soluble fractions (Nordtug, Olsen, Altin, Meier, et al., 2011). Average volumetric oil droplet size was 10.9 μm. *Modified from Hansen et al. (2019). Developmental effects in fish embryos exposed to oil dispersions – The impact of crude oil micro-droplets. Marine Environmental Research, 150, 104753.*

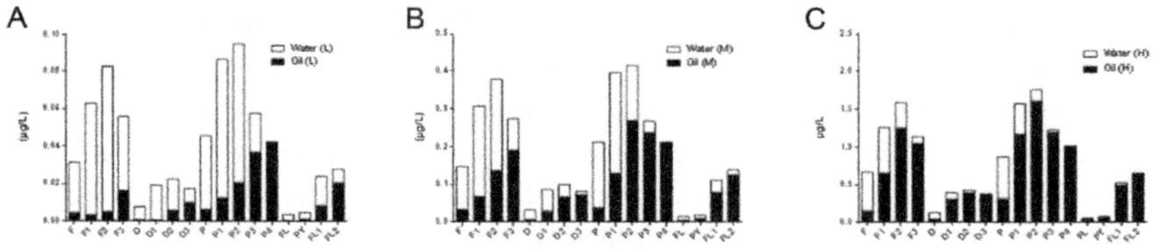

FIGURE 6.2 Distribution of polycyclic aromatic hydrocarbons between oil and water at different concentrations of oil dispersions [A: low (L) concentration, 0.12 mg, B: medium (M) concentration, 0.68 mg/L, and C: high (H) concentration 3.5 mg/L]. Data are created by comparing polycyclic aromatic hydrocarbon concentrations in inline-created dispersions of and their corresponding inline-filtered water-soluble fractions in a flow-through system (Nordtug, Olsen, Altin, Meier, et al., 2011). Average residence time of the dispersions was about 3 h. *Plots were based on raw data from Hansen, B.H., Salaberria, I., Read, K.E., Wold, P.A., Hammer, K.M., Olsen, A.J., ... Bardal, T. (2019). Developmental effects in fish embryos exposed to oil dispersions – The impact of crude oil micro-droplets. Marine Environmental Research, 150, 104753.*

PAH profile in the added oil. For the lowest concentration, the least alkylated PAH homologs are nearly depleted from the oil and, only the more alkylated high K_{ow} components (log $K_{ow} > 5.5$) remains in the oil at approximately their original concentration. This resembles the oil weathering observed after an oil spill. However, in a dispersion with small oil droplets, this process is much faster due the increased interphase area between oil and water.

Dissolved oil components are readily bioavailable and regarded as the main drivers of acute toxicity in marine organisms (Carls et al., 2008; Nordtug, Olsen, Altin, Overrein, et al., 2011; Olsvik et al., 2010; Olsvik et al., 2011). Components with high K_{ow} display higher acute toxicity per mass unit due to their higher potential for bioaccumulation. Since the solubility of these components is also low, they will be retained in the oil phase much longer than the lighter components when the oil is diluted in the water. Consequently, their equilibrium water concentration according to Raoult's law may be relatively unchanged during the dilution process while the lighter components become depleted. The relative contribution of heavier components to toxicity will therefore increase with dilution. Although this causes the mass-based acute toxicity to increase as indicated by a reduction in EC/LC_{50}, the overall acute toxicity of the mixture is usually reduced or unchanged due to the decrease in total concentration of the WSF (Faksness, Altin, Nordtug, Daling, & Hansen, 2015).

6.4.2 Stability of oil dispersions

Preparations of high-energy WAFs (HE-WAFs) were adopted by many laboratories to assess oil dispersion toxicity following the Deepwater Horizon blowout, where chemical dispersants were used to disperse oil during the continuous and highly turbulent release (Sandoval, Ding, & Gardinali, 2017).

The procedure used for creating the test solutions is similar to the WAF system described above, but with rapid stirring using a blender to disperse oil into seawater followed by a 1 hour settling period to allow surfacing of some of the larger oil droplets before collecting the exposure solution. The size distribution of oil droplets created at similar turbulence energy varies greatly between oil qualities, and the oil content in the solution will also be highly variable due to different rise velocity of the droplets. Thus, the result from comparing the toxicity of different oils depends highly on the metrics used to describe the exposure concentration. This may be illustrated by comparing neat oil with the same oil treated with chemical dispersant. Dispersants are "soap-like" chemicals that reduce the interfacial tension between oil and water, and thus reduce the droplet size distribution at a given mixing energy. For example, when comparing similar oil loadings with and without chemical dispersant, the chemically dispersed oil often shows a similar or higher toxicity than the untreated oil due to a more-efficient distribution of oil in the CEWAF. If, on the other hand, the toxicity is related to the total concentration of oil components chemically dispersed oil may appear significantly less toxic than the untreated oil. This is caused by difference in oil droplet size distribution in the two solutions where small droplets in the chemically treated WAF are retained in the water while larger droplets in the untreated oil are rapidly rising to the surface.

The theoretical surfacing velocity of oil droplets in stagnant water (no turbulence) can be calculated by Stoke's law.

$$v_s = \frac{2}{9}\frac{(\rho_p - \rho_f)}{\mu} gR^2 \tag{6.2}$$

ϑ_s is the particles' settling velocity (m/s) (vertically downward if $\rho_p > \rho_f$, upward if $\rho_p < \rho_f$), g is the gravitational acceleration (m/s^2), ρ_p is the mass density of the particles (kg/m^3), and ρ_f is the mass density of the fluid (kg/m^3). μ is the dynamic viscosity (N s/m^2), R is the radius of the spherical object (in m).

According to Stoke's law, the surfacing velocity of small oil droplets increases fourfold for each doubling of the droplet diameter. Due to the differences in surfacing velocity more oil will be retained in the water phase in the CEWAF after a defined settling period. Because oil components contained in oil droplets are not as bioavailable as dissolved components, the droplets contribute directly to the recorded concentration but less to the toxicity. In consequence, components contained within droplets included in sample collected for analyses contribute to the total concentration without being bioavailable. Consequently, the effect concentration will be overestimated causing an underestimation of the toxicity that increases with increasing oil content.

In natural water, the oil droplets will be affected by vertical turbulence and the Reynolds number (Rhodes, 2008), and the rise velocity is expected

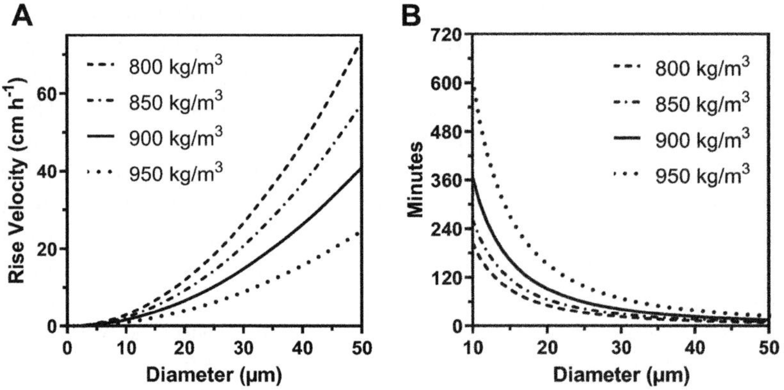

FIGURE 6.3 Terminal rise velocity of oil droplets with different sizes and densities according to Stokes law. (A) Rise velocity. (B) Time to migrate. 10 cm.

to be slower than predicted by Stoke's law. Static or semistatic experimental system normally do not include turbulence, and it is evident from Fig. 6.3 that it is impossible to maintain a controlled and constant exposure situation with dispersions in these systems.

Chemical dispersants enhance the formation of small oil droplets by reducing the oil–water interfacial tension. This will most certainly increase the rate of dissolution due to the increased oil surface. However, due to the limited amounts of dispersants used (application rates 2%–4% of the oil mass), it is not expected that the dispersant will have a large effect on the equilibrium distribution of oil components between oil and water. This is illustrated in Fig. 6.4 showing the distribution of PAHs between oil and water related to log K_{ow} in dispersions with similar oil droplet size distribution and concentration with and without 4% dispersant. The distribution is similar in the two dispersions but suggesting a slightly higher dissolved fraction of the lighter PAHs in the presence of dispersant. Jaggi et al. (2020) showed a moderate but consistently higher dissolved fraction of BTEX components in the water after adding dispersant.

6.5 Characterization of exposure

The standard methodologies to characterize exposure in toxicity tests utilizing crude oil includes liquid–liquid extraction of water samples using dichloromethane (DCM) (USEPA, 1996) followed by chemical analyses using separation of components based on boiling point with gas chromatography (GC). This is most commonly done using GC-flame ionization detection (GC-FID) and GC-mass spectrometry (GC-MS). Using GC-FID, total hydrocarbon (THC), often referred to as total extractable (organic) material

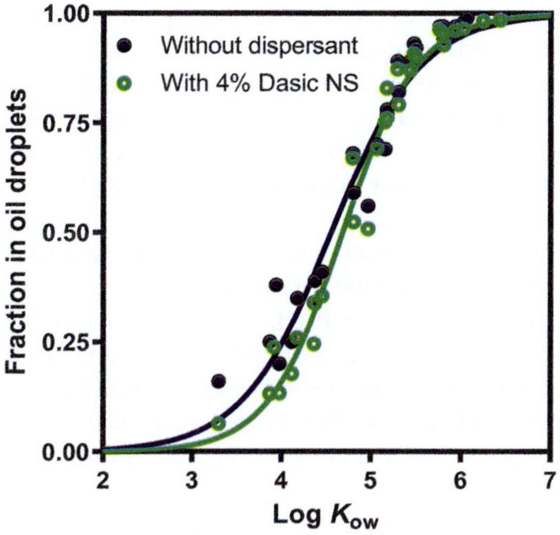

FIGURE 6.4 Partitioning of polycyclic aromatic hydrocarbons between oil and water related to log K_{ow} in dispersions of oil with and without chemical dispersant. Data created by comparing polycyclic aromatic hydrocarbon concentrations in inline-created dispersions of (20 mg oil/L) and their corresponding inline-filtered water-soluble fractions (Nordtug, Olsen, Altin, Meier, et al., 2011; Nordtug, Olsen, Altin, Overrein, et al., 2011). Average volumetric oil droplet size was about 14 μm in both dispersions. Dasic NS (4%) was mixed in to one of the oil batches. *Data replotted from Olsvik et al. (2012). Is chemically dispersed oil more toxic to Atlantic cod (Gadus morhua) larvae than mechanically dispersed oil? A transcriptional evaluation. BMC Genomics, 13.*

and total petroleum hydrocarbons, with carbon numbers typically ranging C_{10} to C_{40} can be quantified (United States-EPA, 2013). This method measures only one total signal from all components extractable by DCM in combination and thus provides limited information about what components are in the mixture. Although most oil components are extractable by DCM, however, some are not and thus notation "THC" may be a bit misleading. As such the method will also not discriminate between petrogenic and biogenic components. This method usually has a high limit of detection, but it is useful as a measure for the total amount of petrogenic material in exposure solutions. GC-MS is more sensitive than GC-FID as it specifically monitors selected masses or ions and provides quantifications of (usually) 22 unsubstituted PAHs and 22 alkylated PAHs (USEPA, 2007). By adding tandem MS (GC-MS/MS), detection limits are reduced even further. For quantifying lighter and more volatile petrogenic components, like BTEX, purge-and-trap GC-MS (United States-EPA, 2006) is widely used.

Common for the mentioned analytical methods is that they are unable to separate between oil compounds present in the water phase (dissolved components) and particulate phase (oil droplets). To do this, either the method for generating oil exposures need to be highly controlled and documented based on particle measurements in combination with analytical chemistry (Nordtug, Olsen, Altin, Meier, et al., 2011) or modeling efforts are needed to postprocess data on chemical composition to estimate droplet concentrations (Hansen, Parkerton, et al., 2019; Redman, 2015; Redman et al., 2007). Importantly, a prerequisite for both approaches is analyses of representative samples of exposure media.

Crude oils from different reservoirs vary markedly in chemical composition, and thus their potential for aquatic toxicity differs markedly. For acute toxicity, using lethality as endpoint, total PAH, or THC is widely used as exposure metrics. This is not unproblematic as all measured (and unmeasured) oil components are not equitoxic on a mass basis; however, quantitative structure–activity relationship (QSAR) models are being used successfully to predict LC_{50}s of mixtures where concentrations of components are based on a molar basis.

6.6 Bioavailability of oil components

6.6.1 The impact of biomass used in toxicity testing

Provided that materials used in the exposure are reasonably inert such as glassware, lipophilic components in static exposure systems partition between between the biomass and water. This process is time and biomass dependent. For nonpolar petrogenic component, there is a correlation between log K_{ow} and bioconcentration factors [the ratio between body concentration and water concentration at equilibrium during constant exposure; bioconcentration factor (BCFs)] (Di Toro & Mcgrath, 2000; McGrath & Di Toro, 2009). Thus, the partitioning between water and biomass can be roughly estimated at equilibrium based on the water-soluble concentration of components and the lipid content of the biomass. This will indicate the worst case since for highly lipophilic components, the uptake relative to the BCF is slow and during a conventional acute study, the final body residue may be much lower than at equilibrium. Highly lipophilic components in general have low solubilities and high BCFs. Thus, in static exposure systems, uptake in the biomass may cause a depletion of these components and underestimation of their toxicity.

6.6.2 Impact of exposure duration and kinetics of uptake and depuration

After an oil spill, the concentration of dissolved components (and oil droplets) decreases due to dilution. The exposure profile for organisms in the recipient greatly depends on the size of the spill. Whereas a small surface oil spill will create a transient increase in concentration (spiked exposure), a large oil spills may create plumes of contaminated water where the dilution mostly occurs in the marginal zone of the plume. This causes different exposure conditions with regard to composition of dissolved oil components and exposure durations. In the case of small oil spills, BTEX may dominate the contribution to toxicity, whereas during a large spill, the extended exposure time and oil weathering will increase the toxic contribution of larger and less-soluble components, such as PAHs. This has important implications for the way toxicity experiments should be conducted.

It is well known that size of organisms affects the rate of uptake and that lipid content is an important determinator for the BCF (Hendriks, Van Der Linde, Cornelissen, & Sijm, 2001). In principle, this implies that in order to characterize the toxicity of a specific oil, the duration of toxicity tests should be adjusted to the size of the test organism. For example, comparing two different developmental stages of the Arctic copepods *Calanus hyperboreus*: the CIII stage (approximate weight 1 mg, 5% lipid content) and the fifth copepodite (CV) stage (approximate weight 10 mg, 30% lipid content), it takes twice as long to reach the same body concentration of C1-phenanthrenes with log K_{ow} of 5.1 (Fig. 6.5). It is also evident that the uptake is much faster in the smaller copepods and that the body concentration for the larger lipid rich copepod is far from being in equilibrium with the water concentration at the end of the exposure period.

The complex mixture of components with their inherently different toxicokinetics represents a challenge in describing the toxicity of crude oil. To

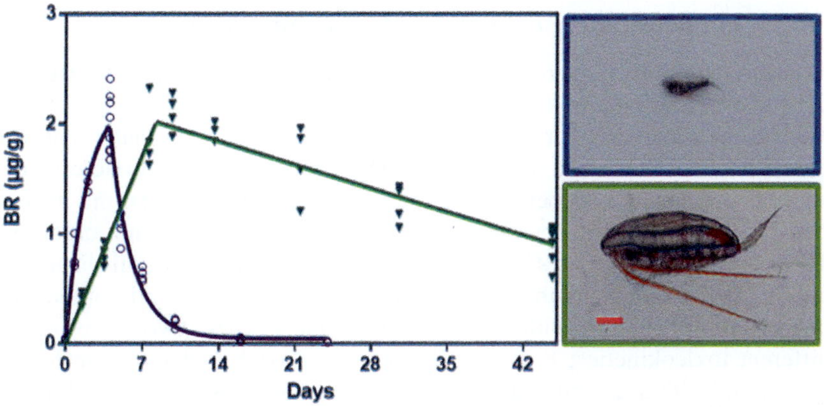

FIGURE 6.5 Left: The combined effects of size and lipid content on body residue. The graphs show kinetics of uptake and depuration of C2-phenanthrene in two developmental stages of the Arctic copepod *Calanus hyperboreus*; CIII (blue, Ĭ1 mg, 5% lipid) exposed to WSF of oil for 4 days and CV (green, Ĭ10 mg, 30% lipid) exposed for 8 days. Body concentrations have been normalized to water concentrations corresponding to 1 mg/L oil. Right: Images of *Calanus hyperboreus*. Top: CIII copepodite with very small lipid content. Bottom: CV copepodite with high lipid content. A scale bar given as red line of 1 mm indicates the scale for both images. *Based on Øverjordet et al. (2018). Toxicokinetics of crude oil components in Arctic copepods. Environmental Science & Technology, 52, 9899–9907.*

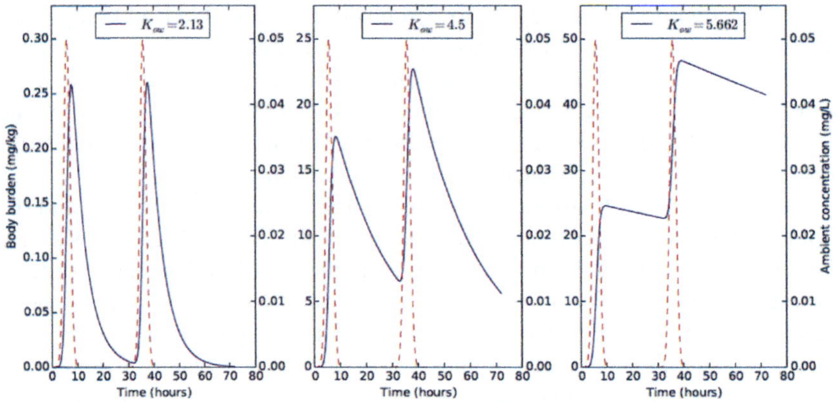

FIGURE 6.6 Single-component body burden resulting from a double exposure spikes, for three oil components with different K_{OW}. The high K_{OW} compound has a very slow rate of depuration, leading to a "pumping" effect where the body residue is increased with each exposure spike encountered, even if these are well separated in time.

visualize differences in toxicokinetics, Fig. 6.6 shows the theoretical body residue of three individual components with different log K_{ow} (2.13, 4.5, and 5.662) during spiked exposures to identical exposure concentrations.

6.6.3 Route of biological uptake of oil components

Exposure solutions prepared from crude oils are highly complex mixtures of mostly hydrocarbons consisting of parent and substituted alkanes, olefins, monocyclic aromatic hydrocarbons (MAHs), heterocyclic aromatic hydrocarbons, and PAHs (Melbye et al., 2009). Their ability to cause toxic effects to aqueous organisms is dependent on their bioavailability, that is, the measure by which various substances in the environment may enter living organisms. There are two main routes of uptake of oil components from oil dispersions into aqueous organisms: through the water phase (dissolved oil components) or through interactions with dispersed oil droplets. These two routes display different toxicokinetics; however, the internal dose by which acute toxicity occurs are within a predictable range (2–8 mmol/kg wet weight) (McCarty & Mackay, 1993). Whereas dissolved components are taken up through passive diffusion over respiration surfaces, droplets may be taken up through direct apical contact, filtration, and ingestion. Filtration and ingestion of oil droplets have been shown in marine filter-feeding organisms, like planktonic crustaceans (Almeda, Baca, Hyatt, & Buskey, 2014; Hansen et al., 2009; Hansen, Altin, Olsen, & Nordtug, 2012; Hansen, Altin, et al., 2017; Hansen, Olsen, et al., 2018), tunicates (Lee, Köster, & Paffenhöfer, 2012), and heterotroph algae (Almeda, Connelly, & Buskey, 2014). Adhesion of oil droplets to the chorion of fish eggs also appears to facilitate transfer of oil components through the chorion to reach the embryo causing toxicity to early life stages of fish (Hansen, Salaberria, et al., 2019; Hansen, Sorensen, et al., 2018; Sørensen et al., 2017; Sørensen et al., 2019; Sørhus et al., 2015).

6.6.4 Body residue as exposure descriptors

The uptake of oil components by organisms is a prerequisite for toxic effects to occur, and the body residue needed to cause acute toxicity is described as the critical body residue (CBR) method for aquatic systems by McCarty and Mackay (1993). For risk assessment processes related to oil spills, this model is valuable when important environmental processes are understood, such as (1) the fate of components released into the aquatic system, (2) accumulation of these components as residues in organisms or specific tissues within organisms, and (3) the relation between body or tissue residues and effects is determined in toxicity tests and bioassays.

The link between internal exposure concentrations and adverse biological responses is currently the most poorly understood aspect. However, by shifting from relating toxicity to water concentrations (as in an LC_{50} test) to relating toxicity to an internal concentration (CBR), several advantages are obtained:

(1) Bioavailability of spilled components is explicitly considered.
(2) Accumulation kinetics are considered, which reduces the confounding effect of exposure duration when interpreting results.
(3) Mixture toxicity may be more readily assessed.
(4) Experimental verification can be readily sought in the lab (and field).

It should also be noted that to a first approximation, the CBR theory implies that narcotic toxicants will produce the same mortality from equal molar body residues (measured in mol/kg), but the water concentration required to produce a given body residue will depend on the BCF of the toxicant.

6.7 Selection of toxicity endpoints

Toxic effects of oil can also occur acutely after a short exposure time, a period after an acute exposure (subacutely or delayed) or after prolonged or chronic exposure. A wide range of different toxic endpoints have been studied in aqueous organisms exposed to oil and oil components, ranging from low levels of biological organization (molecular and cellular effects) to effects on population level. Molecular and cellular endpoints provide valuable insight with the aim of understanding the underlying mechanisms of toxic action and may be used as potential early-warning signals for effects on higher levels of biological organization. An example is the description of the underlying events causing cardiotoxicity in fish larvae exposed to oil dispersions, where a calcium channel in the developing heart of an embryo appears to be the primary target for tricyclic aromatic hydrocarbons. This causes disturbances in cardiac development, which may ultimately result in reduced survival as the larvae develop (Incardona, 2017).

Because of their ecological relevance, the most used endpoints in toxicity tests are, however, lethality, retarded growth and development, and reduced reproduction. The most widely used toxicological endpoint in toxicity tests is lethality, also often referred to as immobilization and/or narcosis. Acute toxicity tests apply short-term exposure times (hours-days) to fixed concentrations, and the acute toxicity threshold is reported as the concentration where 50% of the test population has died at the end of the test period (the LC_{50} value) (Rand, 1995). Although lethality is an endpoint that may occur only at concentrations relevant for a very short period of time during an acute oil spill, there are several advantages of this endpoint. Lethality has, as mentioned, an unquestionable ecological relevance. It is also a definite endpoint (alive vs dead), so there is limited necessity for expert evaluation or advanced analyses. Furthermore, LC_{50} data may also be in QSARs as for nonpolar petrogenic components, acute toxicity correlates with K_{ow} (Di Toro, Mcgrath, & Hansen, 2000). Oil components with higher log K_{ow} have greater toxic potential (i.e., lower LC_{50} values) than those with lower log K_{ow}. Assuming additivity on a molar basis for all components in the exposure solutions, QSARs can be used to predict LC_{50} values for the mixture. These advantages make LC_{50} data possible to implement into environmental models. In addition, several report ratios between LC_{50} and sublethal effects-so-called acute-to-chronic ratios-which also can be utilized for environmental models used for risk and damage assessment (Ahlers et al., 2006; Vestel et al., 2016). Importantly, standard laboratory toxicity testing alone is incapable of providing toxicity data above the individual level; however, if done sensibly and by combining the data using numerical models, risks of toxicological effects on higher level of biological organization may be predicted (Carroll et al., 2018; Nepstad, Hansen, & Skancke, 2020).

6.8 Method to generate parameterized toxicity data for input to risk assessment models

To cope with some of the problems related to toxicity experiments with oil dispersions, SINTEF developed and introduced a test system for toxicity testing of dispersed crude oil (Nordtug, Olsen, Altin, Meier, et al., 2011). The intention of this system was to be able to control the concentration and size of oil droplets during the exposure and to provide parameterized data to risk and damage assessment models DREAM (Dose related Risk and Effect Assessment Model) (Nepstad et al., 2020; Reed & Rye, 2011) and OSCAR (Oil Spill Contingency and Response) (Reed et al., 2000).

6.8.1 Standardization of exposure parameters

The design of the SINTEF dispersion system is based on repeated turbulence in a system with multiple nozzles in series. In this system, for a certain nozzle aperture diameter, the mean droplet size and the droplet size distribution were a function of flow rate and the number of nozzles. The final system (dispersion generator) consisted of four nozzles (ID = 0.5 mm) in series as shown in Fig. 6.7A.

Controlled input of oil and water is obtained by pumping oil and water into the system by predefined ratios. Oil is added through a capillary into the generator after the first nozzle by a syringe pump, and ratio between the flow of oil and seawater defines the concentration of the parent dispersion (Fig. 6.7B). Typically, a water flow of 120–200 mL/min is used. An advantage of the system is that, compared to traditional WAF preparations, very little oil is used. Since all the added oil is dispersed, loss of oil from the dispersion is limited to attachment to surfaces in the flow system. Some of the oil will be lost due to attachment to surfaces and surfacing in the exposure

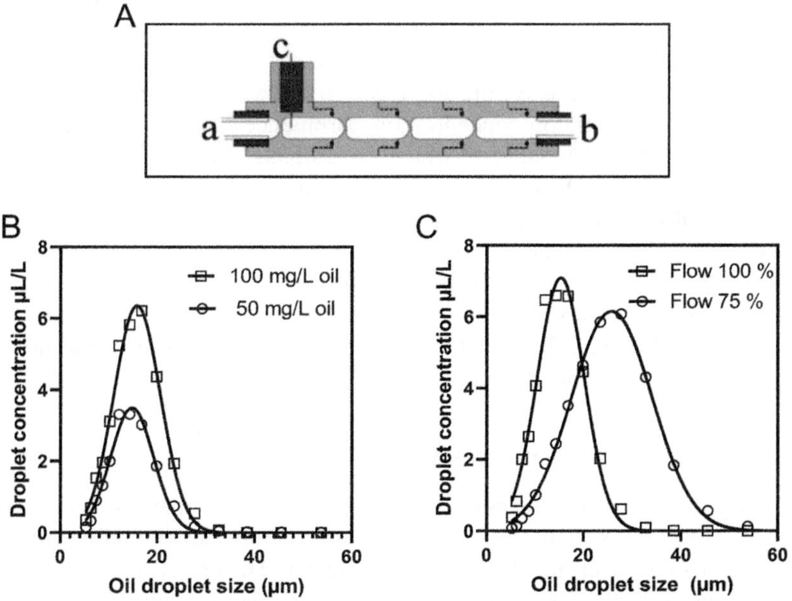

FIGURE 6.7 Creating dispersions by repeated turbulence. (A) The dispersion generator consists of a series of nozzles where water is pushed through (from a to b) creating repeated turbulence. Oil is added through a capillary (c) in front of the first nozzles and subjected to repeated turbulence breaking up the oil. (B) The oil concentration can be accurately controlled by adjusting the addition of oil relative to the water flow through the system. The average droplet size distribution can be controlled by varying the inlet water flow rate. (C) The effect of reducing the water flow by 25% causes an approximate doubling of the average oil droplet size.

solution. The loss to surfaces depends on the oil quality, oil droplet size distribution, flow velocity and the volume/surface ratio of the exposure containers including supply lines. Typically, 100% is dispersed at the outlet of the dispersion generator whereas more than 50% is normally maintained in suspension in the exposure solution.

Dispersions with mean droplet sizes (based on volume) ranging 10–40 μm are produced by adjusting the water flow and thus the turbulence through the generator (Fig. 6.7C). Larger droplets (> 40 μm) have a surfacing velocity that makes them very difficult to keep in suspension in laboratory systems (Brakstad, Nordtug, & Throne-Holst, 2015).

Since this is a flow through system, it eliminates some of the problems related to decreasing exposure concentrations caused by uptake in biomass and attachment of components to surfaces. Thus, the exposure concentrations can be kept constant over time. To maintain the oil droplets in suspension, lengths of tubing is kept to a minimum.

Concentration series are established by an array of computer controlled 3-way magnetic valves alternating between continuous supply lines of the parent dispersion and clean seawater both driven by gravity. The timing of each of the valves can thus be set up to achieve any concentration between that of the parent dispersion and clean seawater by adjusting the time ratio between opening toward the two supply lines.

As discussed previously, oil droplets will surface over time, and thus, to obtain a steady dispersion concentration, turbulence is needed during experiments. The preferred manner is which oil droplets are kept in suspension is using a flow through system with continuous generation of dispersion (Nordtug, Olsen, Altin, Meier, et al., 2011), however for static closed-bottle experiments, this can be applied using rotation of the exposure bottles (Brakstad et al., 2015). For flow through systems, it is important that the inlet for oil dispersions into the exposure system is located close to the bottom of the system and the outlet is at the top due to the surfacing of droplets over time. This prevents the build-up of oil sheen on the water surface.

6.8.2 Application of the method

By using the SINTEF dispersion system, toxicity of oil dispersions has been studied using this system on a range of different species including microorganisms (Brakstad et al., 2018; Ribicic, Netzer, Hazen, et al., 2018), copepods (Hansen et al., 2009; Hansen et al., 2012; Hansen, Jager, et al., 2016; Nordtug et al., 2015; Olsen, Nordtug, Altin, Lervik, & Hansen, 2013; Øverjordet et al., 2018; Hansen, Tarrant, et al., 2017), mussels (Andreassen et al., 2013), and fish (Hansen, Salaberria et al., 2019; Hansen, Sorensen, et al., 2018; Laurel et al., 2019; Nordtug, Olsen, Altin, Overrein, et al., 2011; Olsvik et al., 2010; Olsvik et al., 2011; Olsvik et al., 2012; Sørensen et al., 2019; Sørhus et al., 2015; Sørhus et al., 2016). The system has also been used for studies on oil biodegradation and associated microorganisms (Brakstad et al., 2018; Ribicic, Netzer, Hazen, et al., 2018; Ribicic, Netzer, Winkler et al., 2018; Størdal et al., 2015), formation and fate of oil-related aggregates in seawater at different temperatures (Henry, Netzer, Davies, & Brakstad, 2020).

A good example of the application, and the main reason it was developed, was to assess the contribution of oil droplets to dispersion toxicity. It has been widely accepted that dissolved components are the major drivers for toxicity. However, to test this hypothesis, we used the SINTEF dispersion system to generate and compared toxicities of dispersions and their corresponding filtered WSFs. Several experiments were run on copepods, fish embryos, and fish larvae. For experiments on fish larvae, limited evidence was found to state that oil droplets contribute to dispersion toxicity (Nordtug, Olsen, Altin, Overrein, et al., 2011; Olsvik et al., 2011). However, for fish eggs, adhesion of oil droplets onto the chorion was observed and associated with increased uptake of oil components and a significant increase in toxicity (Hansen, Sorensen, et al., 2018; Sørensen et al., 2017; Sørhus et al., 2015; Sørhus et al., 2016).

Another example was an attempt to evaluate the contribution of dispersants to dispersed oil toxicity. By comparing toxicity of oil dispersions of oil alone and oil-added dispersant, we were able to prove that the main driver for dispersion toxicity was the oil, not the dispersant (Hansen, Lie, et al., 2016; Olsvik et al., 2012). This contrasts with much of the literature, and the reason for this apparent inconsistency was that using the SINTEF dispersion system, we were able to maintain comparable concentrations and oil droplet size distributions between the two treatments. This is in sharp contrast to comparing WAF and CEWAF applications to address the same objective. This example shows the strengths of parameterization of oil exposure experiments.

6.9 Conclusions

Oil is a highly complex mixture and undoubtably toxic when introduced to the environment. However, estimating the potential environmental impact from an oil spill is challenging and suffers from the multitude of experimental methods and exposure metrics used to determine bioavailability and toxicity. With the development of chemical analysis and modeling tools since the establishment of the CROSERF standard, there is a need to revalidate and standardize toxicity methods that are consistent with the development of environmental exposure models. Several new approaches have been suggested that may contribute to this development (Hodson et al., 2019; Nordtug, Olsen, Altin, Meier, et al., 2011; Redman & Parkerton, 2015). In estimating the environmental impact of large oil spills, there is no alternative to modeling. There is thus a need to improve the representation and parameterization of oil toxicity in such a way that data can serve as an input to numerical models considering the fate of the oil in space and time as well as predicting dynamic changes in exposure.

References

Adams, J., Charbonneau, K., Tuori, D., Brown, R. S. & Hodson, P. V. (2017). Review of methods for measuring the toxicity to aquatic organisms of the water accommodated fraction (WAF) and chemically-enhanced water accommodated fraction (CEWAF) of petroleum. *Canadian Science Advisory Secretariat*, 2017/064. xi + 110 p.

Ahlers, J., Riedhammer, C., Vogliano, M., Ebert, R.-U., Kühne, R., & Schüürmann, G. (2006). Acute to chronic ratios in aquatic toxicity-Variation across trophic levels and relationship with chemical structure. *Environmental Toxicology and Chemistry, 25*, 2937–2945.

Almeda, R., Baca, S., Hyatt, C., & Buskey, E. J. (2014). Ingestion and sublethal effects of physically and chemically dispersed crude oil on marine planktonic copepods. *Ecotoxicology, 23*, 988–1003.

Almeda, R., Connelly, T. L., & Buskey, E. J. (2014). Novel insight into the role of heterotrophic dinoflagellates in the fate of crude oil in the sea. *Scientific Reports*, 4.

Andreassen, I., Hammer, K. M., Krause, D., Olsen, A. J., Altin, D. & Nordtug, T. (2013) Removal of oil droplets by benthic filter feeders. In *36th AMOP technical seminar on environmental contamination and response* (pp. 412–422).

Aurand, D. & Coelho, G. M. (1996). In *Proceedings of the fourth meeting of the chemical response to oil spills: Ecological effects research forum.* Ecosystem Management & Associates, Inc, Technical Report 07-03.

Brakstad, O. G., Davies, E. J., Ribicic, D., Winkler, A., Brönner, U., & Netzer, R. J. P. B. (2018). Biodegradation of dispersed oil in natural seawaters from Western Greenland and a Norwegian fjord. *Polar Biology, 41*(2), 1–16.

Brakstad, O. G., Nordtug, T., & Throne-Holst, M. (2015). Biodegradation of dispersed Macondo oil in seawater at low temperature and different oil droplet sizes. *Marine Pollution Bulletin, 93*, 144–152.

Carls, M. G., Holland, L., Larsen, M., Collier, T. K., Scholz, N. L., & Incardona, J. P. (2008). Fish embryos are damaged by dissolved PAHs, not oil particles. *Aquatic Toxicology, 88*, 121–127.

Carroll, J., Vikebø, F., Howell, D., Broch, O. J., Nepstad, R., Augustine, S., ... Juselius, J. (2018). Assessing impacts of simulated oil spills on the Northeast Arctic cod fishery. *Marine Pollution Bulletin, 126*, 63–73.

Cline, P. V., Delfino, J. J., & Rao, P. S. C. (1991). Partitioning of aromatic constituents into water from gasoline and other complex solvent mixtures. *Environmental Science & Technology, 25*, 914–920.

Daling, P. S., Leirvik, F., Almås, I. K., Brandvik, P. J., Hansen, B. H., Lewis, A., & Reed, M. (2014). Surface weathering and dispersibility of MC252 crude oil. *Marine Pollution Bulletin, 87*, 300–310.

Di Toro, D. M., & Mcgrath, J. A. (2000). Technical basis for narcotic chemicals and polycyclic aromatic hydrocarbon criteria. II. Mixtures and sediments. *Environmental Toxicology and Chemistry, 19*, 1971–1982.

Di Toro, D. M., Mcgrath, J. A., & Hansen, D. J. (2000). Technical basis for narcotic chemicals and polycyclic aromatic hydrocarbon criteria. I. Water and tissue. *Environmental Toxicology and Chemistry, 19*, 1951–1970.

Echols, B. S., Smith, A., Gardinali, P., & Rand, G. (2016). An evaluation of select test variables potentially affecting acute oil toxicity. *Archives of Environmental Contamination and Toxicology*, *70*, 392–405.

Eganhouse, R. P., Dorsey, T. F., Phinney, C. S., & Westcott, A. M. (1996). Processes affecting the fate of monoaromatic hydrocarbons in an aquifer contaminated by crude oil. *Environmental Science & Technology*, *30*, 3304–3312.

Faksness, L.-G., Altin, D., Nordtug, T., Daling, P. S., & Hansen, B. H. (2015). Chemical comparison and acute toxicity of water accommodated fraction (WAF) of source and field collected Macondo oils from the Deepwater Horizon spill. *Marine Pollution Bulletin*, *91*, 222–229.

Guggenheim, E. (1937). The theoretical basis of Raoult's law. *Transactions of the Faraday Society*, *33*, 151–156.

Hansen, B. H., Altin, D., Nordtug, T., Øverjordet, I. B., Olsen, A. J., Krause, D., ... Størseth, T. R. (2017). Exposure to crude oil micro-droplets causes reduced food uptake in copepods associated with alteration in their metabolic profiles. *Aquatic Toxicology*, *184*, 94–102.

Hansen, B. H., Altin, D., Olsen, A. J., & Nordtug, T. (2012). Acute toxicity of naturally and chemically dispersed oil on the filter-feeding copepod *Calanus finmarchicus*. *Ecotoxicology and Environmental Safety*, *86*, 38–46.

Hansen, B. H., Jager, T., Altin, D., Øverjordet, I. B., Olsen, A. J., Salaberria, I., & Nordtug, T. (2016). Acute toxicity of dispersed crude oil on the cold-water copepod *Calanus finmarchicus*: Elusive implications of lipid content. *Journal of Toxicology and Environmental Health, Part A*, *79*, 549–557.

Hansen, B. H., Lie, K. K., Størseth, T. R., Nordtug, T., Altin, D., & Olsvik, P. A. (2016). Exposure of first-feeding cod larvae to dispersed crude oil results in similar transcriptional and metabolic responses as food deprivation. *Journal of Toxicology and Environmental Health, Part A*, *79*, 558–571.

Hansen, B. H., Nordtug, T., Altin, D., Booth, A., Hessen, K. M., & Olsen, A. J. (2009). Gene expression of GST and CYP330A1 in lipid-rich and lipid-poor female *Calanus finmarchicus* (Copepoda: Crustacea) exposed to dispersed oil. *Journal of Toxicology and Environmental Health-Part a-Current Issues*, *72*, 131–139.

Hansen, B. H., Olsen, A. J., Salaberria, I., Altin, D., Øverjordet, I. B., Gardinali, P., ... Nordtug, T. (2018). Partitioning of PAHs between crude oil microdroplets, water, and copepod biomass in oil-in-seawater dispersions of different crude oils. *Environmental Science & Technology*, *52*, 14436–14444.

Hansen, B. H., Parkerton, T., Nordtug, T., Størseth, T. R., & Redman, A. (2019). Modeling the toxicity of dissolved crude oil exposures to characterize the sensitivity of cod (*Gadus morhua*) larvae and role of individual and unresolved hydrocarbons. *Marine Pollution Bulletin*, *138*, 286–294.

Hansen, B. H., Salaberria, I., Read, K. E., Wold, P. A., Hammer, K. M., Olsen, A. J., ... Bardal, T. (2019). Developmental effects in fish embryos exposed to oil dispersions – The impact of crude oil micro-droplets. *Marine Environmental Research*, *150*, 104753.

Hansen, B. H., Sorensen, L., Carvalho, P. A., Meier, S., Booth, A. M., Altin, D., ... Nordtug, T. (2018). Adhesion of mechanically and chemically dispersed crude oil droplets to eggs of Atlantic cod (Gadus morhua) and haddock (*Melanogrammus aeglefinus*). *Science of the Total Environment*, *640–641*, 138–143.

Hansen, B. H., Tarrant, A. M., Salaberria, I., Altin, D., Nordtug, T., & Øverjordet, I. B. (2017b). Maternal polycyclic aromatic hydrocarbon (PAH) transfer and effects on offspring of copepods exposed to dispersed oil with and without oil droplets. *Journal of Toxicology and Environmental Health, Part A*, *25*, 1–14.

Hendriks, A. J., Van Der Linde, A., Cornelissen, G., & Sijm, D. T. H. M. (2001). The power of size. 1. Rate constants and equilibrium ratios for accumulation of organic substances related to octanol-water partition ratio and species weight. *Environmental Toxicology and Chemistry*, *20*, 1399–1420.

Henry, I. A., Netzer, R., Davies, E. J., & Brakstad, O. G. (2020). Formation and fate of oil-related aggregates (ORAs) in seawater at different temperatures. *Marine Pollution Bulletin*, *159*, 111483.

Hodson, P. V., Adams, J., & Brown, R. S. (2019). Oil toxicity test methods must be improved. *Environmental Toxicology and Chemistry*, *38*, 302–311.

Incardona, J. P. (2017). Molecular mechanisms of crude oil developmental toxicity in fish. *Archives of Environmental Contamination and Toxicology*, *73*, 19–32.

Jaggi, A., Snowdon, R. W., Radović, J. R., Stopford, A., Oldenburg, T. B., & Larter, S. R. (2020). *Partitioning of organics between oil and water phases with and without the application of dispersants. Deep Oil Spills*. Springer.

Jaggi, A., Snowdon, R. W., Stopford, A., Radović, J. R., Oldenburg, T. B., & Larter, S. R. (2017). Experimental simulation of crude oil-water partitioning behavior of BTEX compounds during a deep submarine oil spill. *Organic Geochemistry*, *108*, 1–8.

Laurel, B. J., Copeman, L. A., Iseri, P., Spencer, M. L., Hutchinson, G., Nordtug, T., ... Boyd, D. T. (2019). Embryonic crude oil exposure impairs growth and lipid allocation in a keystone Arctic forage fish. *iScience*, *19*, 1101–1113.

Lee, L. S., Hagwall, M., Delfino, J. J., & Rao, P. S. C. (1992). Partitioning of polycyclic aromatic hydrocarbons from diesel fuel into water. *Environmental Science & Technology*, *26*, 2104–2110.

Lee, L. S., Rao, P. S. C., & Okuda, I. (1992). Equilibrium partitioning of polycyclic aromatic hydrocarbons from coal tar into water. *Environmental Science & Technology*, *26*, 2110–2115.

Lee, R. F., Köster, M., & Paffenhöfer, G.-A. (2012). Ingestion and defecation of dispersed oil droplets by pelagic tunicates. *Journal of Plankton Research.*

McCarty, L. S., & Mackay, D. (1993). Enhancing ecotoxicological modeling and assessment. *Environmental Science and Technology*, *27*, 1719–1728.

McGrath, J. A., & Di Toro, D. M. (2009). Validation of the target lipid model for toxicity assessment of residual petroleum constituents: Monocyclic and polycyclic aromatic hydrocarbons. *Environmental Toxicology and Chemistry*, *28*, 1130–1148.

Melbye, A. G., Brakstad, O. G., Hokstad, J. N., Gregersen, I. K., Hansen, B. H., Booth, A. M., ... Tollefsen, K. E. (2009). Chemical and toxicological characterization of an unresolved complex mixture-rich biodegraded crude oil. *Environmental Toxicology and Chemistry*, *28*, 1815–1824.

Nepstad, R., Hansen, B. H., & Skancke, J. (2020). North Sea produced water PAH exposure and uptake in early life stages of Atlantic cod. *Marine Environmental Research*, 105203.

Nordtug, T., Olsen, A. J., Altin, D., Meier, S., Overrein, I., Hansen, B. H., & Johansen, Ø. (2011). Method for generating parameterized ecotoxicity data of dispersed oil for use in environmental modelling. *Marine Pollution Bulletin*, *62*, 2106–2113.

Nordtug, T., Olsen, A. J., Altin, D., Overrein, I., Storøy, W., Hansen, B. H., & De Laender, F. (2011). Oil droplets do not affect assimilation and survival probability of first feeding larvae of North-East Arctic cod. *Science of the Total Environment*, *412*, 148–153.

Nordtug, T., Olsen, A. J., Salaberria, I., Øverjordet, I. B., Altin, D., Størdal, I. F., & Hansen, B. H. (2015). Oil droplet ingestion and oil fouling in the copepod *Calanus finmarchicus* exposed to mechanically and chemically dispersed crude oil. *Environmental Toxicology and Chemistry*, *34*, 1899–1906.

NRC. (2005). *Oil spill dispersants: Efficacy and effects*. Washington D.C: The National Academies Press.

Olsen, A. J., Nordtug, T., Altin, D., Lervik, M., & Hansen, B. H. (2013). Effects of dispersed oil on reproduction in the cold water copepod *Calanus finmarchicus* (Gunnerus). *Environmental Toxicology and Chemistry, 32*, 2045–2055.

Olsvik, P. A., Hansen, B. H., Nordtug, T., Moren, M., Nolen, E., & Lie, K. K. (2011). Transcriptional evidence for low contribution of oil droplets to acute toxicity from dispersed oil in first feeding Atlantic cod (*Gadus morhua*) larvae. *Comparative Biochemistry and Physiology C: Toxicology & Pharmacology, 154*, 333–345.

Olsvik, P. A., Lie, K. K., Nordtug, T., & Hansen, B. H. (2012). Is chemically dispersed oil more toxic to Atlantic cod (*Gadus morhua*) larvae than mechanically dispersed oil? A transcriptional evaluation. *BMC Genomics, 13*, 702.

Olsvik, P. A., Nordtug, T., Altin, D., Lie, K. K., Overrein, I., & Hansen, B. H. (2010). Transcriptional effects on glutathione S-transferases in first feeding Atlantic cod (*Gadus morhua*) larvae exposed to crude oil. *Chemosphere, 79*, 905–913.

Øverjordet, I. B., Nepstad, R., Hansen, B. H., Jager, T., Farkas, J., Altin, D., ... Nordtug, T. (2018). Toxicokinetics of crude oil components in Arctic copepods. *Environmental Science & Technology, 52*, 9899–9907.

Picel, K. C., Stamoudis, V. C., & Simmons, M. S. (1988). Distribution coefficients for chemical components of a coal-oil/water system. *Water Research, 22*, 1189–1199.

Rand, G. M. (1995). *Fundamentals of aquatic toxicology: Effects, environmental fate and risk assessment*. Taylor & Francis Group.

Redman, A. (2015). Role of entrained droplet oil on the bioavailability of petroleum substances in aqueous exposures. *Marine Pollution Bulletin, 97*, 342–348.

Redman, A. D., & Parkerton, T. F. (2015). Guidance for improving comparability and relevance of oil toxicity tests. *Marine Pollution Bulletin, 98*, 156–170.

Redman, A., Mcgrath, J. A., Stubblefield, W. A., Maki, A., & Di Toro, D. (2007). Quantifying the concentration of crude oil microdroplets in oil-water preparations. SETAC USA.

Reed, M., & Rye, H. (2011). *The DREAM model and the environmental impact factor: Decision support for environmental risk management. Produced Water*. Springer.

Reed, M., Daling, P. S., Brakstad, O. G., Singsaas, I., Faksness, L. -G., Hetland, B. & Ekrol, N. (2000). OSCAR2000: A multi-component 3-dimensional oil spill contingency and response model.

Rhodes, M. J. (2008). *Introduction to particle technology*. John Wiley & Sons.

Ribicic, D., Netzer, R., Hazen, T. C., Techtmann, S. M., Drabløs, F., & Brakstad, O. G. (2018). Microbial community and metagenome dynamics during biodegradation of dispersed oil reveals potential key-players in cold Norwegian seawater. *Marine Pollution Bulletin, 129*, 370–378.

Ribicic, D., Netzer, R., Winkler, A., & Brakstad, O. G. (2018). Microbial communities in seawater from an Arctic and a temperate Norwegian fjord and their potentials for biodegradation of chemically dispersed oil at low seawater temperatures. *Marine Pollution Bulletin, 129*, 308–317.

Sandoval, K., Ding, Y., & Gardinali, P. (2017). Characterization and environmental relevance of oil water preparations of fresh and weathered MC-252 Macondo oils used in toxicology testing. *Science of the Total Environment, 576*, 118–128.

Shiu, W. Y., Maijanen, A., Ng, A. L., & Mackay, D. (1988). Preparation of aqueous solutions of sparingly soluble organic substances: II. Multicomponent systems–Hydrocarbon mixtures and petroleum products. *Environmental Toxicology and Chemistry, 7*, 125–137.

Singer, M. M., Aurand, D., Bragin, G. E., Clark, J. R., Coelho, G. M., Sowby, M. L., & Tjeerdema, R. S. (2000). Standardization of the preparation and quantitation of water-accommodated fractions of petroleum for toxicity testing. *Marine Pollution Bulletin*, *40*, 1007–1016.

Sørensen, L., Hansen, B. H., Farkas, J., Donald, C. E., Robson, W. J., Tonkin, A., ... Rowland, S. J. (2019). Accumulation and toxicity of monoaromatic petroleum hydrocarbons in early life stages of cod and haddock. *Environmental Pollution*, *251*, 212–220.

Sørensen, L., Sørhus, E., Nordtug, T., Incardona, J. P., Linbo, T. L., Giovanetti, L., ... Meier, S. (2017). Oil droplet fouling and differential toxicokinetics of polycyclic aromatic hydrocarbons in embryos of Atlantic haddock and cod. *PLoS One*, *12*, e0180048.

Sørhus, E., Edvardsen, R. B., Karlsen, Ø., Nordtug, T., Van Der Meeren, T., Thorsen, A., ... Meier, S. (2015). Unexpected interaction with dispersed crude oil droplets drives severe toxicity in Atlantic haddock embryos. *PLoS One*, 10.

Sørhus, E., Incardona, J. P., Karlsen, Ø., Linbo, T., Sørensen, L., Nordtug, T., ... Meier, S. (2016). Crude oil exposures reveal roles for intracellular calcium cycling in haddock craniofacial and cardiac development. *Scientific Reports*, *6*, 31058.

Størdal, I. F., Olsen, A. J., Jenssen, B. M., Netzer, R., Hansen, B. H., Altin, D., & Brakstad, O. G. (2015). Concentrations of viable oil-degrading microorganisms are increased in feces from Calanus finmarchicus feeding in petroleum oil dispersions. *Marine Pollution Bulletin*, *98*, 69–77.

USEPA. (1996). Method 3510C: Separatory funnel liquid–liquid extraction.

United States-EPA. (2006). Method 8260C, 2006. Volatile organic compounds by gas chromatography/mass spectrometry (GC/MS).

USEPA. (2007). Method 8270D: Semivolatile Organic Compounds by GC/MS.

United States-EPA. (2013). EPA Method 8015D (SW-846): Nonhalogenated Organics Using GC/FID. United States-EPA.

Vestel, J., Caldwell, D. J., Constantine, L., Vincent, J., Davidson, T., Dolan, D. G., ... Ryan, J. J. (2016). Use of acute and chronic ecotoxicity data in environmental risk assessment of pharmaceuticals. *Environmental Toxicology and Chemistry*, *35*, 1201–1212.

Chapter 7

Chemical assessments of sources, fate, and impacts of marine oil spills

Jagoš R. Radović
Department of Geoscience, University of Calgary, Calgary, AB, Canada

7.1 Introduction

Anthropogenic oil releases to marine environment occur on daily basis; most of them are small spills, related to routine shipping operations, and activities of offshore and onshore oil and gas facilities. Overall, industrial oil releases amount to roughly half of approximately 1.2 M tonnes of oil that is introduced to the global ocean annually, while the remainder is being sourced from natural seepage (Boehmer-Christiansen, 2008).

Before continuing, it is essential to more precisely define some fundamental geochemical terms that are often discussed interchangeably in oil spill science but can cause a certain degree of confusion with nonexperts; following condensed definitions are based on Overton et al. (2016), readers are advised to refer to that paper and references within for more detailed information.

Petroleum is a general name for a variety of subsurface fluids occurring in geological reservoirs that are commercially produced through engineered drilling and excavation, or released naturally through seepage, as crude oil or natural gas, or both, and sometimes as liquid condensates (gases that condense to liquids at surface pressures) and bitumens (biodegraded, viscous oils). On the other hand, crude oil is a more specific term that encompasses liquid phase reservoir fluids containing up to 50% (by weight) of dissolved natural gas, at reservoir pressures. Gas rich oils are often called "live oils," while crude oils that contain no gas are known as "dead oils"; this distinction is very relevant for different oil spill scenarios—deepwater well blowouts [see the description of the Deepwater Horizon (DWH) incident below] involve releases of live crudes, while in surface spill scenarios, we are typically dealing with degassed, "dead" oils. Finally, the term "petroleum products" involves various types of petrogenic materials that

Marine Hydrocarbon Spill Assessments. DOI: https://doi.org/10.1016/B978-0-12-819354-9.00009-0

are produced by industrial refining, and/or mixing and blending of crude oils and refinery streams.

In the chemical sense, all the above substances comprise hydrocarbons, that is, molecules that contain only hydrogen and carbon atoms, and a fraction of compounds that are termed nonhydrocarbons, which include heteroatom-bearing molecules, namely, species with nitrogen, oxygen, and/or sulfur moieties, and organically bound metals (nickel and vanadium porphyrins) (see Fig. 7.2 and Section 7.2). Typically, hydrocarbons are predominant component of crudes and petroleum products, but in some cases, such as in oils that have been heavily biodegraded during their geological history, and/or subjected to extensive environmental transformations post spillage, nonhydrocarbon components can represent the major fraction of oil (Section 7.3).

Large-scale oil spill incidents occur rarely, but when they do, they typically cause major environmental damage because of the large volume of oil that is released in a relatively short time frame, within a constrained marine area. Historically, catastrophic spills were related either to tanker accidents, such as the sinking of Exxon Valdez (1989) and Prestige (2003), or blowouts on the offshore oil assets, the likes of Ixtoc-1 (1979) and the DWH (2010) disasters in the Gulf of Mexico. The DWH blowout occured from a well at 1500 m depth, causing a leak that lasted for 87 days, releasing 5.3×10^{11} g of oil (defined as molecules with ≥ 6 carbons that are liquid at 1 atm) and 1.7×10^{11} g of natural gas (hydrocarbons with ≤ 5 carbons that are gases at 1 atm) at high pressure (1.5×10^4 kPa) (Reddy et al., 2012; Kujawinski et al., 2020). Ixtoc-1 incident is historic precedent to DWH, during which a similar amount of oil was released from a gushing well offshore the Mexican coast, albeit in much shallower water, at around 60 m depth (Jernelöv and Lindén, 1981).

Shipping safety improved significantly in the past 50 years, reducing the average number of tanker spills per year from nearly 80 in the 1970s, to only six in the last decade (ITOPF, 2020). On the other hand, in recent years, some regional seas have experienced a rise in shipping activity; most notably the Arctic, where, due to global warming and sea-ice reduction, there has been an increase in marine traffic (Ghosh and Rubly, 2015). Such trends are causing a concern related to possible incidental spills, in particular, because of the sensitivity of the Arctic ecosystem, and the lack of spill response infrastructure and resources in this remote area.

Another important trend relevant to oil spill preparedness is the global push of the industry to offshore drilling in ultra-deep waters (>1 km depth). For example, in 2017 most of the crude oil produced in the Gulf of Mexico (GoM) came from ultra-deep wells (Murawski et al., 2020). DWH was a stark example of the challenges involved in responding to deepwater oil releases, and their complex and lasting environmental and human health impacts (Farrington et al., 2016).

Finally, in the past two decades, new types of unconventional crude oil types came to the market, most notably heavy oils and bitumens, being produced and transported in increasing volumes, both on land and the sea (Williams, 2019). Due to their biodegraded nature, these oil types have unique physicochemical properties which pose specific challenges in terms of their environmental fate and impact assessments (Radović et al., 2018).

Once released to marine environment, oil undergoes a complex sequence of processes including physicochemical transformation and partitioning to atmosphere and aqueous phase, as well as direct and indirect interactions with biogenic components of marine environment (Radović et al., 2012; Farrington et al., 2016) (Fig. 7.1). Depending on the oil type and spill scenario, the residence time of oil in the environment can vary widely, from hours/days to decades (Radović et al., 2020a). Knowing the above, it is important to emphasize that oil spill research evolved with every new spill case, advancing our understanding of oil fate and impacts in different marine environments and under different release scenarios. That progress was enabled, in large part, by new analytical technologies and methods allowing a more comprehensive insight into complex biogeochemical variables and processes involved in assessments of oil spill sources (Section 7.2), environmental fate (Section 7.3), and impacts (Section 7.4).

This chapter will showcase a nonexhaustive selection of new insights that were generated in the past 10 years of research that investigated some of the major marine oil spills and will discuss their significance for the assessment of potential future spills.

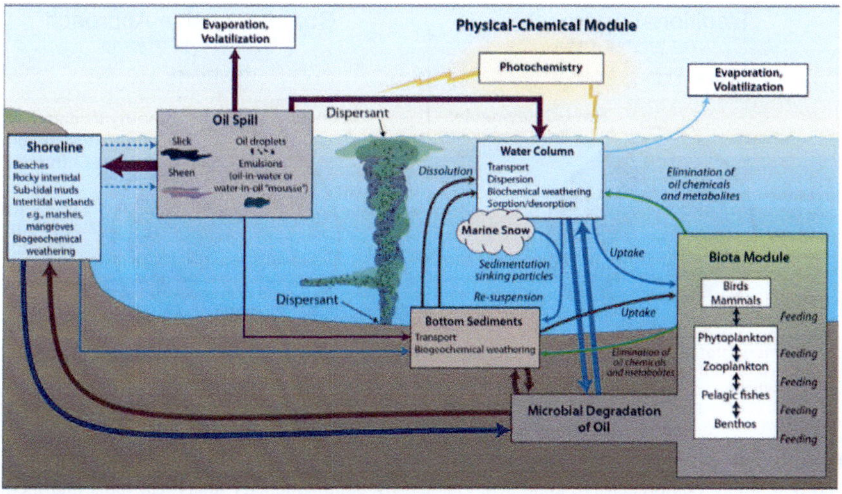

FIGURE 7.1 Schematic overview summarizing the fate of spilled oil in marine environment and relevant physiochemical and biological processes. *From Farrington, J. W., Burns, K. A. & Leinen, M. S. (2016). Synthesis and crosscutting topics.* Oceanography, 29, *204–213.*

7.2 Spill source assessment

Postspill oil detection and source assessment are crucial for planning spill response measures, natural resource damage assessment, and long-term monitoring and understanding of the ultimate fate of spilled oil. Various technologies are used to that end, ranging from remote sensing to in situ sensors and laboratory-based techniques (White et al., 2016).

Traditionally, the workhorse of lab-based chemical oil spill assessments has been one-dimensional capillary gas chromatography (GC), coupled to either flame ionization detector (FID) or quadrupole mass spectrometer (MS) (Fig. 7.2). GC-based protocols are still the most common workflow in oil spill laboratories, due to the robustness and reproducibility of analytical results, which make them suitable for standardization (Daling et al., 2014). Such standard methods rely on (semi)quantitation of well-established oil proxies including *n*-alkanes, polycyclic aromatic hydrocarbons (PAHs), and biomarkers, that is, source-specific molecular fossils of parent organisms, for example, bacteria and plants.

Occurrence and distribution of those diagnostic compounds create a unique chemical "fingerprint," which can be genetically related to a specific oil spill source. For example, Stout used this approach, namely, an adaptation of the European oil spill identification standard (Daling et al., 2014), to compare approximately 1600 oily samples collected over multiple years during natural resources damage assessment investigation of the DWH incident (Stout, 2016). These samples included a large variety of weathered oiled

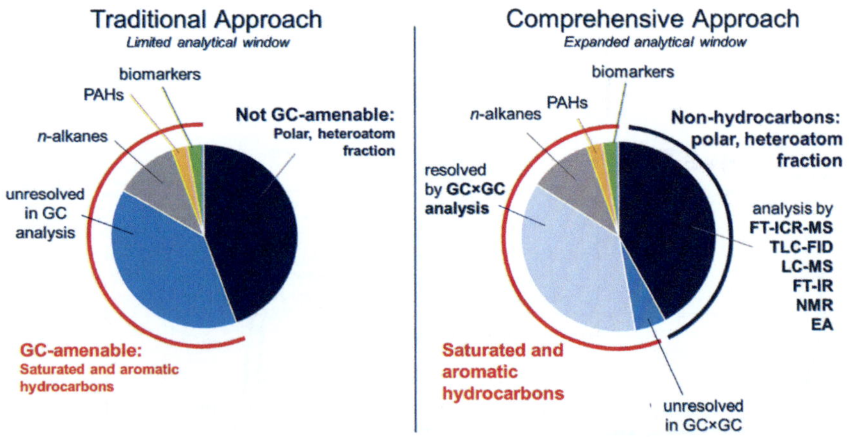

FIGURE 7.2 Comparison of traditional, GC-based approach to chemical oil spill assessments, and new, more comprehensive approach, enabled by complementary analytical tools and methods that expand the analytical window into newly resolved saturated and aromatic hydrocarbons, and non-GC-amenable oil fractions consisting of polar, heteroatom-bearing compounds. *GC*, Gas chromatography.

matrices represented by floating (sheens, slicks, and mousses) and stranded (tar balls, oiled sand, vegetation, and debris) materials collected both during the DHW spill and in the months after the well was successfully capped in July 2010.

In the study, Stout used a three-tiered approach, wherein GC-FID results provided a qualitative assessment of the presence of oil fingerprint. In the second tier, quantitative comparison of 29 diagnostic ratios of PAHs, triterpanes, steranes, and triaromatic steroids derived from GC–MS data was performed. Finally, in the third tier, all the available chemical information was synthesized, and samples were classified into four categories in relation to their similarity to the source oil released from the Macondo well during the DWH blowout. Even in this very complex oil spill assessment scenario, where many concurrent spill sources were present, both from very active offshore oil production and ubiquitous natural seepage in the Gulf region, the GC-based fingerprinting method was able to robustly distinguish Macondo crude oil from other south Louisiana crude oils and make a positive identification to the DWH blowout (Stout, 2016).

Notwithstanding, in certain scenarios the utility of this traditional approach can be significantly limited, for example, in spill cases which involve unconventional oils (e.g., heavy, biodegraded or residual oils), mixtures of different oil sources/types, and/or heavily weathered oil residues. In such cases, volatile and semivolatile oil fractions are either too complex or depleted and degraded, making them less amenable for the GC analysis. In addition, weathered oil and oil spill residues will often be mixed with naturally occurring organic matter present in the water, or coastal and seafloor sediments. These background organic matter species can interfere and "overprint" diagnostic petroleum components, making conventional source assessments more challenging. For example, in the aftermath of DWH, blooms of *Trichodesmium* cyanobacterium often occurred at the same location with the surfaced oil slicks and were a source of interfering *n*-alkanes (*n*-C15, and *n*-C17) that are commonly used in spill fingerprinting to assess the spill source and/or the degree of oil biodegradation (White et al., 2019a).

Conveniently, in the past decade, significant technological advancements of laboratory-based techniques occurred, enabling researchers to overcome some of these challenges (Table 7.1). Developments such as multidimensional chromatography and (ultra)high-resolution mass spectrometry afforded an increased resolving power, broader dynamic range, lower detection limits, and better selectivity of analytical methods. These new technologies have been combined with traditional tools in a complementary way and expanded the analytical window of oil (spill) analysis, allowing detailed molecular characterization of thousands of compounds that were previously part of petroleum "black matter," such as the operationally defined resins and asphaltenes, containing polar, nonhydrocarbon oil fractions (Overton et al., 2016) (Fig. 7.2).

TABLE 7.1 Overview of laboratory-based analytical platforms used to identify and characterize oil spills..

Instrument*	Analytes	Selectivity	Certainty	Sensitivity	Speed	Usability	Availability
GC-FID	C8–C40 hydrocarbons	High	High	High	Weeks	Specialized	Wide
GC–MS	C8–C40 hydrocarbons	High	High	High	Weeks	Specialized	Wide
GC × GC-FID	C8–C40 hydrocarbons	Highest	Highest	High	Months	Specialized	Limited
GC-HTSD	C5–C120 hydrocarbon homologs separated by boiling point	Average	High	High	Days	Specialized	Wide
GC × GC–MS (TOF and HRT)	C8–C40 hydrocarbons	Highest	Highest	High	Months	Specialized	Limited
FT-IR	Bulk oil	Average	Average	<Average	Days	Specialized	Wide
TLC-FID	Fractions of oil separated by polarity	Average	Highest	Average	Days	Nonexpert	Wide
FT-ICR-MS	Hydrocarbons, oxidized hydrocarbons, and polar (NSO compounds)	Highest	High	Highest	Months	Specialized	Limited
NMR	Bulk oil or oil fractions	High	High	Highest	Months	Specialized	Limited
GC-IR-MS	Stable carbon and hydrogen isotopic composition of oil compounds	Highest	Highest	Highest	Weeks	Specialized	Limited
AMS	14C composition of oil and oil compounds	Highest	Highest	Highest	Months	Specialized	Limited
Ramped pyrolysis	Fractions of oil separated by thermochemical stability	Average	High	Average	Weeks	Specialized	Limited

**AMS*, Accelerator mass spectrometry; *FID*, flame ionization detection; *FT-ICR-MS*, Fourier-transform ion cyclotron mass spectrometry; *FT-IR*, Fourier-transform infrared spectroscopy; *GC*, gas chromatography; *GC × GC*, comprehensive two-dimensional gas chromatography; *HRT*, high-resolution TOF; *HTSD*, high-temperature simulated distillation; *MS*, mass spectrometry; *NMR*, nuclear magnetic resonance; *TLC*, thin-layer chromatography; *TOF*, time-of-flight.
Source: Modified from White, H. K., Conmy, R. N., Macdonald, I. R. & Reddy, C. M. (2016). Methods of oil detection in response to the Deepwater Horizon oil spill. *Oceanography*, 76–87.

The power of these new methods for improved spill source assessments has been demonstrated in recent studies of DWH and Ixtoc-1 blowouts in the GoM. Most notably, Aeppli and colleagues leveraged better chromatographic resolution, mass spectral information, and sensitivity afforded by comprehensive two-dimensional GC (GC × GC) coupled to TOF and FID, to identify DWH oil in slicks and weathered coastal residues (sand patties) washing ashore along the US Gulf Coast, and to improve spill forensic assessments by quantifying and decoupling the effects of microbial degradation and photooxidation on major classes of diagnostic biomarkers, such as homohopanes and triaromatic steranes (Aeppli et al., 2014; Aeppli et al., 2012) (Fig. 7.3).

More recently, Nelson et al. (2019) used the power of new high mass resolution (HRT) detector coupled to GC × GC to further expand the inventory of GC-amenable diagnostic markers in DWH and Ixtoc-1 source oils, by identifying extended series of alkylated PAHs, sulfur and nitrogen heteroaromatic species, and nontypical de-A-sterane biomarkers (Fig. 7.4). Such specific and recalcitrant chemical proxies are crucial to identify spill residues over multiyear or multidecade periods.

For example, in 2016 an international research consortium revisited sites in the southern GoM that were impacted by the Ixtoc-1 blowout, including mangrove forests of Yucatan peninsula, coastal systems of Bay of Campeche, and offshore deepwater sites. Among various findings of this research expedition, researchers were able to identify residues of Ixtoc-1 crude oil in samples from the prop roots of mangrove trees of the western Yucatan Peninsula, by combining sensitive GC–MS/MS-MRM analysis of

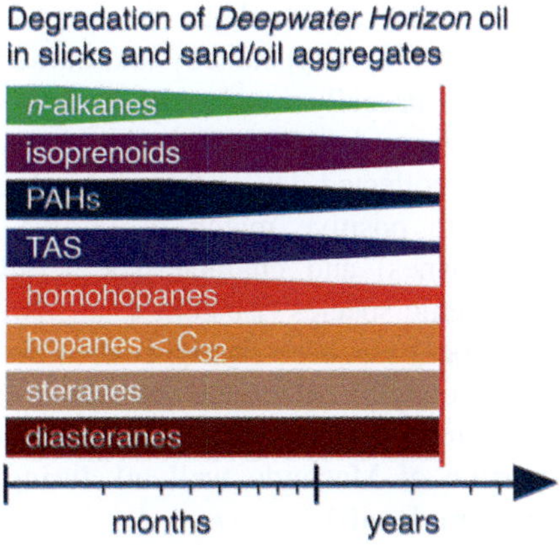

FIGURE 7.3 Assessment of degradation and recalcitrance of major classes of petroleum compounds used in oil spill forensics, based on the data from DWH oils slicks and coastal residues. *DWH*, Deepwater horizon. *From Aeppli, C., Nelson, R. K., Radović, J. R., Carmichael, C. A., Valentine, D. L., & Reddy, C. M. (2014). Recalcitrance and degradation of petroleum biomarkers upon abiotic and biotic natural weathering of Deepwater Horizon oil.* Environmental Science and Technology, 48, *6726–6734.*

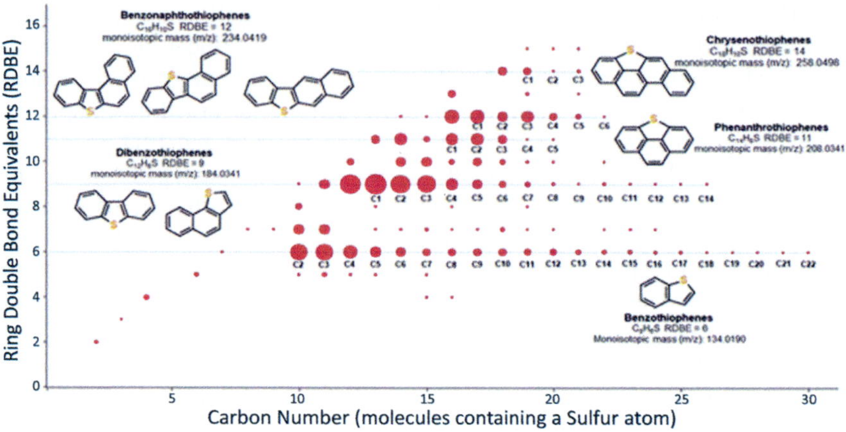

FIGURE 7.4 Plot of the RDBE (rings or double bonds) against carbon number for compounds containing sulfur heteroatoms (sulphur-bearing polycyclic aromatic hydrocarbons, PASHs) detected in Ixtoc-1 crude oil, produced using a petroleomics application for LECO's ChromaTOF software tailored for HRT multidimensional GC × GC data. In this case, PASHs were identified for compounds with RDBE values of 6, 9, 11, 12, and 14 (corresponding to benzothiophenes, dibenzothiophenes, phenanthrothiophenes, benzonaphthothiophenes, and chrysenothiophenes). *GC*, Gas chromatography; *HRT*, high mass resolution; *RDBE*, ring double bond equivalent. *From Nelson, R. K., Gosselin, K. M., Hollander, D. J., Murawski, S. A., Gracia, A., Reddy, C. M., et al. (2019). Exploring the complexity of two iconic crude oil spills in the Gulf of Mexico (Ixtoc I and Deepwater Horizon) using comprehensive two-dimensional gas chromatography (GC × GC).* Energy & Fuels, 33, *3925–3933.*

recalcitrant biomarkers, and nontarget screening with ultrahigh-resolution FT-ICR-MS instrument which revealed the presence of recalcitrant, high-molecular weight (HMW) oxygenated, and/or sulfur-bearing species, indicating a preservation of partially weathered oil in these low energy environments for more than three decades (Radović et al., 2020b). Similarly, deepwater sediment cores collected during this research campaign, contained stratigraphic horizons with positive match to Ixtoc-1 oil, based on GC-amenable proxies (triterpanes) and characteristic FT-ICR-MS fingerprints enriched in HMW petrogenic species (Lincoln et al., 2020). The results suggest a possibility that Ixtoc-1 has been transported to the deep sea by a sedimentation mechanism known as MOSSFA (marine-oil-snow sedimentation and flocculent accumulation), which was also responsible to benthic deposition of significant portion of Macondo well oil during the DWH incident (see Section 7.3.2 for more details).

7.3 Assessment of environmental fate

Weathering is a broad term that encompasses various abiotic and biotic processes occurring in all marine compartments, including surface waters, water column, bottom sediment, and shoreline (Fig. 7.1) (Farrington et al., 2016). These processes deplete and/or transform oil compounds, often leaving residues that can be more recalcitrant than the parent species. From the mass balance perspective most relevant processes that ultimately remove oil from marine system are evaporation, that is, volatilization to the atmosphere, photolysis, and microbial degradation. Certain spill response measures such as in situ burning (ISB) and dispersant application are designed to enhance these natural routes of oil removal, namely, volatilization and (bio)degradation.

Oil on the surface can sink to the seafloor due to agglomeration with mineral and biological particles, that is, phytoplankton (see Section 7.3.2), or densification through the loss of volatile fractions, for example, after ISB. In the case of deepwater submarine oil releases, such as DWH, some portion of the oil may never experience surface weathering processes but will instead be partitioned to deep, near seabed waters and sediment.

Surfaced slicks and floating emulsified oil can be transported for hundreds of kilometers and reach shorelines. There, the impacts and ultimate fate of oil will depend on the shore type, for example, in intertidal wetlands such as saltmarshes and mangroves, organic-rich sediment, and vegetation cover will favor interaction with oil and its retention and long-term preservation (Radović et al., 2020b; Reddy et al., 2002), while more dynamic coastal environments would promote oil removal (Jahns et al., 1991).

Disentangling the complexity of weathering and other postspill processes and interactions is a cumbersome task. First, transformation of oil components changes their properties in a way that makes them less amenable to GC-based analyses—as the weathering progresses, GC chromatograms become less useful, dominated by a characteristic unresolved complex mixture "hump" (Farrington and Quinn, 2015). In addition to weathering changes, spill residues mix with other petrogenic and biogenic signals in the environment, which makes fate assessment even more challenging. As mentioned in the previous section, new analytical approaches and tools helped resolve some of the issues related to studying complex, environmentally altered oil residues, contributing to better understanding of postspill processes and fate.

7.3.1 Re-evaluation of postspill weathering processes

Weathering of oil in the environment is a complex interplay of physiochemical and biological processes, occurring over wide temporal scale, from hours to years, or even decades (Fig. 7.5). Traditionally, it has been considered that immediately after a spill, evaporation acts as a principal spill weathering process, which rapidly removes a large mass fraction of volatile oil compounds to the atmosphere (National Research Council, 2003). Following evaporation, microbial degradation has been considered as an important secondary removal process, occurring over a wider temporal scale, from

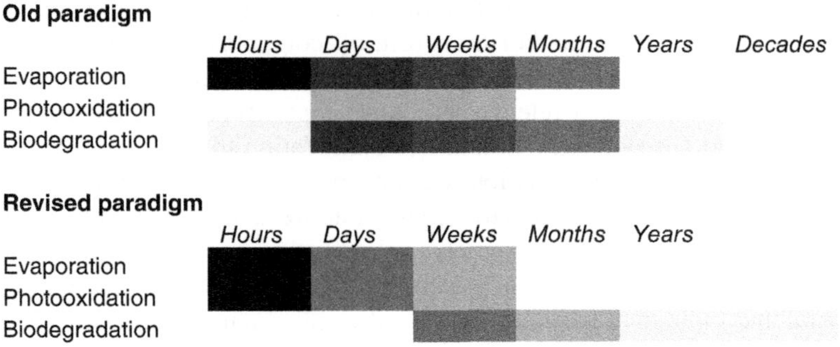

FIGURE 7.5 Relative importance of different postspill weathering processes has been reevaluated in the aftermath of the DWH, by recognizing the significance of photooxidation processes in the initial weathering stages. The darker shades indicate higher importance of a specific weathering process (Ward et al., 2018b). *DWH*, Deepwater horizon. *Based on National Research Council. (2003). Oil in the sea III: Inputs, fates, and effects. National Academies Press; Ward, C. P., & Overton, E. B. (2020). How the 2010 Deepwater Horizon spill reshaped our understanding of crude oil photochemical weathering at sea: A past, present, and future perspective.* Environmental Science: Processes & Impacts, 22, *1125–1138; White, H. K., et al. (2018b). Partial photochemical oxidation was a dominant fate of Deepwater Horizon surface oil.* Environmental Science and Technology, 52, *1797–1805.*

hours/days up to year(s), depending on the biodegradability of specific oil spill components. On the other hand, light-driven processes, for example, photooxidation, have been typically given a low, to negligible importance in the conventional weathering models and viewed only as a minor oil transformation/removal pathway, compared to evaporation and biodegradation (National Research Council, 2003) (Fig. 7.5).

However, in recent years, particularly after the DWH, numerous studies, both in the laboratory and in the field, challenged the previous conceptions related to the importance of photooxidation. These investigations demonstrated that surface-weathered oil residues exhibited a rapid depletion of GC-amenable fractions, coupled to concomitant enrichment of oil fractions eluting with more polar solvents, as would be expected of so-called resin and asphaltene fractions, as conventionally defined in petroleum geochemistry (Aeppli et al., 2012). Distinctly from resins and asphaltenes, the polar fractions of weathered oils had some unique properties, most notably high oxygen content, with enriched isotopic signature characteristic suggesting oxygen sourcing from the molecular oxygen (O_2) and not water molecules (Ward et al., 2019). In addition, by combining some conventional analytical tools (FT-IR), and new high-resolution platforms, such as GC × GC and FT-ICR-MS, it has been demonstrated that the oxygen was preferentially introduced as hydroxyl and carbonyl functionality (e.g., ketones), which increased the complexity of spilled oil by at least twofold, via transformation of parent hydrocarbon species (Aeppli et al., 2012; Ruddy et al., 2014; Radović et al., 2014a) (Fig. 7.6). The

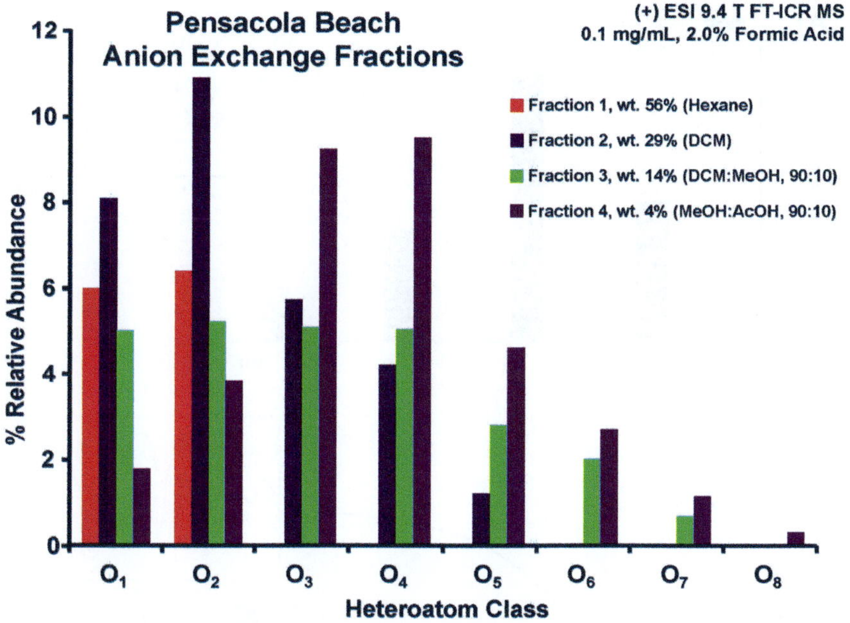

FIGURE 7.6 Heteroatom class distribution of weathered DWH spill showing the increase of highly oxidized species, with more polar character (elute with polar solvents), products of rapid photooxidation of spilled oil. *DWH*, Deepwater horizon. *From Ruddy, B. M., Huettel, M., Kostka, J. E., Lobodin, V. V., Bythell, B. J., Mckenna, A. M., et al. (2014). Targeted petroleomics: Analytical investigation of macondo well oil oxidation products from Pensacola Beach.* Energy & Fuels, 28, *4043–4050.*

magnitude and rate of observed photooxidative changes were comparable to evaporation, it is estimated that roughly 50% of hydrocarbons in surfaced oil slicks were oxidized within a week (Ward and Overton, 2020).

Interestingly, DHW spill is not the only one where such compositional changes caused by light have been observed. For example, laboratory and field studies of weathered heavy fuel oil released during Prestige tanker incident (2003) show analogous production/enrichment of polar, oxygen-bearing species (Radović et al., 2014a; Díez et al., 2007; Fernández-Varela et al., 2006) (Fig. 7.7). The extent of photooxidation after Prestige spill was never robustly modeled and quantified as was the case post-DWH; notwithstanding, the observation of same photochemical changes suggests the importance of photooxidation beyond a specific spill scenario and oil type, warranting further research.

The results of the above-discussed studies also revealed an important aspect of photooxidation pathways during surface spill weathering; in addition to likely direct oxidation of photo-absorbing aromatic species (e.g., PAHs), an important indirect photooxidation mechanism was at play wherein a chromophore, like a PAH or asphaltene, photochemically generates

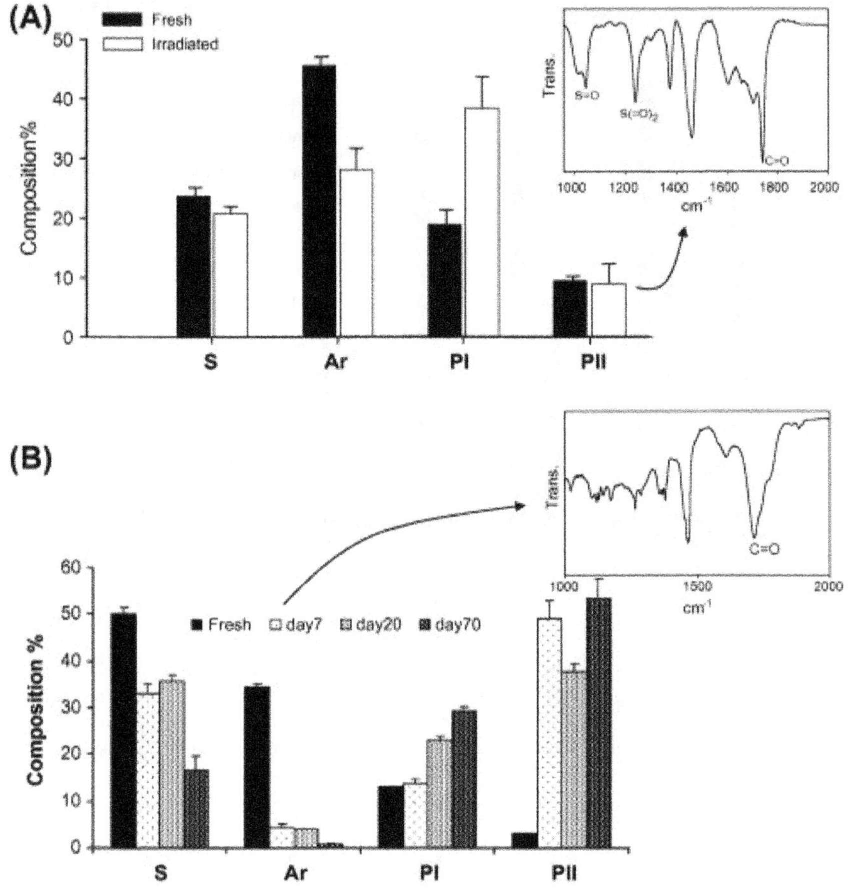

FIGURE 7.7 Photooxidation-driven compositional changes of oil released during the Prestige tanker spill (A), and DWH blowout (B). TLC-FID analysis reveals the enrichment of polar oil fractions (PI and PII), likely oxidation products of parent saturate (S) and aromatic (A) fractions, as indicated by the occurrence of oxygen-bearing functionalities ($C = O$, $S = (O)x$) in the FT-IR spectra. *DWH*, Deepwater horizon; *FID*, flame ionization detector; *FT-IR*, Fourier-transform infrared spectroscopy; *TLC*, thin-layer chromatography. *From Radović, J. R., Aeppli, C., Nelson, R. K., Jimenez, N., Reddy, C. M., Bayona, J. M., et al. (2014a). Assessment of photochemical processes in marine oil spill fingerprinting.* Marine Pollution Bulletin, 79, *268–277.*

reactive oxygen species, like singlet oxygen, peroxy radicals, and hydroxyl radical which end up indirectly transforming oil species/fractions that cannot directly absorb the sunlight, such as saturated hydrocarbons (Hall et al., 2013; Ward and Overton, 2020) (Fig. 7.7).

Introduction of oxygen functionalities into parent hydrocarbons changes their environmental behavior, impacts, and fate. Increased polarity of oxidized products favors their partitioning to the aqueous phase, and possibly bioavailability (Vaughan et al., 2016; Liu and Kujawinski, 2015). Recent experimental observations indicate that similar effects on aqueous distribution of petroleum components can be caused by some of the typical response measures such as dispersant application and ISB (Jaggi et al., 2019; Jaggi et al., 2020). Once in the aqueous phase, it is possible that oxidized products of oil transformation can be more amenable to biodegradation (Harriman et al., 2017; Liu et al., 2020). On the other hand, similar, highly oxidized species have been detected in oily matrices by FT-ICR-MS, for example, in decades-old coastal residues of historic Ixtoc-1 spill, in settings that are less conducive to fast biodegradation, that is, marshy, hypoxic coastal environments, and deepwater sediments (Radović et al., 2020b; Lincoln et al., 2020). Disentangling and quantification of various environmental pathways and ultimate sinks of these newly described oxygen-rich materials will remain a major research focus in the coming years.

From the response perspective, most relevant implication of rapid and extensive photooxidation is the change of oil interfacial properties from hydrophobic to amphiphilic, due to the introduction of polar, oxygen-bearing functionalities into parent hydrocarbon molecules, which can be detrimental to dispersant efficiency, because it promotes oil emulsification (Ward et al., 2018a).

7.3.2 Novel postspill phenomena—marine-oil-snow sedimentation and flocculent accumulation

Some estimates of the DWH spill budget indicate that as much as 23% of spilled oil has not been accounted for, that is, it was not directly recovered, burned, dispersed, evaporated, nor dissolved (Mcnutt et al., 2012). Research conducted in the years following the spill revealed that up to 14% of the spilled oil might have been affected by a process that became known as MOSSFA, which transported the oil to the seafloor (Burd et al., 2020; Daly et al., 2016; Valentine et al., 2014) (Fig. 7.8). MOSSFA was caused primarily by physical aggregation of oil droplets with naturally occurring marine particulate matter, such as marine snow, phytoplankton, microbes, mineral particles, and organic detritus. In addition, biological processes have also contributed to marine-oil-snow (MOS) formation, for example, by ingestion of oil particles by zooplankton and subsequent excretion via fecal pellets, or the production of exopolymeric substances and biofilms by planktonic marine bacteria, which enhanced oil-particulate aggregation (White et al., 2019b; Burd et al., 2020). Physicochemical changes of oil on the surface, such as evaporation and photooxidation, as well as application of dispersants have been considered as variables that could possibly play a role in MOS formation; notwithstanding, comprehensive understanding of their effects is still a subject of ongoing research (Dutta and Harayama, 2000; Passow and Ziervogel, 2016; Daly et al., 2016; Burd et al., 2020; Santschi, 2018; Ward et al., 2018a; Ward and Overton, 2020). It must be noted that interactions of particulates and oil have

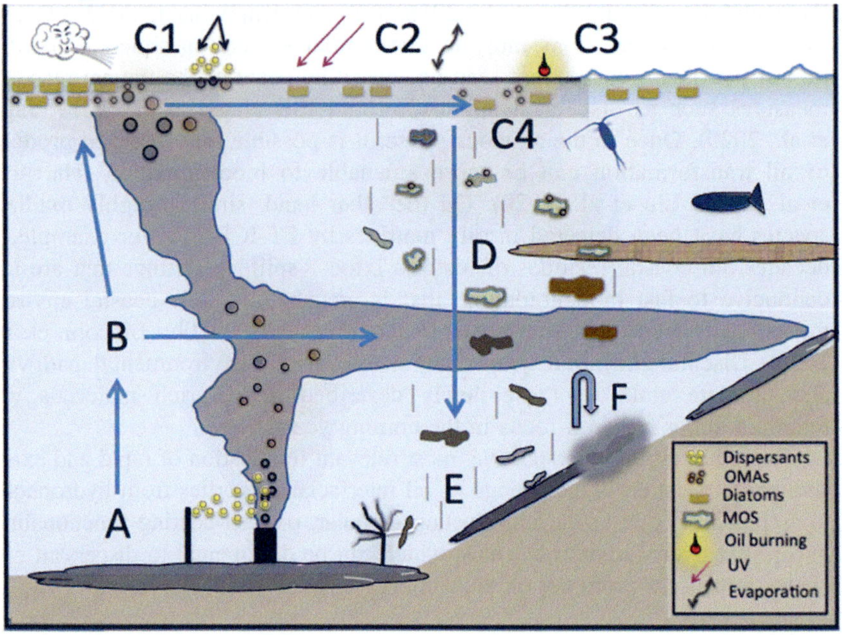

FIGURE 7.8 Conceptual diagram of MOSSFA related processes from the source of oil discharge to the fate of hydrocarbons in sediments. Point (A) shows the release of oil at the wellhead and application of dispersants and (B) represents rising oil droplets and gas bubbles and the formation of a deep oil plume. Points (C1–C4) shows surface processes influencing the formation of MOS: (C1) illustrates wind impacts, a diatom bloom, and application of surface dispersants; (C2) shows oil transformation due to UV light and evaporation; (C3) depicts the role of aerosols and oil burning in creating new material sources; and (C4) shows processes impacting sinking MOS particles in surface waters and as particles sink through (D) a benthic nepheloid layer and deep oil plumes. Point (E) shows benthic sedimentation of MOS and flocculation onto corals, and (F) represents resuspension of oiled sediments due to turbulence. See the text for a detailed explanation of the figure. *MOSSFA*, Marine-oil-snow sedimentation and flocculent accumulation. *From Daly, K. L., Passow, U., Chanton, J. & Hollander, D. (2016). Assessing the impacts of oil-associated marine snow formation and sedimentation during and after the Deepwater Horizon oil spill.* Anthropocene, 13, *18–33.*

been recognized as an important postspill process for many years (Lee, 2002), albeit much of the historic research has been focused on mineral particles such as clays. However, DWH spill provided a unique "natural laboratory" to study oil-particle aggregation and sedimentation on large scale, through a concerted and well-funded (Gulf of Mexico Research Initiative, GoMRI) effort of many research groups, which resulted in major breakthroughs in our understanding and assessment of this complex postspill phenomenon.

Here, again, studies using novel chemical assessment tools and approaches made significant contributions; for example, FT-ICR-MS and NMR were successfully used to characterize oil-marine snow associations in mesocosm experiments

(Hatcher et al., 2018; Radović et al., 2020a). The power of high-resolution mass spectrometry was leveraged to confirm the presence of recalcitrant oil residues in sediment cores collected at 500 m depth in the southern GoM (Lincoln et al., 2020; Schwing et al., 2020), which, in addition to other spill markers, suggests that MOSSFA-like process could have been responsible for extensive sedimentation (up to 25% of oil) that was reported in the aftermath of the 1979 Ixtoc-1 spill (Jernelöv and Lindén, 1981). These findings warrant a long-term monitoring of possible chronic impacts of MOSSFA events on benthic environments, and the assessment of possibility of oil resuspension and redistribution. The likelihood of MOSSFA occurrences in other spill scenarios and in different marine environments, such as northern-latitude seas, also needs to be evaluated (Vonk et al., 2015; Suja et al., 2017).

Finally, several laboratory studies showed that formation of MOS can slow down the biodegradation processes and in concert with other limitations of key biogeochemical gradients (oxygen, nutrients, etc.) can potentially lead to long-term preservation of oil in seafloor sediments (Kostka et al., 2020; Langenhoff et al., 2020; Van Eenennaam et al., 2018; Van Eenennaam et al., 2019) (Fig. 7.9).

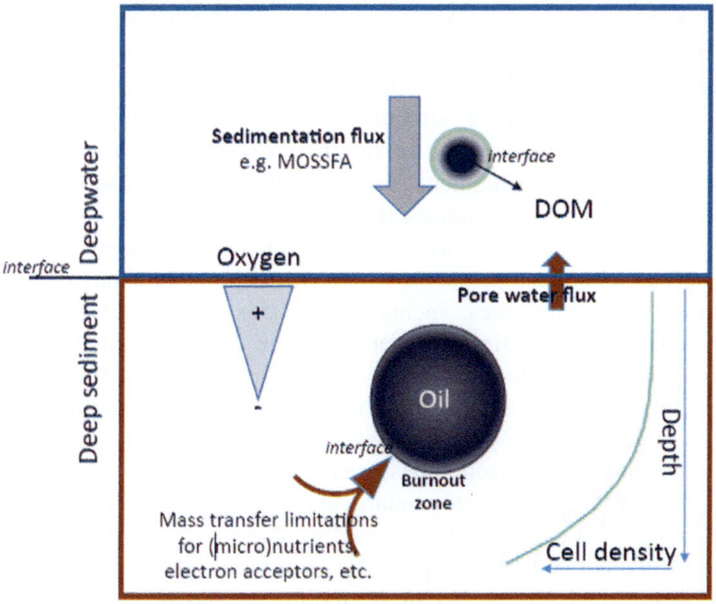

FIGURE 7.9 Conceptual model for oil biodegradation in the deep sea. Depending on the amount of deposited oil, due to physical mass transfer limitations of many key variables such as oxygen, electron acceptors, and others, occurring at interfaces, there is a potential for long-term preservation of oil in the sediment, even when hydrocarbon-degrading microorganisms are present—so-called burnout effect. *From Kostka, J. E., Joye, S. B., Overholt, W., Bubenheim, P., Hackbusch, S., Larter, S. R., et al. (2020). Biodegradation of petroleum hydrocarbons in the deep sea. In:* Deep oil spills. *Springer.*

7.4 Assessments of oil spills impacts

Chemical complexity of oil and its transformations in the environment poses a significant challenge for effective and comprehensive assessment of spill impacts. For example, typical toxicity evaluations are based on a-priori knowledge of types and amounts of chemical substances that are released to the environment. As it was demonstrated in previous sections, in the case of oil spills, such targeted approach is severely limited because of the wide range of poorly characterized chemical species that are present in the spilled oil, and are produced by various weathering processes after its release to the marine environment.

In this case, a nontargeted, exploratory strategy might be more appropriate, which attempts to assess and identify compounds and compounds classes that represent the greatest risk to the environment. Effect-directed analysis (EDA) is one of such approaches, which is based on a broad range analysis that combines chemical fractionation and characterization with response quantification of target effects of interest, typically using bioassays (Brack, 2003). Ideally, these bioassays should have specific endpoints related to contaminant interaction with receptors that mediate toxic effects, for example, binding with aryl-hydrogen (AhR) as a proxy for carcinogenicity, or interaction with androgen and estrogen receptors as an indication of a potential to cause endocrine disruption (Thomas et al., 2009; Radović et al., 2014b). More recently, quantitative structure–activity relationship computational models have emerged as a tool for in silico assessment of receptor binding, which can substitute or augment bioassays, and thus increase the scope and throughput of toxicity assessments (Radović et al., 2014b).

First step in the EDA workflow is chemical fractionation, which combines various laboratory protocols to isolate target compounds over a range of compositional properties, including methods of solid-phase extraction, pressurized liquid extraction, (semi)preparative gas and liquid chromatography, gel permeation chromatography, and many others. Isolated chemical fractions are then analyzed by appropriate instruments, typically by gas and liquid chromatography coupled to mass spectrometry, or more recently, with novel, multidimensional techniques, such as GC × GC-ToF, in an attempt to further expand the window of nontargeted screening (Radović et al., 2012). As an example, GC × GC-ToF was used to characterize fresh and artificially weathered (evaporated, photooxidized) samples of North Sea crude oil and residual heavy fuel oil. In this study, multidimensional output of GC × GC platform (first and second dimension retention times) has been leveraged to build an N-way partial least-squares chemometric model that correlated the chemical composition of different oil subfractions to AhR activity assessed via bioassays (Fig. 7.10). In this nontargeted way, alkyl-substituted three and four-ring aromatic systems, including nitrogen and sulfur heteroaromatics, have been identified in the active fractions by the intensity-weighting of their

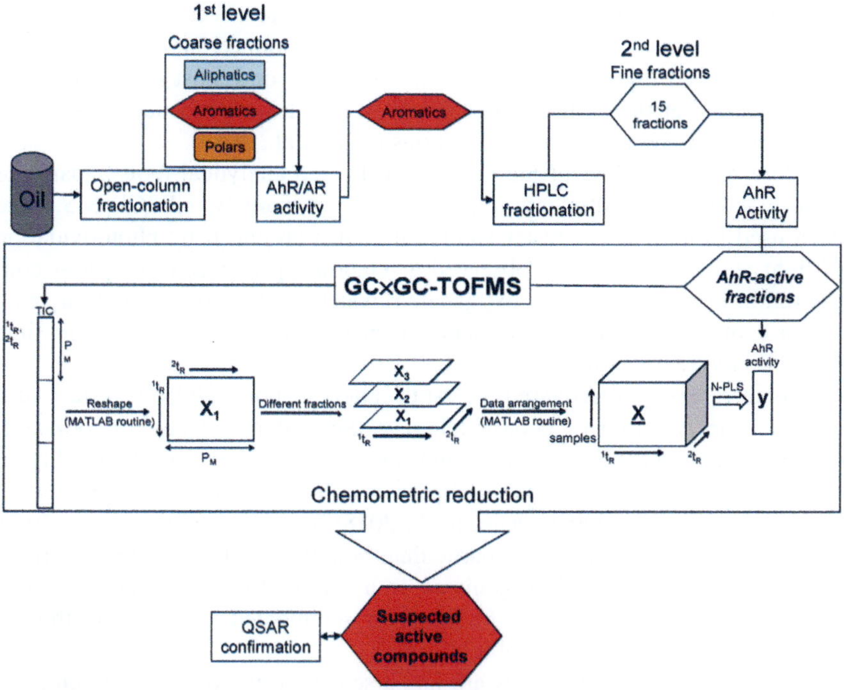

FIGURE 7.10 Flowchart of EDA that combined chemical fractionation of oil, GC × GC analyses, and bioassays with chemometric modeling to identify oil components with toxic potential. *EDA*, Effect-directed analysis; *GC*, gas chromatography. *From Radović, J. R., Thomas, K. V., Parastar, H., Díez, S., Tauler, R., & Bayona, J. M. (2014b). Chemometrics-assisted effect-directed analysis of crude and refined oil using comprehensive two-dimensional gas chromatography–time-of-flight mass spectrometry.* Environmental Science and Technology, 48, *3074–3083.*

contributions to the observed toxic effects. This type of approach is relevant because typical toxicity assessment protocols in postspill monitoring focus on a suite of nonsubstituted PAHs (Allan et al., 2012), while more complex, alkylated aromatics are not considered, as they cannot be easily resolved and measured using conventional, one-dimensional chromatography. However, alkylated species often exhibit more pronounced toxic effects than their unsubstituted counterparts (Muusse et al., 2012; Vrabie et al., 2012).

In addition to complex aromatic species, a major open research need in oil spill impact assessments is the newly identified oxidation products (Section 7.3.1). As discussed previously, the introduction of polar, oxygen-bearing functional groups into parent hydrocarbons changes their physico-chemical properties in a way that can increase environmental mobility (e.g., solubility) of otherwise hydrophobic oil components, which in turn can enhance the bioavailability of spilled material (Liu and Kujawinski, 2015; Vaughan et al., 2016).

Earlier toxicity studies attempting to evaluate the effect of spill weathering on oil toxicity investigated fractions of evaporated and photooxidized oils and showed that the polar fraction was the second most important toxic component, after the aromatic, despite its overall low abundance in comparison to the saturate and aromatic fractions (Rial et al., 2013).

More recently, some of the new, sophisticated analytical tools capable of molecular-level characterization of polar species, namely, FT-ICR-MS, have been used to look at the water-soluble organics produced by photooxidation of light and heavy oils and observed a relationship between the initial concentration increase of higher oxygenated, water-soluble species and an acute increase of toxicity, which gradually decreased after longer exposure time ($>$24 hours) (Zito et al., 2019).

Another newly discovered postspill phenomenon, MOSSFA, has also attracted the attention of ecotoxicologists. Several recent studies looked at the effects of MOSSFA in micro and mesocosm studies revealing a higher toxicity of MOS aggregates, in comparison to only oil treatments, in certain benthic species (Van Eenennaam et al., 2018; Van Eenennaam et al., 2019). Their authors hypothesize that particulate-bound oil could possibly be more available to certain benthic organisms, in a way that could initiate the introduction and bio-accumulative transport of sedimented oil from benthos to higher elements of marine trophic chains.

The above-described studies are just a selection of what is a nascent, and rapidly growing field in oil spill science, attempting to push the boundaries of risk and impact assessments to previously overlooked chemical components of fresh and/or environmentally altered oil (Fingas, 2017).

7.5 Conclusions and recommendations

Sections 7.2 and 7.3 highlighted the complexity involved in understanding chemical composition of oil spills, both due to the plethora of crude oil types, petroleum products, and mixtures thereof, and/or caused by postspill biotic and abiotic transformations. New instrumental platforms and workflows that combine novel technologies with more conventional oil spill assessment tools, such as one-dimensional gas chromatography, have proven their analytical power and value-add for better understanding of poorly resolved oil fractions. These include complex alkylated species, and polar, heteroatom-rich and/or HMW compounds. However, there is still much work to be done to improve these new assessment protocols, by standardizing them across different laboratories and instruments and by increasing their throughput via improved sample preparation protocols (e.g., online sample cleanup and preconcentration) and/or chemometric approaches to quickly process and interpret big datasets that are outputted by high-resolution and/or multidimensional instruments. Next, renewed appreciation of postspill processes such as photooxidation and oil-particulate interactions (MOSSFA) is just the first step to full understanding of coupling of abiotic and biotic spill degradation/transformation processes and their influence on environmental (re)distribution of petroleum compounds, ultimate fate and bioavailability, and toxicity to marine organisms. In regard to the latter, Section 7.4 showcased some initial results of in vitro and mesocosm toxicity studies that are only a glimpse into important new areas of future research to better quantify observed impacts and extrapolate those to ecosystem scales.

Finally, all of these new research avenues have important implications for the improvement of response strategies such as ISB and dispersant application, as well as better preparedness for future large spills in challenging environments, for example ultra-deep waters or northern-latitude seas.

Acknowledgments

JRR acknowledges generous support to oil spill research in which he participated during the past 10 years, including project grants from the Spanish Ministry for Science and Innovation (MCIN) (projects Ref. CTM2008−02718-E/MAR and CTM2008−02721-E/MAR), and the Gulf of Mexico Research Initiative, GoMRI (projects Deep-C and C-IMAGE), as well as his fellowship from the Spanish National Research Council (CSIC) and the European Social Fund (ESF). Christopher M. Reddy is acknowledged for his contribution to Fig. 7.2. The author also thanks the book editor Oleg Makarynskyy for the kind invitation and helpful editorial revisions.

References

Aeppli, C., Carmichael, C. A., Nelson, R. K., Lemkau, K. L., Graham, W. M., Redmond, M. C., et al. (2012). Oil weathering after the Deepwater Horizon disaster led to the formation of oxygenated residues. *Environmental Science and Technology*, *46*, 8799−8807.

Aeppli, C., Nelson, R. K., Radović, J. R., Carmichael, C. A., Valentine, D. L., & Reddy, C. M. (2014). Recalcitrance and degradation of petroleum biomarkers upon abiotic and biotic natural weathering of Deepwater Horizon oil. *Environmental Science and Technology*, *48*, 6726−6734.

Allan, S. E., Smith, B. W., & Anderson, K. A. (2012). Impact of the Deepwater Horizon oil spill on bioavailable polycyclic aromatic hydrocarbons in Gulf of Mexico coastal waters. *Environmental Science and Technology*, *46*, 2033−2039.

Boehmer-Christiansen, S. (2008). Estimates of oil entering the marine environment from sea-based activities, reports and studies no. 75. *Energy & Environment*, *19*, 765.

Brack, W. (2003). Effect-directed analysis: A promising tool for the identification of organic toxicants in complex mixtures? *Analytical and Bioanalytical Chemistry*, *377*, 397−407.

Burd, A. B., Chanton, J. P., Daly, K. L., Gilbert, S., Passow, U., & Quigg, A. (2020). The science behind marine-oil snow and MOSSFA: Past, present, and future. *Progress in Oceanography*, *187*, 102398.

Daling, P., Faksness, L., Hansen, A., Kienhuis, P., & Duus, R. (2014). Improved methodology for oil spill identification of waterborne petroleum and petroleum products. In: *Proceedings of 4th international conference on oils and environment.*

Daly, K. L., Passow, U., Chanton, J., & Hollander, D. (2016). Assessing the impacts of oil-associated marine snow formation and sedimentation during and after the Deepwater Horizon oil spill. *Anthropocene*, *13*, 18–33.

Díez, S., Jover, E., Bayona, J. M., & Albaigés, J. (2007). Prestige oil spill. III. Fate of a heavy oil in the marine environment. *Environmental Science and Technology*, *41*, 3075–3082.

Dutta, T. K., & Harayama, S. (2000). Fate of crude oil by the combination of photooxidation and biodegradation. *Environmental Science and Technology*, *34*, 1500–1505.

Farrington, J. W., Burns, K. A., & Leinen, M. S. (2016). Synthesis and crosscutting topics. *Oceanography*, *29*, 204–213.

Farrington, J. W., & Quinn, J. G. (2015). "Unresolved complex mixture" (UCM): A brief history of the term and moving beyond it. *Marine Pollution Bulletin*, *96*, 29–31.

Fernández-Varela, R., Gómez-Carracedo, M., Fresco-Rivera, P., Andrade, J., Muniategui, S., & Prada, D. (2006). Monitoring photooxidation of the Prestige's oil spill by attenuated total reflectance infrared spectroscopy. *Talanta*, *69*, 409–417.

Fingas, M. (2017). Polar compounds in oils and their aquatic toxicity. *International Oil Spill Conference Proceedings*, *2017*, 2017036.

Ghosh, S., & Rubly, C. (2015). The emergence of Arctic shipping: Issues, threats, costs, and risk-mitigating strategies of the Polar Code. *Australian Journal of Maritime & Ocean Affairs*, *7*, 171–182.

Hall, G. J., Frysinger, G. S., Aeppli, C., Carmichael, C. A., Gros, J., Lemkau, K. L., et al. (2013). Oxygenated weathering products of Deepwater Horizon oil come from surprising precursors. *Marine Pollution Bulletin*, *75*, 140–149.

Harriman, B. H., Zito, P., Podgorski, D. C., Tarr, M. A., & Suflita, J. M. (2017). Impact of photo-oxidation and biodegradation on the fate of oil spilled during the Deepwater Horizon incident: Advanced stages of weathering. *Environmental Science and Technology*, *51*, 7412–7421.

Hatcher, P. G., Obeid, W., Wozniak, A. S., Xu, C., Zhang, S., Santschi, P. H., et al. (2018). Identifying oil/marine snow associations in mesocosm simulations of the Deepwater Horizon oil spill event using solid-state 13C NMR spectroscopy. *Marine Pollution Bulletin*, *126*, 159–165.

ITOPF. (2020). *Oil tanker spill statistics 2019*. ITOPF.

Jaggi, A., Radović, J. R., Snowdon, L. R., Larter, S. R., & Oldenburg, T. B. (2019). Composition of the dissolved organic matter produced during in situ burning of spilled oil. *Organic Geochemistry*, *138*, 103926.

Jaggi, A., Snowdon, R. W., Radović, J. R., Stopford, A., Oldenburg, T. B., & Larter, S. R. (2020). *Partitioning of organics between oil and water phases with and without the application of dispersants. Deep oil spills*. Springer.

Jahns, H. O., Bragg, J. R., Dash, L. C., & Owens, E. H. (1991). Natural cleaning of shorelines following the Exxon Valdez spill. *International Oil Spill Conference Proceedings*, *1991*, 167–176.

Jernelöv, A., & Lindén, O. (1981). Ixtoc I: A case study of the world's largest oil spill. *Ambio*, 299–306.

Kostka, J. E., Joye, S. B., Overholt, W., Bubenheim, P., Hackbusch, S., Larter, S. R., et al. (2020). *Biodegradation of petroleum hydrocarbons in the deep sea. Deep oil spills*. Springer.

Kujawinski, E. B., Reddy, C. M., Rodgers, R. P., Thrash, J. C., Valentine, D. L., & White, H. K. (2020). The first decade of scientific insights from the Deepwater Horizon oil release. *Nature Reviews Earth & Environment*, *1*, 237–250.

Langenhoff, A. A. M., Rahsepar, S., Van Eenennaam, J. S., Radović, J. R., Oldenburg, T. B. P., Foekema, E., et al. (2020). Effect of marine snow on microbial oil degradation. In

S. A. Murawski, C. H. Ainsworth, S. Gilbert, D. J. Hollander, C. B. Paris, M. Schlüter, & D. L. Wetzel (Eds.), *Deep oil spills: Facts, fate, and effects*. Cham: Springer International Publishing.

Lee, K. (2002). Oil–particle interactions in aquatic environments: influence on the transport, fate, effect and remediation of oil spills. *Spill Science & Technology Bulletin*, *8*, 3–8.

Lincoln, S. A., Radović, J. R., Gracia, A., Jaggi, A., Oldenburg, T. B. P., Larter, S. R., et al. (2020). Molecular legacy of the 1979 Ixtoc 1 oil spill in deep-sea sediments of the Southern Gulf of Mexico. In S. A. Murawski, C. H. Ainsworth, S. Gilbert, D. J. Hollander, C. B. Paris, M. Schlüter, & D. L. Wetzel (Eds.), *Deep oil spills: Facts, fate, and effects*. Cham: Springer International Publishing.

Liu, Y., & Kujawinski, E. B. (2015). Chemical composition and potential environmental impacts of water-soluble polar crude oil components inferred from ESI FT-ICR MS. *PLoS One*, *10*, e0136376.

Liu, Y., White, H. K., Simister, R. L., Waite, D., Lyons, S. L., & Kujawinski, E. B. (2020). Probing the chemical transformation of seawater-soluble crude oil components during microbial oxidation. *ACS Earth and Space Chemistry*, *4*, 690–701.

Mcnutt, M. K., Chu, S., Lubchenco, J., Hunter, T., Dreyfus, G., Murawski, S. A., et al. (2012). Applications of science and engineering to quantify and control the *Deepwater Horizon* oil spill. *Proceedings of the National Academy of Sciences of the United States of America*, *109*, 20222–20228.

Murawski, S. A., Hollander, D. J., Gilbert, S., & Gracia, A. (2020). Deepwater oil and gas production in the Gulf of Mexico and related global trends. In S. A. Murawski, C. H. Ainsworth, S. Gilbert, D. J. Hollander, C. B. Paris, M. Schlüter, & D. L. Wetzel (Eds.), *Scenarios and responses to future deep oil spills: Fighting the next war*. Cham: Springer International Publishing.

Muusse, M., Langford, K., Tollefsen, K. E., Cornelissen, G., Haglund, P., Hylland, K., et al. (2012). Characterization of AhR agonist compounds in roadside snow. *Analytical and Bioanalytical Chemistry*, *403*, 2047–2056.

National Research Council. (2003). *Oil in the sea III: Inputs, fates, and effects*. National Academies Press.

Nelson, R. K., Gosselin, K. M., Hollander, D. J., Murawski, S. A., Gracia, A., Reddy, C. M., et al. (2019). Exploring the complexity of two iconic crude oil spills in the Gulf of Mexico (Ixtoc I and Deepwater Horizon) using comprehensive two-dimensional gas chromatography (GC × GC). *Energy & Fuels*, *33*, 3925–3933.

Overton, E. B., Wade, T. L., Radović, J. R., Meyer, B. M., Miles, M. S., & Larter, S. R. (2016). Chemical composition of Macondo and other crude oils and compositional alterations during oil spills. *Oceanography*, *29*, 50–63.

Passow, U., & Ziervogel, K. (2016). Marine snow sedimented oil released during the Deepwater Horizon spill. *Oceanography*, *29*, 118–125.

Radović, J. R., Aeppli, C., Nelson, R. K., Jimenez, N., Reddy, C. M., Bayona, J. M., et al. (2014a). Assessment of photochemical processes in marine oil spill fingerprinting. *Marine Pollution Bulletin*, *79*, 268–277.

Radović, J. R., Jaggi, A., Silva, R. C., Snowdon, R., Waggoner, D. C., Hatcher, P. G., et al. (2020a). *Applications of FTICR-MS in oil spill studies. Deep oil spills*. Springer.

Radović, J. R., Oldenburg, T. B. P., & Larter, S. R. (2018). Chapter 19—Environmental assessment of spills related to oil exploitation in Canada's oil sands region. In S. A. Stout, & Z. Wang (Eds.), *Oil spill environmental forensics case studies*. Butterworth-Heinemann.

Radović, J. R., Rial, D., Lyons, B. P., Harman, C., Viñas, L., Beiras, R., et al. (2012). Post-incident monitoring to evaluate environmental damage from shipping incidents: Chemical and biological assessments. *Journal of Environmental Management, 109*, 136–153.

Radović, J. R., Romero, I. C., Oldenburg, T. B., Larter, S. R., & Tunnell, J. W. (2020b). *40 Years of weathering of coastal oil residues in the Southern Gulf of Mexico. Deep oil spills.* Springer.

Radović, J. R., Thomas, K. V., Parastar, H., Díez, S., Tauler, R., & Bayona, J. M. (2014b). Chemometrics-assisted effect-directed analysis of crude and refined oil using comprehensive two-dimensional gas chromatography–time-of-flight mass spectrometry. *Environmental Science and Technology, 48*, 3074–3083.

Reddy, C. M., Arey, J. S., Seewald, J. S., Sylva, S. P., Lemkau, K. L., Nelson, R. K., et al. (2012). Composition and fate of gas and oil released to the water column during the *Deepwater Horizon* oil spill. *Proceedings of the National Academy of Sciences of the United States of America, 109*, 20229–20234.

Reddy, C. M., Eglinton, T. I., Hounshell, A., White, H. K., Xu, L., Gaines, R. B., et al. (2002). The West Falmouth oil spill after thirty years: the persistence of petroleum hydrocarbons in marsh sediments. *Environmental Science and Technology, 36*, 4754–4760.

Rial, D., Radović, J. R., Bayona, J. M., Macrae, K., Thomas, K. V., & Beiras, R. (2013). Effects of simulated weathering on the toxicity of selected crude oils and their components to sea urchin embryos. *Journal of Hazardous Materials, 260*, 67–73.

Ruddy, B. M., Huettel, M., Kostka, J. E., Lobodin, V. V., Bythell, B. J., Mckenna, A. M., et al. (2014). Targeted petroleomics: Analytical investigation of Macondo well oil oxidation products from Pensacola Beach. *Energy & Fuels, 28*, 4043–4050.

Santschi, P. H. (2018). Marine colloids, agents of the self-cleansing capacity of aquatic systems: Historical perspective and new discoveries. *Marine Chemistry, 207*, 124–135.

Schwing, P. T., Hollander, D. J., Brooks, G. R., Larson, R. A., Hastings, D. W., Chanton, J. P., et al. (2020). The sedimentary record of MOSSFA events in the Gulf of Mexico: A comparison of the Deepwater Horizon (2010) and Ixtoc 1 (1979) oil spills. In S. A. Murawski, C. H. Ainsworth, S. Gilbert, D. J. Hollander, C. B. Paris, M. Schlüter, & D. L. Wetzel (Eds.), *Deep oil spills: Facts, fate, and effects*. Cham: Springer International Publishing.

Stout, S. A. (2016). Oil spill fingerprinting method for oily matrices used in the Deepwater Horizon NRDA. *Environmental Forensics, 17*, 218–243.

Suja, L. D., Summers, S., & Gutierrez, T. (2017). Role of EPS, dispersant and nutrients on the microbial response and MOS formation in the subarctic northeast Atlantic. *Frontiers in Microbiology, 8*, 676.

Thomas, K. V., Langford, K., Petersen, K., Smith, A. J., & Tollefsen, K. E. (2009). Effect-directed identification of naphthenic acids as important in vitro xeno-estrogens and anti-androgens in North Sea offshore produced water discharges. *Environmental Science and Technology, 43*, 8066–8071.

Valentine, D. L., Fisher, G. B., Bagby, S. C., Nelson, R. K., Reddy, C. M., Sylva, S. P., et al. (2014). Fallout plume of submerged oil from *Deepwater Horizon*. *Proceedings of the National Academy of Sciences of the United States of America, 111*, 15906–15911.

Van Eenennaam., Rahsepar, J. S., Radović, S., Oldenburg, J. R., Wonink, T. B. P., Langenhoff, J., et al. (2018). Marine snow increases the adverse effects of oil on benthic invertebrates. *Marine Pollution Bulletin, 126*, 339–348.

Van Eenennaam., Rohal, J. S., Montagna, M., Radović, P. A., Oldenburg, J. R., Romero, T. B. P., et al. (2019). Ecotoxicological benthic impacts of experimental oil-contaminated marine snow deposition. *Marine Pollution Bulletin, 141*, 164–175.

Vaughan, P. P., Wilson, T., Kamerman, R., Hagy, M. E., Mckenna, A., Chen, H., et al. (2016). Photochemical changes in water accommodated fractions of MC252 and surrogate oil created during solar exposure as determined by FT-ICR MS. *Marine Pollution Bulletin, 104*, 262–268.

Vonk, S. M., Hollander, D. J., & Murk, A. J. (2015). Was the extreme and wide-spread marine oil-snow sedimentation and flocculent accumulation (MOSSFA) event during the Deepwater Horizon blow-out unique? *Marine Pollution Bulletin, 100*, 5–12.

Vrabie, C. M., Sinnige, T. L., Murk, A. J., & Jonker, M. T. O. (2012). Effect-directed assessment of the bioaccumulation potential and chemical nature of Ah receptor agonists in crude and refined oils. *Environmental Science and Technology, 46*, 1572–1580.

Ward, C. P., Armstrong, C. J., Conmy, R. N., French-Mccay, D. P., & Reddy, C. M. (2018a). Photochemical oxidation of oil reduced the effectiveness of aerial dispersants applied in response to the Deepwater Horizon spill. *Environmental Science and Technology Letters, 5*, 226–231.

Ward, C. P., & Overton, E. B. (2020). How the 2010 Deepwater Horizon spill reshaped our understanding of crude oil photochemical weathering at sea: A past, present, and future perspective. *Environmental Science: Processes & Impacts, 22*, 1125–1138.

Ward, C. P., Sharpless, C. M., Valentine, D. L., Aeppli, C., Sutherland, K. M., Wankel, S. D., et al. (2019). Oxygen isotopes (δ18O) trace photochemical hydrocarbon oxidation at the sea surface. *Geophysical Research Letters, 46*, 6745–6754.

Ward, C. P., Sharpless, C. M., Valentine, D. L., French-Mccay, D. P., Aeppli, C., White, H. K., et al. (2018b). Partial photochemical oxidation was a dominant fate of Deepwater Horizon surface oil. *Environmental Science and Technology, 52*, 1797–1805.

White, H. K., Conmy, R. N., Macdonald, I. R., & Reddy, C. M. (2016). Methods of oil detection in response to the Deepwater Horizon oil spill. *Oceanography*, 76–87.

White, H. K., Marx, C. T., Valentine, D. L., Sharpless, C., Aeppli, C., Gosselin, K. M., et al. (2019a). Examining inputs of biogenic and oil-derived hydrocarbons in surface waters following the Deepwater Horizon oil spill. *ACS Earth Space Chemistry, 3*, 1329–1337.

White, H. K., Marx, C. T., Valentine, D. L., Sharpless, C., Aeppli, C., Gosselin, K. M., et al. (2019b). Examining inputs of biogenic and oil-derived hydrocarbons in surface waters following the Deepwater Horizon oil spill. *ACS Earth Space Chemistry, 3*, 1329–1337.

Williams, N. (2019). *Canadian company ships solid oil sands bitumen to Chinese refinery*. Reuters.

Zito, P., Podgorski, D. C., Johnson, J., Chen, H., Rodgers, R. P., Guillemette, F., et al. (2019). Molecular-level composition and acute toxicity of photosolubilized petrogenic carbon. *Environmental Science and Technology, 53*, 8235–8243.

Chapter 8

Spill impact and response analyses

Jake Nelson[1,2]
[1]*Oak Ridge Institute for Science and Education, National Energy Technology Laboratory, Oak Ridge, TN, United States,* [2]*Department of Geosciences, Auburn University, Auburn, AL, United States*

8.1 Introduction

Over the past several decades, many hydrocarbon and chemical spills of varying size, duration, and severity have occurred. Readers are likely familiar with some of the most notable spills, including the Exxon Valdez in Prince William Sound, Alaska; the Prestige spill off the coast of Galicia, Spain; the Deepwater Horizon in the Gulf of Mexico (GOM); the Montara in the Timor Sea; and most recently, the Sanchi Tanker spill in the East China Sea. Among other things, these spills highlight the importance of developing models and methods to assess the potential damage that future spills may cause. The empirical evidence of harm caused by spills provide critical information to responders, decision-makers, and research communities seeking to enhance spill response in the future. In particular, the past spill events offer an opportunity to refine, enhance, and develop risk, impact, and response analyses capable of addressing the myriad factors related to the deleterious effects on the ocean and coastal landscape. With the continued advancement of risk, impact, and response analyses, the literature does seem to be converging on how to address the challenges within each of these respective areas of research (Nelson & Grubesic, 2018a; Sepp Neves, Pinardi, & Martins, 2016; Sepp Neves, Pinardi, Navarra, & Trotta, 2020) and furthermore, how to enhance response efforts (Galt and Payton, 1999). Taken together, these three facets—risk, impact, and response—are critical for enhancing overall spill preparedness.

Galt and Payton (Galt & Payton, 1997) posed a series of questions that framed a "roadmap" for illustrating how risk assessment and impact may be used in conjunction with response analysis to mitigate the effects of a spill. The questions, which are still valid today, ask:

- What the (hydrocarbon/chemical) slick is posing a threat to?
- Where is it likely to go?

Marine Hydrocarbon Spill Assessments. DOI: https://doi.org/10.1016/B978-0-12-819354-9.00001-6

- Who is likely to get hit by the slick?
- What are the likely impacts?
- What can be done to limit/prevent the impacts?

Over the course of a spill event these questions may be asked several times, and each time the answer could be different compared to the previous time it was asked. This variation is, of course, due to the nature of the spilled substance/slick behavior and the associated response process. Both the spill and the threatened resources will evolve as the slick changes through time and moves through space. This dynamic underpins the importance of developing representative risk and impact assessments. Even more importantly, it underpins the need to develop robust and flexible frameworks that re-evaluate these questions as they are asked. That being said, it is also important to keep in mind that these are models of a specific situation, and that all models must contend with some amount of uncertainty (Gasparotti, 2010; Sebastiao & Soares, 2006; Shuohui, Xuejing, Shuang, & Xuan, 2006). However, researchers and practitioners continue to advance these models to make them as representative as possible for both the environment and the spilled substance (Bauer and Rose, 2015; Sebastião and Soares, 2007; Zhang, Easton, & Steiner, 1997). Taking advantage of new and improved data describing the environment (see Chapter 1: Baseline Data for Spill Assessments: Ambient Conditions, Socioeconomic Data, Sensitivity Maps), the development of methodological standards, and advances in computing, the fields of risk, impact, and response analysis are being made more robust.

Taken together, risk, impact, and response analysis provide a foundational framework for spill preparedness. Although they inform one another, they are distinguished by their methodological and theoretical underpinnings. Risk is defined as the combination of the probability of event occurrence and magnitude/severity of consequences (SRA, 2015). In other words, risk is measured as the chance of a hazard occurring combined with impact. As the definition implies, risk calculations require some measure of impact. Impact, then, is a measure of vulnerability to the adverse effects of a spill combined with the degree in which the spilled substance (or slick) has come into contact with those vulnerable areas. Risk and impact are thus inextricably linked, yet conceptually different in the way they are determined.

In many ways, response analysis involves a thorough understanding of both risk and impact. Consider offshore oil exploration and production activities. In the United States the Bureau of Safety and Environmental Enforcement, under the Federal Water Pollution Control Act[1] (Copeland, 1999), requires the operator of a well to submit plans that detail how the organization will respond to a worst-case discharge event from an offshore facility (BSEE, 2018). In Australia, these powers are vested in the National

1. More commonly known in the United States as the Clean Water Act.

Offshore Petroleum Safety and Environmental Management Authority, which came into existence after the 11-week long Montara spill event. Preparing the contingency/response plans involves, among other things, an assessment of risk and potential impact to enhance the preparedness of the operator and agencies involved in responding to an occurrence. Furthermore, the International Association of Oil and Gas Producers notes the importance of contingency planning to support response, which involves an understanding of the level of *risk* and potential *impact* to "key ecological and socioeconomic sensitivities which may be threatened under realistic planning scenarios" (IOGP, 2015). Simply put, an accurate response analysis depends on a robust measure of both risk and impact.

The remainder of this chapter will detail the models and methods used to develop and apply risk, impact, and response analyses to various offshore spill settings. Section 8.2 will address some foundational elements of these topics, including the theories and frameworks underlying the analyses. Specifically, each subsection will highlight the more common frameworks used in risk, impact, and response analysis methods with reference to some of the seminal texts in each of the fields. By the end of Section 8.2, the reader should have a good understanding of the interworkings of each model and, more importantly, why these models are critical for enhancing response preparedness. Section 8.3 will move to specific applications and examples of risk, impact, and response analyses. Details on the development and use of these applications are provided to help in developing a working knowledge of these methods, and understand what these methods offer interested practitioners. The last section will touch on some of the promising future developments for this field.

8.2 Impact, risk, and response analysis—theory and practice

Within the oil spill literature risk and impact are closely associated, especially when there is an explicit consideration of a modeled hydrocarbon/chemical spill. The numerical modeling of spills provides researchers and practitioners with a mechanism for establishing how likely it is that oil will come into contact with off- and/or onshore, pelagic, and benthic resources (e.g., ecosystems), and also socioeconomic and cultural infrastructure and assets. Determining that likelihood depends on the calculation of a risk metric. This risk metric can be conceptualized as the probability of oil occurring in a particular location combined with the impact/s that the spilled substance may cause within the environment. Responders in the field and the researchers in the laboratory commonly use these results to determine where response resources should be allocated (Li, Cai, Lin, Chen, & Zhang, 2016).

Spill transport and weathering models provide the means for estimating the likelihood that a spilled substance may beach at a location at a certain decay stage, and advances in data collection and the ability to store data in easily accessible formats allow for the determination of what may be impacted. This cooperative structure of risk embracing both the probability and impact of a spill does not mean that the assessment of one cannot be performed without the assessment of the other. Indeed, many impact studies of hydrocarbon spills are performed without regard to oiling probability, and risk analyses are completed without regard for the severity of oiling. However, for all intents and purposes, a true risk analysis should consider both. In doing so, the recorded impacts can effectively be scaled by the probability that they may actually come to fruition, making for a more informed analysis.

8.2.1 Impact assessment

Impact calculations are performed in two distinct ways. The first can be classified as ex-post which involves the real world, observable effects measured in the field. Ex-post approaches take place following a spill event. The second approach can be classified as ex-ante. This class of approaches relies on modeled slick trajectories and substance weathering, and outputs that assess where the slick is going, the concentration (in terms of either volume, mass, or concentration), and what the slick may come into contact with. The second approach is commonly performed under the auspices of preparedness and impact forecasting for evaluating potential scenarios. Where the latter is concerned, the calculation of potential impact can take many different forms. Generally speaking, however, the impact is some function of the *severity* of impact and the probability of oil/substance reaching a certain onshore location. Severity considers the vulnerability of the environmental or socioeconomic assets that could be adversely affected by the substance and combines it with the degree of oiling (e.g., how much oil has arrived in a particular location). Mathematically, this can be defined as:

$$impact = f(v, d) \tag{8.1}$$

where impact is measured by the vulnerability (v) to adverse oiling effects and the degree of oiling (d). A formal calculation of impact is not provided due to the abundance of algorithms and methods that researchers can use for its calculation. However, there is a general process and logic to the calculations which will be detailed below.

8.2.1.1 Vulnerability

Important for practitioners to keep in mind is that the impact value is often data-dependent and will only be as representative as the data used in the calculation of the associated parts. One of the first steps in calculating impact is to determine vulnerability. Vulnerability is a measure of susceptibility to damage (from spilled substance) and the capacity to adapt to the stresses caused by damage (Adger, 2006). One of the most commonly used measures of vulnerability for hydrocarbon and chemical spills are Environmental Sensitivity Index (ESI) values for shoreline segments (Gundlach & Hayes, 1978; Jensen, Ramsey, & Holmes, 1990). As the name implies, this system of shoreline classification provides a measure of sensitivity to oil for different segments of the shoreline. It considers the geomorphology, exposure to wind, ability to recover, and how easy it is to clean via anthropogenic methods. Researchers often use ESI values as a stand-in or as an additional measure of environmental vulnerability in impact assessments (Al Shami, Harik, & Alameddine, 2017; Canu, Solidoro, & Bandelj, 2015; Castanedo, Juanes, & Medina, 2009). Combined with additional environmental datasets, vulnerability can be represented at an increasingly high fidelity with flexibility in how the datasets are combined for impact estimations.

Researchers in the laboratory and responders in the field need to also be aware of the vulnerability related to the socioeconomic sectors of impacted areas (Nelson, Grubesic, Sim, & Rose, 2018). Communities that are reliant on the ocean for their economic livelihood are much more vulnerable to the effects of a hydrocarbon/chemical substance spill than, say, communities that rely on the onshore production, manufacturing, or construction industries. Again, adding datasets to the vulnerability calculation which are specific to the assets within the study area will provide a better estimation of overall vulnerability.

Another advantage of using multiple data sets is the ability to analyze "what-if" scenarios by applying different weighting schemes to the data sets (Fattal, Maanan, & Tillier, 2010). The simple presence or absence of a particular data set is no doubt useful, but can be refined with weights that reflect the vulnerability of a particular asset in relation to other assets in the study area. Leveraging expert knowledge to assign the weights is one of the more common approaches (Cai, Yan, Ni, & Wang, 2015), and how these weights are assigned is often situation dependent and will vary for different study areas. To this extent, the importance of expert and stakeholder input in this process cannot be stressed enough.

As an example, a portion of the study area may contain public beaches, hotels, and marinas as well as boat ramps and water intakes. Each of these can be classified into one larger socioeconomic category or could be broken down further into tourism and recreation (beaches, hotels, marinas) and infrastructure (boat ramps, water intakes) classifications. Weights can then be applied to each class which would reflect expert opinion on the sensitivity of the resources or some known, empirically derived values.

A prime example of this operationalization is demonstrated in Fattal et al. (2010). In their oil spill impact analysis model, they created a global measure for vulnerability by combining scores for environmental vulnerability and socioeconomic vulnerability. Environmental vulnerability was a composite score consisting of shoreline exposure, coastal morphology, habitat sensitivity, intensity of pollution, and marine weather. Each of those subsectors was rated on a scale from 1 to 3 reflecting the intensity. Socioeconomic vulnerability was another composite score making use of a weighting scheme, which included heritage, human activities (fishing, tourism, industry, and agriculture), infrastructures, and crisis management. In this example, human activities were weighted the highest, infrastructure and crisis management were weighted less but equal to each other, and heritage did not have a weight assigned. Weights were based on the cost of these resources relative to one another.

8.2.1.2 Degree of oiling

The degree of oiling can be modeled or observed. For the former, oil spill modeling software packages provide a host of useful metrics to determine the degree of oiling (see Chapters 3, 4, and 5 for details on spill modeling). The estimation of how much oil (or its most harmful components, or other toxic substance) has accumulated in a location is important to consider when calculating impact because it serves as a way to scale the impacts by the amount of substance an area received. Scaling or normalizing also allows for comparison across multiple spill scenarios (Nelson and Grubesic, 2018), and the ability to determine which spill event qualifies as the worst case (Fingas, 2016; Nelson & Grubesic, 2021).

Degree of oiling may be conceptualized in several ways, but is commonly captured using discrete units of analysis across the study area (Nelson and Grubesic, 2018b), or with continuous surfaces reflecting the amount of oil (Canu et al., 2015). The resolution of the former determines how refined the understanding of the resulting impacts will be. Larger units of analysis can provide an excellent "bird's eye" view of the study area but could obfuscate local heterogeneity in impacts and oiling at smaller neighborhood or community levels (Fattal, Maanan, & Tillier, 2010; Nelson, Grubesic, Sim, Rose, & Graham, 2015).

Degree of oiling by continuous surface is not as common in risk and impact studies, but it may have several advantages. Parcels of spilled substance in the form of Lagrangian particles are often used to represent a slick (where each individual particle has a unique set of characteristics that degrade and/or change over the duration of a spill), although in reality, a slick may be an amorphous, semiviscous material that leaves trails of residue as it moves across the water. Oil slicks form tails where thickness is larger in some areas and smaller in others. Thus when oil makes landfall, concentrations can be higher in some areas and lower in others, tapering off as one moves away from the immediate landfall area (Sammarco, Kolian, & Warby, 2013). As a result, transforming the parcels/particles into a continuous surface may provide a more realistic representation of a slick (Nelson and Grubesic, 2017b).

Determining the degree of oiling and slick concentrations are some of the first steps for real-time oil spill response. Following a significant oiling event such as the Deepwater Horizon (DWH), the on-scene incident commander will dispatch response crews to affected shorelines to assess the physical characteristics of the area, how much oil has been deposited, and then to decide on the appropriate remediation techniques. This process is referred to as a Shoreline Cleanup Assessment Technique (SCAT) (NOAA, 2013). SCAT was initially developed following the Exxon Valdez oil spill in 1989 to provide responders with a formalized method of documenting a spill's impacts. It is now a regular part of oil spill response in such countries as the United States and Australia. Readers are encouraged to check the National Oceanic and Atmospheric Administration's (NOAA) website for more information on SCAT.[2]

8.2.1.3 Determining impact

Once the degree of oiling and vulnerability have been established, researchers turn their attention to the calculation of impact and its geographic distribution. Magnitude of impact is directly related to how much oil has arrived at a location/habitat and what type of asset the oil is interacting with. As mentioned previously, impact can be modeled in a number of ways and depends largely on how weights are assigned and how the degree of oiling is incorporated (if at all). This combination of factors should not change the final interpretation of what areas will be most impacted. Instead, magnitude of impacts will reflect the value of the variables that are used to describe each of the component parts.

When comparing multiple spill scenarios, it is helpful to scale the impact values to a common minimum and maximum impact. This is advantageous for several reasons. First, it allows for direct comparison across spills and gives decision-makers the ability to easily visualize which scenarios are most impactful. Second, a common impact scale makes the outcome of different weighting schemes directly observable and comparable (Nelson & Grubesic, 2021). Both of these metrics are tantamount when determining how to allocate response resources to staging areas or dispatch them to the spill itself.

8.2.2 Risk

In its most general form, risk is a function of the probability that an event will occur. It is the consequences of an activity and associated uncertainties (SRA, 2015). There are three questions one can ask when it comes to evaluating the risk level for any type of spill scenario (Kaplan and Garrick, 1981):

(a) What can go wrong?
(b) What is the probability of it going wrong?
(c) What are the consequences if it does go wrong?

2. https://response.restoration.noaa.gov/oil-and-chemical-spills/oil-spills/resources/shoreline-cleanup-and-assessment-technique-scat.html.

Where oil spill risk analysis is concerned, the associated analysis is used as a means for developing deployment strategies and to minimize the risk of accidental discharge to improve safety in the marine environment (Gasparotti, 2010). The evaluation of risk takes the following general form:

$$risk = f(i, r) \tag{8.2}$$

where risk is some function of impact (i, which can also be considered the consequence or degree of harm), combined with the probability (r, the chance that a particular area may come into contact with the spilled substance) that the event will occur. There is some variation in how each of these two components are determined as well as the spatial scale that they are evaluated at. Since risk involves the level of impact, the spatial and temporal scales between the two (impact and risk) should match. A few of the different approaches used to conceptualize and operationalize the assessment of the probability of oiling are detailed next.

8.2.2.1 Probability of oiling

There are two common approaches for evaluating risk, both having to do with likelihood. The first focuses on determining the probability of an event actually occurring, while the second proceeds by assuming the event has occured and determines the probability that an area will come into contact with a slick. When taking the approach of the former, researchers may use historic spill event information, the density of marine traffic, or the number of hydrocarbon-related projects/drilling rigs/oil platforms in an area. The idea is that if an area has experienced spills in the past or has more hydrocarbon-related activity, the likelihood of a future spill will be greater (Olita, Cucco, & Simeone, 2012). As an example, Fernández-Macho (2016) performed a risk analysis for all of the coastal areas of European countries using data on spills from 1970 to 2014. The researcher used location and release information combined with the geomorphological condition of the coastline and the prevailing ocean currents to develop risk indices for each of the coastal European countries. The database used for the analysis is maintained by the International Tanker Owners Pollution Federation (ITOPF, 2019) and is freely available, which provides a great resource for risk modeling. Other examples of risk assessment likelihood from past spill events include the work by Lee and Jung (2015) on tanker incidents in offshore Korea, and Liu, Meng, and Xing (2015) who identified large tanker spills in the Chinese Sea over the past 29 years.

The other approach to addressing the probability of oil spill occurrence is to use the prevalence of hydrocarbon-related infrastructure and/or marine traffic (Lan, Liang, & Bao, 2015). For example, in addition to the use of past spill events Liu et al. (2015) identified the density of oil platforms and ship traffic using remotely sensed imagery. Similarly, Mokhtari, Hosseini, and

Danehkar (2015) used distance to oil wells, coastlines, ship traffic, and oil facilities to identify the most at-risk oiling locations in the Persian Gulf. The final map was a continuous surface illustrating the mean probability of an oil spill occurring across their study area. Finally, Olita et al. (2012) used a vessel tracking dataset with 35,000 records to normalize their marine oil spill risk assessment model. Given the average number of ships that passed through their study area, the researchers assumed a potential oil spill amount. After combining this information with ocean currents, the researchers were able to derive risk maps for the shoreline.

In all of the above examples, risk is a function of how much of a given measure exists for the study location. Now that oil spill models are more widely available and computational memory and processing speed are growing, the more common approach to evaluating oil spill risk involves ensemble-type modeling approaches and advanced oil spill simulations (Sepp Neves et al., 2016). In these models, hundreds or thousands of spill events are simulated. The simulations may take place at a single location or be geographically dispersed. For instance, Boer, den, Azevedo, and Vaz (2014) simulated 3500 spills in a coastal lagoon using VOILS, an advanced spill model based on Eulerian–Lagrangian transport and unstructured horizontal grids (Azevedo, Oliveira, Fortunato, Zhang, & Baptista, 2014). The results of the analysis were used to derive a hazard probability map based on several oil spill locations across a lagoon. In a similar approach, 2,018 spills were simulated by Guillen, Rainey, and Morin (2004) across 91 different starting locations in the GOM over a simulated 9-year period to identify seasonal risk variation along with aggregated risk information for the shoreline. Other researchers have utilized more common spill models including MEDSLIK-II and GNOME to gain valuable insight into the likely locations of oil landfall (Al Shami et al., 2017; Barker, 1999).

Other methods for calculating risk without having to rely on ensemble approaches also exist. These alternatives are advantageous if the researchers do not have access to powerful computers or are not capable of running oil spill simulations in ensemble mode—which usually means some sort of command-line implementation or custom computer code. One example is Monte Carol approaches. With this approach, only a few spills need to be simulated. Using the mean and standard deviations from those simulations, a probability distribution can be derived without having to rerun the entire oil spill model multiple times (Nelson and Grubesic, 2017b). This provides an unbiased estimate of oiling likelihood along with associated error measures.

All of the metrics used in risk analysis continue to evolve. Those highlighted above are some of the most widely used and efforts are underway to develop common standards and approaches for risk estimation (Sepp Neves, Pinardi, & Martins, 2015). Returning to our original risk analysis equation, once the probability of oiling has been determined, it is combined with the impact score (detailed in Section 8.2.1 above) to complete the risk analysis. Most of the time, this is multiplicative [see, e.g., Lan et al. (2015)], additive (Fernández-Macho, 2016), or a mix of both (Azevedo, Fortunato, & Epifânio, 2017).

8.3 Response analysis

In practice in both the United States and Australia, impact analyses play an important role in the creation of plans that guide response efforts during a spill event. In the United States, response plans are organized in a hierarchical structure reflecting the geographic and operational extent of the response requirements. The Area Contingency Plan (ACP) covers the largest geographic extent and provides a governing structure for all agencies within the geographic bounds of the ACP. The ACP aids in coordinating where and how response equipment should be allocated. The Geographic Response Plan (GRP) is more specific, both in operational and geographic scope. For instance, within the United States GOM as well as other coastal areas, the ACP is divided into smaller districts that set the bounds for individual GRPs (Fig. 8.1). Within these smaller districts, sensitive resources can be identified by area experts and stakeholders as priorities for protection. The impact

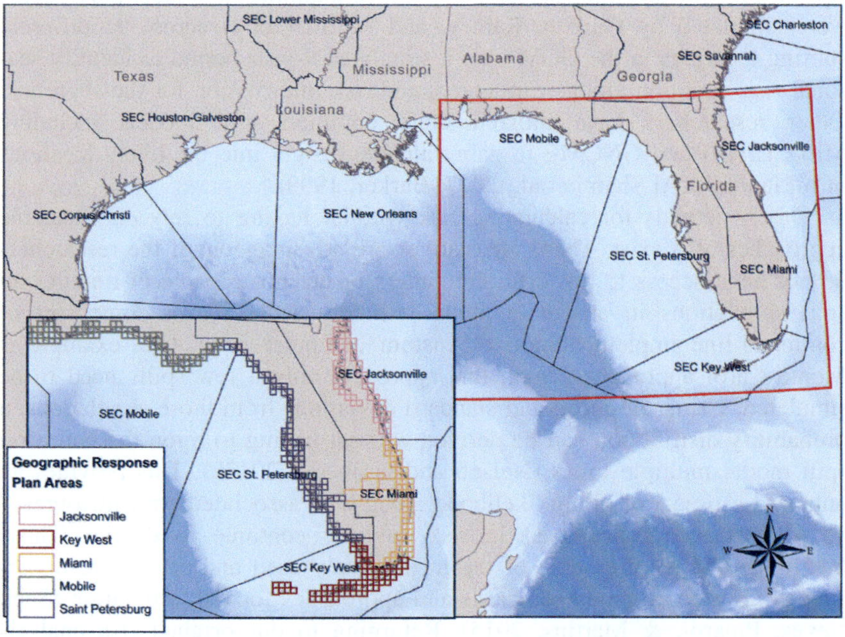

FIGURE 8.1 An example of Geographic Response Plan boundaries (inset map) and the specific Area Contingency Plans (black "SEC" outlines) in the Gulf of Mexico.

assessments described in the previous section offer valuable guidance at this stage of the response planning process.

In addition to identifying the sensitive coastal resources, GRPs guide where and how response resources should be allocated, such as containment booms and skimmers. However, during a spill event, these plans can change based on the spill location, movement of the hydrocarbon/chemical slick, available personnel and equipment, and new information describing the areas in need of protection. As one example, during the early stages of the DWH, new information on land topography and locations of sensitive receptors had to be incorporated into response plans before command units could deploy response equipment with certainty. Over the course of the spill, response teams had to consider the different environmental conditions and reassess the feasibility of the booming strategies in a dynamic manner. Critical throughout the process was a continual assessment of booming effectiveness and strategy (USCG, 2011). In several instances, certain boom types were found to be ineffective and had to be changed.

The dynamic process of planning and executing these response plans for rapid response to spills cannot be overstated. Sensitive assets, both onshore and offshore, are immediately at risk of being harmed when a spill begins. As the slick moves through space and evolves/weathers over time, the areas of protection may also change. This observation during the DWH highlighted how dynamic the overall response operation needed to be (Liu, MacFadyen, Ji, & Weisberg, 2013). Recent research to enhance response strategies in order to accomodate spill dynamics have recently been underway, and include spatial optimization and geospatial analysis.

8.3.1 Response strategies

There are three overarching areas of response operation defined in the literature. The first is classified as *strategic* operations (Iakovou, Ip, Douligeris, & Korde, 1996). Analyses at the strategic level seek to optimize where valuable and often scarce response resources should be located to ensure an optimal response. The second type of response operation is *tactical* (Grubesic, Wei, & Nelson, 2017). Tactical response operations focus on the optimal allocation of resources from their storage locations but extend even more broadly to include decisions on how the equipment should be used once it is at the spill site (e.g., how long equipment stays on scene, what type of equipment should be used, and where exclusion booms should be allocated). The last area of research (sometimes collapsed into the tactical response) is the *operational* response. These types of response analyses are performed at much higher detail and with a panoptic approach. They consider the minute details of optimizing the distribution of booms and skimmers, the time required for equipment to be in place, and the type and size of personnel teams required for response.

As noted by Iakovou, Ip, Douligeris, & Korde (1996), decisions at the operational level are complicated and heavily dependent on the response team availability, structure, and the stockpile of equipment at the storage locations proximal to the spill site. Decisions at the operational level are built on knowing where resources are located, how much of that resource is available, and how those available resources should be deployed with respect to the spill. Iakovou, Ip, Douligeris, & Korde (1996) acknowledge that for major oil spills, incident commanders may choose to engage all available resources, regardless of whether that particular type of equipment is necessary for response. While it is of course better to be overprepared than underprepared, Iakovou and colleagues go on to note the existence of several examples when equipment was transported to a spill site but never used. This is wasteful, both in terms of time and money, and can lead to an ineffective response operation. To avoid these wasteful scenarios from occurring, researchers have been developing optimization models and decision support systems to aid responders in their decisions to strategically place equipment at storage sites and to optimally mount a tactical response.

8.3.2 Spatial optimization

Solutions for determining where response equipment should be located and the decision about how that set of equipment should be dispatched to the site of the spill has, generally, been approached through spatial optimization. Spatial optimization has gained much traction in the field of geography and, similar to mathematics or economics in which it is often applied, consists of three main components: an objective, decisions to be made, and constraints (Tong and Murray, 2012). The objective is the goal to be achieved—either maximizing some kind of benefit or minimizing costs—and is represented through single or multiple objective functions. Decision variables, like their name implies, correspond to the explicit policy decision to be made. In the context of spill response, decision variables can be where staging areas or response equipment should be located, the specific areas that require protection, or how much of a particular type of equipment should be dispatched or stored. Finally, constraints determine the conditions that must be satisfied in accordance with the problem under study. Often, this has to do with budget (constrained by a maximum available budget) but can also focus on time (how much time responders have to arrive to a spill or affected site), and/or some allowed level of environmental impact.

Within a spatial optimization problem, these three components—objective, decisions, and constraints—are combined to outline the geographic problem of interest. In addition, the geographic nature of these problems means that each of the component parts has spatially interdependent relationships. In other words, space is explicitly considered by these models.

As a basis for these models, the following generic formulation of an optimization problem from Tong and Murray (2012) may be adapted:

$$\max:\ g(k) \tag{8.3}$$

Subject to:

$$f_i(k) \le a_i \forall_i \tag{8.4}$$

$$k\ conditions \tag{8.5}$$

where k is a vector of decision variables, a_i places limitations on what the value for all ($\forall_i$) functions (f_i) can be, and Eq. (8.5) denotes the requirements that the model must meet to be solved. The decision variables can represent a number of items, such as minimizing costs or distance traveled, maximizing the number of protected/boomed sensitive receptors, maximizing the total amount of oil/chemical substance that can be cleaned, or minimizing total impact [Eq. (8.3)]. f_i and g are both functions of k. These are the constraints on the model, effectively providing limits to the objective function. As an example, if the objective is to minimize the total impact of a spill, constraints could be the number of boats available, the total length of containment boom, or the number of personnel. Given multiple objectives and constraints, the model would provide optimal combinations of available resources that work to minimize the impact over time. These optimal combinations can be visualized using a Pareto optimal curve (Fig. 8.2). Choices along

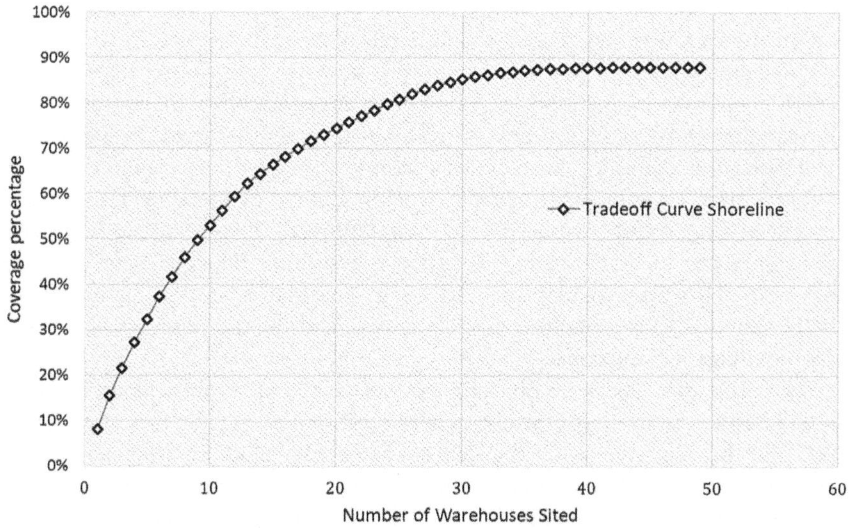

FIGURE 8.2 Pareto optimal curve balancing the percentage of resources covered and the number of locations to achieve that coverage.

the curve are optimal with respect to the amount of resources required to achieve the resulting impacts.

Fig. 8.2 was created with an optimization model seeking to maximize response coverage of sensitive environmental resources with a travel time constraint of 1 hour (Grubesic, Wei, & Nelson, 2018). The objective of the model was to identify where along the coast warehouses with equipment stockpiles should be placed to minimize damage to sensitive environmental resources within a 1-hour time horizon. The reader will note that the curve begins to plateau at about 35 warehouses. This means that the installation of additional warehouses beyond 35, no matter where they are located, is redundant in the sense that the max amount of assets can be covered with 35 strategically placed warehouses. Any more than 35 will not improve the overall coverage percentage. Furthermore, any point on this curve can be considered an optimal solution given the spatial configuration of the warehouses and the constraints.

There is an implicit trade-off being made between coverage and warehouses in this example. Researchers and practitioners can therefore use the Pareto optimal curve to make valuable decisions regarding spill response. Perhaps the most valuable output of these spatial optimization models is the spatial configuration of the 35 selected storage sites that maximize the coverage illustrated by the curve. Details on the process leading up to the curve and the steps specific to the two major areas of response operations—strategic and tactical—are detailed below.

8.3.3 Strategic

Before moving into the strategic optimization of spill response, it is important to acknowledge a key caveat: during a worst-case spill event like the DWH, incident commanders and on-scene coordinators will operationalize all available resources regardless of whether they will be used. In the case of the DWH, for example, the spill was simply too large and required significant resources to combat it. In fact, over the course of the spill, hundreds of vessels of opportunity were enlisted and deployed. These were private vessels contracted to assist crews in response activities. This, of course, is not the norm. However, readers should be aware that a spill as large as the DWH can easily overwhelm response crews, and contracting private vessels may be a part of a response plan.

The strategic aspect of oil spill response generally takes place before a spill. Planners must first consider where spills might occur, their frequency, size, and the sensitive resources that could be impacted. Contingency analyses using spill transport and weathering/decay modeling packages (such as those described in Chapter 5: Oil Spill Modelling Assessment) will help to identify these at-risk areas and characterize the dynamics of potential spills (French-McCay, 2004; Nelson & Grubesic, 2018; Spaulding, 2017).

In addition to using spill models, researchers may use historical spill accounts or known shipping routes as potential spill locations (Belardo, Harrald, Wallace, & Ward, 1984; Verma, Gendreau, & Laporte, 2013). Once the likely areas of spill occurrence have been identified, planners can then decide, given the information about potential spills, where to place response resources and how much of each resource should be located within each stockpile.

Strategic response spatial optimization models begin with a set of locations where equipment could, in theory, be placed. This may be near existing boat ramps, within docking areas, marinas, or beaches, whichever makes sense for the context under investigation. Decisions about which locations are feasible for storage or staging depend on the local landscape and existing infrastructure. Before beginning the modeling process, researchers and practitioners should identify which criteria for inclusion should be of decisive value before determining the initial set of storage or staging locations.

Next, researchers must choose what criteria will drive location siting decisions. In other words, the model asks: where should this equipment be sited given the protection requirements? In the case of Verma et al. (2013), the researchers used historical spill locations and existing marine traffic routes to operationalize their two-stage stochastic programming model. They chose the historical spill sites and shipping routes as the most likely places where a future spill could occur. Consequently, equipment was sited so that it would adequately cover these areas. One of the objectives of the model was to identify the potential storage sites that enabled responders to arrive at any of the identified spill sites within 6 hours, given a specified oversea transportation speed. In an older, yet important piece, Belardo, Harrald, Wallace, & Ward, 1984 used a similar bundle of information to identify storage equipment locations. They used historical spill events and local shipping traffic and also considered weather factors.

Two constraints have already been mentioned regarding Eq. (8.5). These are the limits that determine what the model can and cannot do. In Verma et al. (2013), the researchers were choosing to site equipment in locations where responders were able to arrive within a 6-hour time frame. The time frame was the constraint placed on the model, and any solution had to meet that (or tighter) time constraint to be considered optimal. They also included monetary constraints by specifying a total amount of money that the cost of response had to adhere to. When evaluating possible solutions, the model only identified geographic locations that met those two constraining conditions. Additionally, other constraints can be added that limit the total amount of equipment that can be sited. Models that impose equipment constraints must determine what equipment should be allocated in order to maximize overall response coverage or minimize potential harm. These decisions can be driven by the relative sensitivity of the surrounding environment or frequency and magnitude of potential spills, as they were in Verma et al. (2013).

If it is the former, more resources are likely to be placed in areas with larger numbers of sensitive assets. If it is the latter, more resources will be sited in areas with higher spill frequency or where the potential spills have historically been large in magnitude. This may seem intuitive, but the power of the optimization model is that the resulting output will be optimized in such a way that the exact number of resources required to protect sensitive resources or to cleanup spills of varying sizes will be identified. Using these modeled results can lead to a more efficient and economical response operation.

8.3.4 Tactical

Compared to strategic spill response models, tactical response can become quite complicated. There are additional levels of detail and uncertainties from both practical and theoretical points of view. Tactical response planning focuses on which marine transportation resources are required, what equipment should be placed on the vessels responding to the spill, and where those resources should come from. Thus optimization models face temporal, resource, and monetary constraints but may also contend with where the most efficient set of resources should depart from, given the known allocation of resources at each of the equipment storage sites or staging areas.

Similar to strategic optimization, tactical optimization begins by defining an objective function—either minimizing or maximizing some aspect of the spill response. One example of such an approach is given by Psaraftis and Ziogas (1985). The focus of their model is on the development of a decision algorithm to dispatch equipment sets to a spill location. It is based on the aggregate cleanup capability needed within discrete stages of the cleanup operation. The idea is that during the course of a spill, different sets of equipment will be required depending on the discharge rate, type of oil/pollutant, prevailing ambient conditions, and movement of the slick after the initial spill. The notion of time-variant equipment sets is also addressed in Wilhelm, Srinivasa, and Wilhelm (1997) which seeks to minimize response time while explicitly considering the number of components available at each staging area, the storage capacity of vessels, and the number of components used in the formation of equipment sets.

As a more recent example, Gkonis, Kakalis, Ventikos, Ventikos, & Psaraftis (2007) develop a tactical decision approach that considers costs with respect to time and, in a similar way to Wilhelm et al. (1997), take into account the changes to oil due to weathering over time. Damage potential is included as a categorical variable (high, medium, and low) following each spill simulation which is curiously nonspatial with respect to spread and impacted areas. It does follow earlier work by considering the optimal set of equipment required and where it is stored. Even more recent are the works by Grubesic et al. (2017, 2018), which were developed for use with new spill models capable of simulating individual spill parcels at a fine-grained spatiotemporal resolution. Moreover, recent work has extended the tactical approach to determine both the optimal set of equipment to be dispatched and how exclusion booms should be distributed to protect sensitive receptors (Grubesic et al., 2018; Zhong & You, 2011). Total costs to deploy and maintain the boom, as well as the transportation of the boom are important considerations in these models (Etkin, 2004; Schmidt Etkin, 2009), but so too is the identification of the sensitive receptors.

There are two critical components of tactical response models that are worth elaborating on. The first is equipment sets which refers to the mix of equipment required to respond to a spill. Wilhelm and Srinivasa (1997) detail this aspect well. They note that cleanup consists of a variety of individual cleanup *components* that are owned by public and private companies and stored at known locations. In the event of a spill, these individual pieces of equipment are dispatched to *staging areas* where they are assembled to form the *response system* or *response sets*. A response system is thus the location of a single or several staging areas that contain a specific set of components. These locations are what is generally evaluated by the optimization models. As one example, sets can comprise booms, pumps, and the storage capacity of the responding vessels. It may also include the location of that equipment. Given the set of equipment, the objective of the model is to minimize response cost. Cost is often assumed based on a general damage functions or ancillary data describing an assumed cost of cleaning up oil/pollutant given a particular coastal setting (Etkin, 2004).

The second critical aspect worth elaborating on is the idea behind cases or scenario evaluation. Evaluating different cases gives responders the ability to compare and contrast the outcomes of different response options across several different scenarios. In Psaraftis and Ziogas (1985), scenarios consisted of do-nothing (also referred to as benign neglect), optimal (response and damage are minimized using any possible combination of cleanup equipment), and limited where certain constraints of the model are restricted (such as cleanup resources, storage capacity, or travel time). The different response options are then compared to determine the difference in overall cost and impact. Similarly in Grubesic et al. (2018), scenarios consist of benign neglect, best-case scenario, and restricted equipment sets to illustrate the differences in final impact.

8.4 Applications

In the previous sections, the methodological and theoretical foundations for spill risk, impact, and response analyses commonly used by researchers were provided. The following sections will present examples of their application, focusing on each of the individual components and their operationalization. These applications are meant to provide the reader with a general framework and a guide for analysis. In practice, some of the steps involved in implementing these analyses will depend on the availability of data, the study area, access to software, and importantly, the research question of interest. They are also just one specific example. Readers should be aware that there are indeed other methods for approaching these questions, but the general framework for how the individual pieces of information come together is generalizable in the sense that they build upon fundamental aspects found throughout the spill risk, impact, and response literature. Most of these fundamentals have briefly been covered in the preceding sections and are also covered in more detail elsewhere (e.g., Nelson and Grubesic, 2018).

8.4.1 Estimating spill risk and impact: an application in the Gulf of Mexico

For governments, responders, and other stakeholders, the accurate quantification of potential impact stemming from pollutant spills is critical for improving response effectiveness and spill preparedness. In the United States, offshore oil development—both exploration and production—has varied from year to year, often as a function of the political administration. However, the GOM has remained one of the largest offshore production areas in the United States. This is because both exploration and production in the GOM are an important part of the United States energy budget (EIA, 2018) and support thousands of industry-related jobs in GOM coastal communities. There are thousands of offshore oil wells and hundreds of associated platforms which is why many oil spill impact models have used the GOM as a study area. Furthermore, the DWH disaster occurred in the GOM, along with several other smaller spills in the recent past. Because of this activity, massive efforts have been made to analyze, collect, and create data sets representing the systems in this area. As mentioned previously, this particular approach to risk and impact is data driven. The active oil industry and data availability for the GOM allow for a comprehensive illustrative example.

8.4.1.1 Units of analysis

A spill impact assessment requires a mechanism for analyzing areas within the larger study area. By dividing the study area into discrete units of analysis, the coincidence of a floating or deposited pollutant and sensitive assets can be assessed and subsequently analyzed on an individual basis. Researchers may do this in a number of ways, including the use of administrative units, response areas, or arbitrary divisions created by the researcher. For this analysis, we will be using a 2 km × 2 km grid created using a GIS software package. The grid stretches over the entire coastline of the GOM and is wide enough to include any barrier islands that may lie just off the mainland shore, as well as the entirety of bays that cut inland. The objective at this stage is to design the grid in such a way that all the underlying data sets used to model the vulnerability of the coastal areas are captured. If data are sparse, larger units of analysis should be used. With higher density data, smaller units of analysis are advised as they allow for the estimation of impacts at finer scales. The more refined the data, the more heterogeneous the resulting impacts may be. Thus, without a proper analysis of resolution, impacts at the meso- or microscale could be obfuscated and provide an inaccurate sense of impact and its location.

8.4.1.2 Scenario specification

As detailed in Eq. (8.1), the impact calculation is based on the vulnerability of the coastal and marine assets that may be affected by a spill. This value is combined with the degree of oiling. For this example, vulnerability is defined as the total number of sensitive assets within a given unit of analysis where a unit of analysis is one individual grid cell (2 km × 2 km). The degree of oiling is determined by the BLOSOM[3] oil spill model. Because spill modeling has been covered at length in Chapter 5, Oil Spill Modelling Assessment, only a brief summary of the spill simulation is provided.

Due to variations in the dispersion of oil throughout the water column, 30 separate simulations were performed and averages were taken. All settings remained the same across simulations. The spills began at 27.74°N, −89.36°W in the offshore GOM, 90 miles south of Louisiana on May 1 (Fig. 8.3). This spill occurred at a depth of 4169.9 ft. The release of crude oil occurred over a 7-day period at a rate of about 400 barrels per day. The total amount of oil released over the 1-week blowout was 1935.17 barrels (bbl) after weathering [using the evaporation model detailed in Tkalin (1986)] (Fig. 8.4). The model then followed the transport of the oil through space and time as it traveled across the water column over the course of 50 days. Updated location and oil slick characteristic information were recorded and output in shapefile format every 24 simulated hours. In total, there were 50 individual shapefiles detailing the evolution of the spilled oil for each of the 30 simulations. By the end of each of the simulations, the majority of the oil had degraded (weathered). An average of 390.15 bbls beached and 1105.56 bbls remained in the offshore environment. At this point, the vulnerability and degree of oiling were calculated for each grid cell.

8.4.1.3 Vulnerability calculation

Vulnerability is defined as the number of sensitive receptors/assets that could be negatively affected by the presence of spilled hydrocarbon or other pollutants/chemical substances. The calculation of vulnerability can be as simple as summing the number of sensitive assets that occur within an individual unit of analysis. This approach would assume that all receptors are equally

3. https://edx.netl.doe.gov/blosom/.

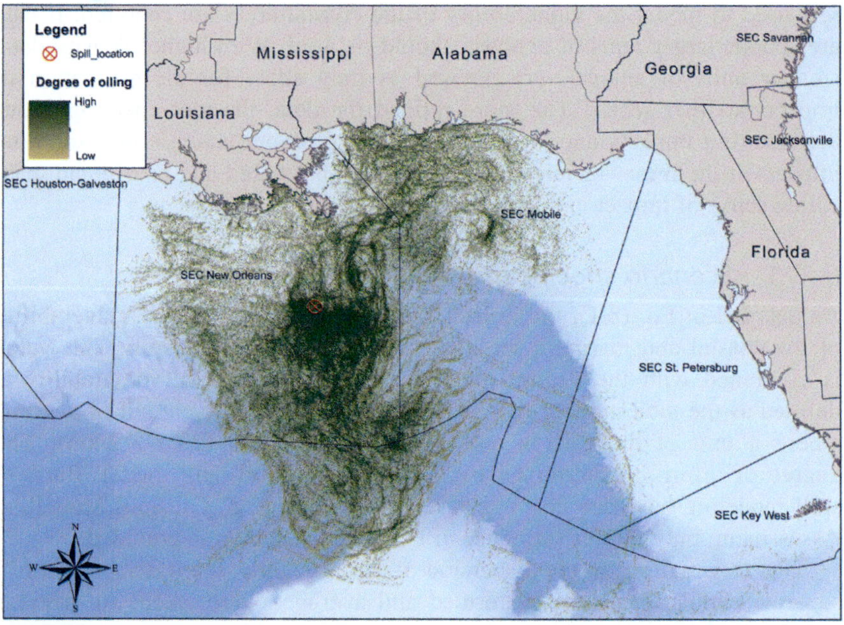

FIGURE 8.3 The oil spill simulation location and extent in the Gulf of Mexico. Darker colors of green indicate more oil passed through that location over the duration of the simulation.

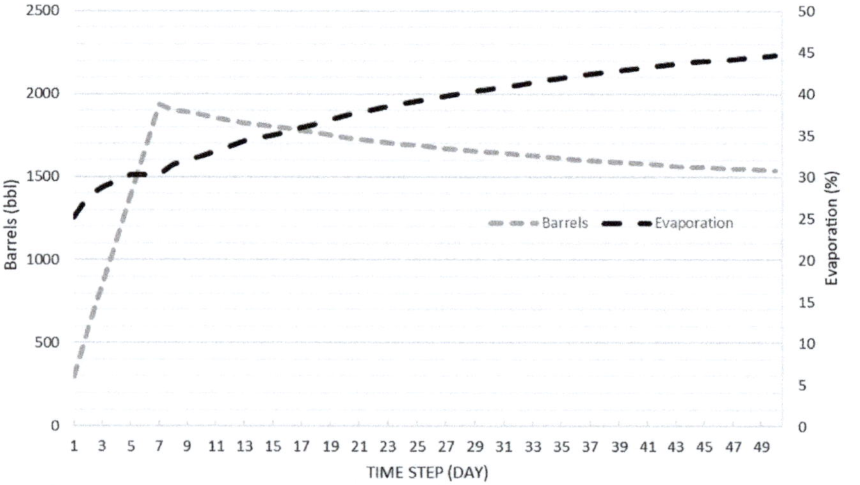

FIGURE 8.4 Association between the amount of oil in the marine environment and the amount of oil weathering over time.

important, and that the locations with the highest number of receptors are the ones that will be the most impacted. However, this may not always be true as some assets will likely be more sensitive to the effects of oiling, or furthermore, are more critical to protect because of their economic importance. For the latter situation, the user can start by determining the number of assets impacted by the spill and then apply weights to the different types of assets in the region.

For this example, weights were applied to the different categories of sensitive assets within each of the larger asset categories—environmental and socioeconomic. After simulating the spill and determining where oil made landfall, the vulnerability calculation began by determining the relative weights for each of the subcategories within the broader environmental and socioeconomic categories (Table 8.1).

For demonstration purposes, the subcategories were weighted as follows. Recreation/Tourism (RT) is highest as this area depends heavily on the revenue brought in by the tourism industry. Essential fish habitat (EFH) is weighted slightly lower but high enough to reflect the importance to the seafood industry and the relative sensitivity of these species to the detrimental effects of oiling. Ecologic impacts (EC) are weighted the same as EFH because of their sensitivity to oil, slow recovery time, and importance to the surrounding ecosystems. Infrastructure (IF) is weighted the lowest because of resilience to the effects of oiling and low sensitivity to negative effects. Shut in of oil operations could result from a major spill but will only have a minimal impact (relatively speaking) when compared to the other categories; it was therefore assigned the lowest of all weights. Final equations for each subcategory impact are detailed below:

$$SE_{vul} = 3\left(\sum RT_1, RT_2, \ldots RT_n\right) + \left(\sum IF_1, IF_2, \ldots IF_n\right) \tag{8.6}$$

$$ENV_{vul} = 2(EFH_1, EFH_2, \ldots EFH_n) + 2\left(\sum EC_1, EC_2, \ldots EC_n\right) \tag{8.7}$$

where RT are the individual assets (n) representing recreation and tourism, IF are the individual assets of infrastructure, EFH are the essential fish habitats, and EC are the ecological variables within a grid cell. Again, the type and number of assets within a grid cell will likely change based on data availability. But as previously mentioned, with more data comes the ability to refine weights and better understand and assess the heterogeneity across the coastal landscape.

Once the vulnerability score for each affected unit of analysis are calculated, the total vulnerability score is obtained by adding together the final scores for the broader categories (Fig. 8.5):

$$FV = SE_{vul} + ENV_{vul} \tag{8.8}$$

TABLE 8.1 Break down of the data set type, the subsector, and the sector that the data set is grouped into.

Data set	Subsector	Sector
Beach access	Recreation/Tourism	Socioeconomic (SE)
Marinas		
Boat ramps		
Drinking-water intake		
Parks		
Piers		
Tourism/rec business		
Migratory pelagic	Essential fish habitat	Environmental (ENV)
Red drum		
Reef fish		
Spiny lobster		
Albacore tuna		
Sharpnose shark		
Big eye tuna		
Blacknose shark		
Blacktip shark		
Tiger shark		
White marlin		
Yellowfin tuna		
Bluefin tuna		
Coral reef/hard bottom habitat	Ecological	ENV
Artificial reef locations		
Critical wildlife areas		
Sea turtle nesting beaches		
Wildlife refuge		
Oyster habitat		
Environmental Sensitivity Index		
Marine protect areas		
Seafood processing plant	Infrastructure	SE

(*Continued*)

TABLE 8.1 (Continued)

Data set	Subsector	Sector
Airports		
Coastal roads		
Refineries		
LNG facilities		
Oil platforms		
Oil pipelines		
Oil wells		

Used in the weighting scheme for vulnerability.

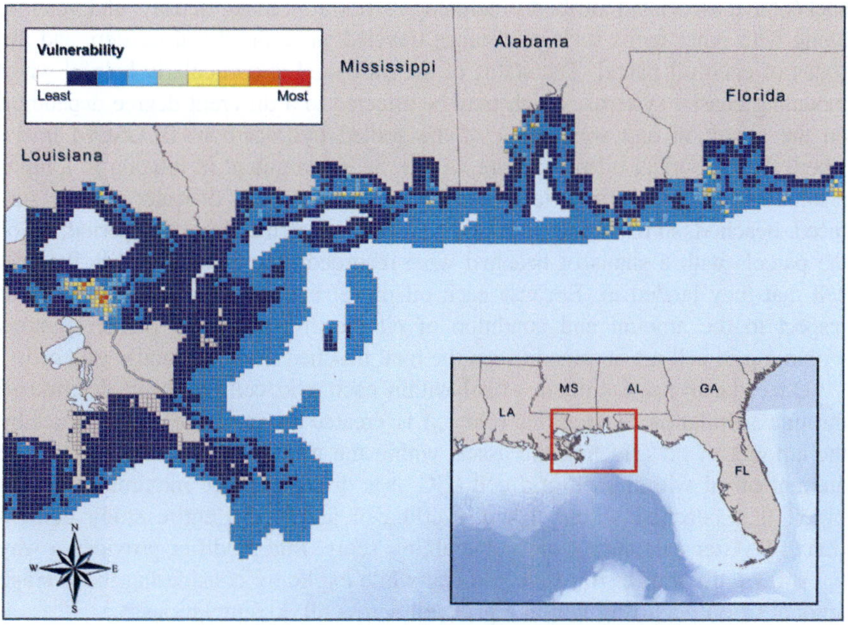

FIGURE 8.5 The spatial distribution of vulnerability using the total vulnerability score calculated for each of the grid cells in the study area delineated by the red line in the inset.

The result will be a range of values that can vary from cell to cell, making it difficult to compare in later calculations. We can transform these values to something more comparable by scaling them to a different range but the relative difference between the values will remain the same. For this analysis, the values were rescaled between 1 and 5, with 5 being the most vulnerable area and 1 being the least vulnerable:

$$\left(\frac{\max_{new} - \min_{new}}{\max_{old} - \min_{old}}\right)^{*}(FV - \max_{old}) + \max_{new} \tag{8.9}$$

where $\max_{new}$ is 5, $\min_{new}$ is 1, $\max_{old}$, and $\min_{old}$ are the maximum and minimum values that result from the calculation of final vulnerability, and FV is the vulnerability score for each of the individual units of analysis (grid cells in this case). With that completed, all of the vulnerability scores across all units of analysis will have a value between 1 and 5. The final step in the calculation of impact is the combination of vulnerability with degree of oiling.

8.4.1.4 Degree of oiling

The oil spill trajectory model BLOSOM is flexible with respect to the type of data outputs and the information the outputs contain. The model will create text files or shapefiles that record, among other things, the location of the individual oil parcels in time and space. Each of the individual oil parcels that make up the slick/s have associated attributes including its position, mass, density, and volume, along with other items such as distance traveled since the last time step, and the status of each oil parcel. The status of the oil parcel is particularly helpful when modeling coastal systems, which may be affected to a different degree depending on the condition and weathering of the spilled hydrocarbon. BLOSOM tracks where the individual oil parcels are within the environment in relation to a high-resolution bathymetry raster layer. Parcels can have a status of water column, surfaced, beached, sunk, or out of bounds. At the end of the simulated period, all of the parcels with a status of beached were recorded and associated with the grid cell that they landed in. Because each oil parcel has unique characteristics with respect to the amount and condition of oil they represent, all parcel amounts within a grid cell are summed to get the total beached oil in a specific grid cell.

Once the average amount of oil within each grid cell has been determined through simulation, a modifier (Oil_{mod}) is created to scale the total impact by the amount of oil affecting the assets within the grid cell. For this analysis, the amount of oil within each grid cell (GC_{oil}) is divided by the maximum amount of oil that beached within a single grid cell across the entire study domain ($\max_{oil}$). After calculating the vulnerability score, this modifier provides a way to estimate the impact for each grid cell while explicitly considering the average amount of oil occurring within a grid cell across all 30 simulations.

$$oil_{\mathrm{mod}} = \frac{GC_{oil}}{\max_{oil}} \tag{8.10}$$

8.4.1.5 Impact calculation

The impact calculation is a combination of degree of oiling and vulnerability score. The approach taken here was fairly simple and consisted of capturing the total amount of oil that beached within each of the grid cells (average from 30 simulations), transforming those values to range between 0 and 1 and then multiplying it by the vulnerability score, which was also transformed to range between 1 and 5. Thus the final impact score will be in a range between 0 and 5.

$$impact = oil_{\text{mod}} \times FV \tag{8.11}$$

Having calculated the impacts, the results can be mapped to visually determine where the most and least impacted areas are located (Fig. 8.6). This is advantageous for several reasons. First, a responder can change the weights of their model to reflect different protection priorities and compare the resulting impact locations. Second, responders can tailor their contingency plans based on where the highest impacts are located. Third, these impacts, and the process of modeling them, can be used as valuable inputs for response operations models which will be detailed in the next section.

8.4.1.6 Risk analysis

Once the potential impact for each of the grid cells are determined, the risk indicates how likely it is that oiling will occur for a particular location. Within the actual offshore environment and in the laboratory, there is some variation where

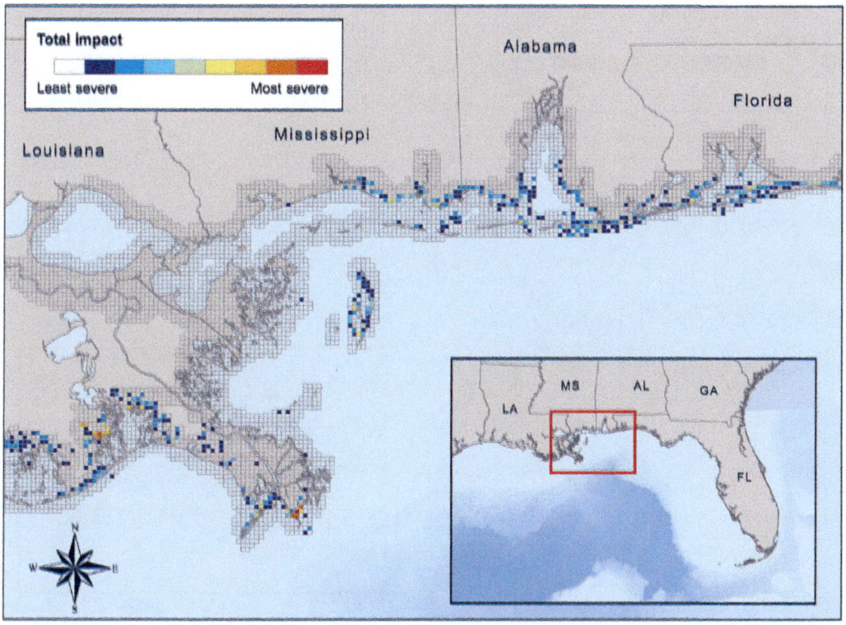

FIGURE 8.6 The spatial distribution of impact after vulnerability is combined with degree of oiling.

oil will make landfall. Most oil spill models are deterministic, but have built-in randomness in the dispersion algorithms to better reflect what might be experienced in a real-world setting (Spaulding, 2017). As a result, two simulations run by the same model under the same settings will likely have different outcomes with respect to where the slick makes landfall.

The likelihood of occurrence is based on where the slick is most likely to make landfall. That is, the 30 simulations were used to calculate a probability of occurrence based on how often that grid cell had been affected by oil. As was done for the impact analysis, the oil parcels that made landfall were joined to the grid following each simulation. For the simulations where oil landed in the grid cell a value of 1 was recorded. For the days where no oil landed in the grid cell a value of 0 was recorded. Dividing the final count of the number of simulations where oil was present (between 0 and 30) by the total number of simulations (30) gave a probability of occurrence for the grid cell.

After calculating the probability of occurrence for each grid cell the final risk metric can be calculated. The use of Eq. (8.2) combines the impact calculated in the previous section with the probability of occurrence. This yields the final distribution of total oil spill risk across the study area (Fig. 8.7). The highest risk areas are those in red, the lowest risk areas are in blue, and the areas where there is no calculated risk do not have a fill. Given this spill scenario, the most at-risk areas are on the south end of Louisiana.

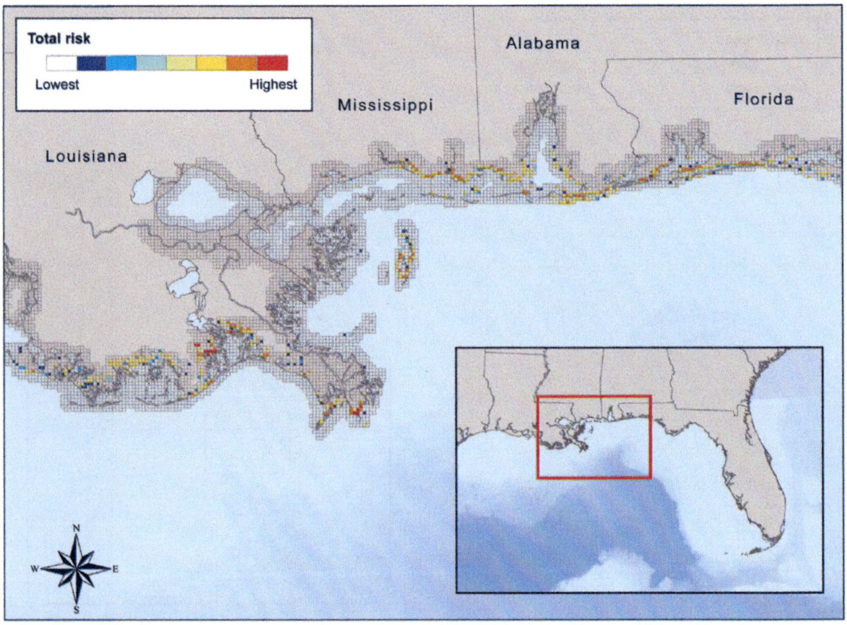

FIGURE 8.7 Final risk map for the study is determined by combining the impact and probability of oil occurrence.

8.4.2 Response analysis: the tactical analysis and coordination for oil spill suite

The response analysis can take on many forms including strategic, tactical, and operational. The goal of each response type varies slightly in the focus of response. As detailed in Section 8.3.3 strategic response focuses on where response resources should be allocated before a spill, tactical response is concerned with how to best allocate a finite amount of resources to the spill, and operational response focuses on the efficient coordination of the numerous facets of the response effort overall. To satisfy all three response forms, the Geoinformatics and Policy Analytics Lab at the University of Texas at Austin has created the Tactical Analysis and Coordination for Oil Spills (TACOS) suite, which is a platform that combines spatial optimization and geospatial analysis to inform response operations.

The suite is informed by the outputs of an oil spill model of the user's choosing and uses the behavior of the spill through time and space to determine how to allocate response equipment in the most efficient way. The tool also gives response coordinators the ability to vary the aggressiveness of the response by specifying the percentage of oil targeted for cleanup during each time step. The current iteration of the tool starts at 75% and increases in 5% increments to 95% and also includes an option for benign neglect, or no cleanup at all. After a spill has been analyzed to completion, the model estimates the resulting impacts. Some of the functionality built into the TACOS suite gives responders the ability to compare different response strategies and resulting impacts in a dynamically linked web mapping system. In doing so, responders can evaluate how to best respond to a spill by visually comparing the resulting impacts and the response resources required to meet the user-defined response targets.

The current iteration of the tool utilizes a built-in spatial optimization model for tactical response, which mimics the one detailed in Section 8.3.4. The associated impact analysis built into TACOS is calculated using an approach similar to the one detailed in the previous section. The combination of these two approaches highlights the importance of having accurate impact assessments, without which the response may not be as efficient or well informed. In the following sections, several of the submodels within the TACOS suite are detailed, with close attention paid to the underlying workflow. Ultimately, the TACOS suite provides a framework for aiding in response analysis through the use of analytic tools; the suite is continually being enhanced.

8.4.2.1 Oil Spill Cleanup Operations Model

Within the TACOS modeling suite is a set of equations that evaluate modeled oil spills and how to optimally dispatch resources to the spill in consideration of the surrounding environment. The Oil Spill Cleanup Operations Model (OSCOM) evaluates the behavior of the plume/slick and determines how to optimally dispatch vessels for on-water cleanup operations. The critical factors required to perform the analysis are the amount of oil in the water column and on the water surface, the total capacity of the equipment at each staging area and of the boats responding to the spill, the speed that the boats can travel, and the oil spill cleanup target. Each of these parameters is represented as unique variables within the optimization model and can be varied depending on the characteristics of the problem scenario.

Two optimization models, which are a variation on the general form detailed in Section 8.3.2, are described in more detail in the following sections. They include an equipment dispatch location model, which has not yet been incorporated into the TACOS suite, and the Exclusion Boom Allocation Model (EBAM), which is geared towards optimally dispatching booms to sensitive coastal resources given the potential impact that the oil may cause.

8.4.2.1.1 Determining response equipment dispatch locations

For each modeled time step, the TACOS model reevaluates where the optimal location to dispatch response resources is located given the behavior of the spilled substance, how much oil/pollutant is in the environment, the overall cleanup target, and the capacity of the response vessels at each of the dispatch locations. A more detailed description of this model can be found in Grubesic et al. (2017). The model does this by solving the optimization model for each day of the spill using the indices below:

I = set of staging areas indexed by i

J = set of oil spill locations indexed by j or i

Parameters:

Γ = oil cleanup target

P_e = operating capacity of cleanup equipment in the response vessel

V_j = volume of spilled oil at site j

D_{ij} = cost/time to dispatch vessel from staging area i to soil site j

Ω_j = set of spill locations that are within the containment area of site j

N_i = available number of vessels in i

Decision variables:

x_{ij} = number of vessels dispatched from area i to spill site j

$$u_j = \begin{cases} 1; & \text{if spill site } j \text{ is cleaned by vessels} \\ 0; & \text{otherwise} \end{cases}$$

The TACOS model currently treats all boat ramp locations as potential staging areas (indexed by I). This data have been preloaded into the modeling suite but is dynamic in the sense that they can change based on user knowledge of the exact locations of equipment staging areas. Additionally, the type and amount of equipment at each staging location can also be changed. For the test case, however, the amount and type of equipment has been randomly distributed across each of the staging locations.

When the objective [Eq. (8.12)] is to minimize the total time that it takes to dispatch resources to the spill, then the total "cost" of response should be taken into account. Cost is conceptualized as the total distance traveled by all vessels responding to the spill. In other words, costs are the accumulated distance traveled by all vessels, to and from the spill, but could be modified to represent monetized values related to the operating costs of vessels. The constraints (8.13) and (8.14) limit the total capacity of cleanup equipment at each staging area and the total capacity per vessel, respectively. The amount of oil being cleaned by the vessels cannot exceed the amount set in Eq. (8.14). The constraint (8.15) makes sure that only areas in which a vessel has been sent are cleaned and Eq. (8.16) specifies that only the amount of oil specified by the user is removed (Grubesic et al., 2017).

$$\min \sum_{j \in J} \sum_{j \in I} x_{ij} D_{ij} \tag{8.12}$$

Subject to:

$$\sum_{j \in J} x_{ij} \leq N_i, \forall_i \in I, t \in T \tag{8.13}$$

$$\sum_{i \in I} x_{ij} P_e \leq \sum_{l \in \Omega_j} V_l, \forall j \in J \tag{8.14}$$

$$\sum_{i} \sum_{l \in \Omega_j} x_{ij} \geq u_j, \forall j \tag{8.15}$$

$$\sum_{j} u_j V_j \geq \Gamma \tag{8.16}$$

The above model is solved for each day of the spill, resulting in a graphical representation of the plume/slick that identifies which areas have been "cleaned." The cleaned area and amount is considered during every subsequent iteration of the model. It is also used to determine the most optimal location to dispatch equipment from, given the equipment constraints at each specified staging area and the user-specified cleanup target that the model must meet. As the cleanup target gets more aggressive (more oil removed from the environment), more resources are required to meet the target, especially as the amount and extent of the oil plume/slick increases over the duration of the spill event. As a plume/slick moves through space and time, the most efficient (with respect to travel distance and time) staging location will also change. This gives response decision-makers the geospatial intelligence necessary for an efficient and effective response to spills.

The first step in a response analysis is to simulate the path of the slick and to determine the level of coastal impact. The TACOS model takes an approach to impact calculation similar to the one detailed in the impact section above. Briefly, a grid is preloaded into TACOS with the locations of sensitive assets. This grid is then compared to where the oil is expected to make landfall. The OSCOM model uses the amount of oil within those grid cells in combination with the number of sensitive coastal assets in the area to get an overall impact score for each individual grid cell.

Next, responders can set a baseline response option within TACOS and then compare it with other (more or less aggressive) response strategies. In Fig. 8.8, the benign neglect option (Psaraftis & Ziogas, 1985) is used as the baseline. The model allows the user to visually compare the difference in impact under different response options.

Noticeable in Fig. 8.8 are the patches of dark blue across the impact grid. These indicate that the alternative response options will have fewer impacts than the baseline option at those locations. The deeper the shade of blue, the fewer the impacts compared to baseline. This makes sense given the choice of benign neglect as the baseline. A point worth noting is the variation—or lack thereof—in blue color outlined by the red rectangles. The similar blue color across all alternative response strategies tells us that differences in impact will not be substantial. That is, the impacts will remain relatively stable no matter how aggressive the response is. With this information in mind, a response coordinator may opt for a less aggressive response (a 75% or 85% cleanup target) that is likely to be cheaper yet achieve a similar or the same level of protection.

Assume for the sake of this example that the response option of choice is in between the most aggressive and least aggressive, such as an 80% cleanup target. The response coordinators can now direct their attention to assessing where response equipment should be dispatched from. Again, the underlying

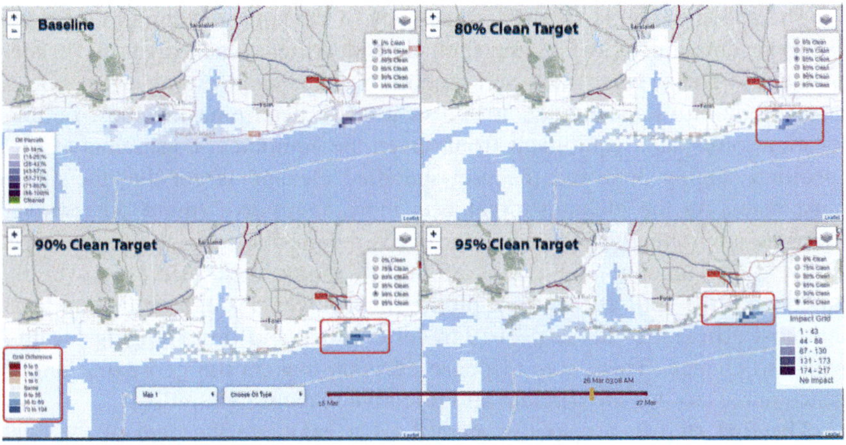

FIGURE 8.8 Example of the user interface for the online TACOS model showing the differences in impact resulting from selected response options.

optimization model determines these locations based on the movement of the spilt oil and whether there are enough resources located at each staging area to achieve the 80% cleanup target. If the cleanup target cannot be achieved with just the resources at one location, the model will identify the next closest location to "borrow" equipment from. It does this for every time step of the spill. Fig. 8.9 shows the location of the optimal staging equipment changing with the slick movement. In addition, the response teams will continually remove oil from the environment, which is also considered by the optimization model ensuring that only the equipment required to meet the cleanup target is dispatched.

8.4.2.2 Exclusion Boom Allocation Model

EBAM is another tactical response optimization model that determines where and how to best allocate exclusion booms to sensitive coastal areas. Like OSCOM, the EBAM considers several important pieces of geospatial information including the location of the staging areas and length of boom at each staging area, where and to what degree the potential impacts on the coastal environment will be (given a simulated oil spill), and allocates boom from staging areas accordingly. An overview of this model is provided here but a more detailed explanation of the inputs, outputs, and model development can be found in Grubesic et al. (2018). One may see from the

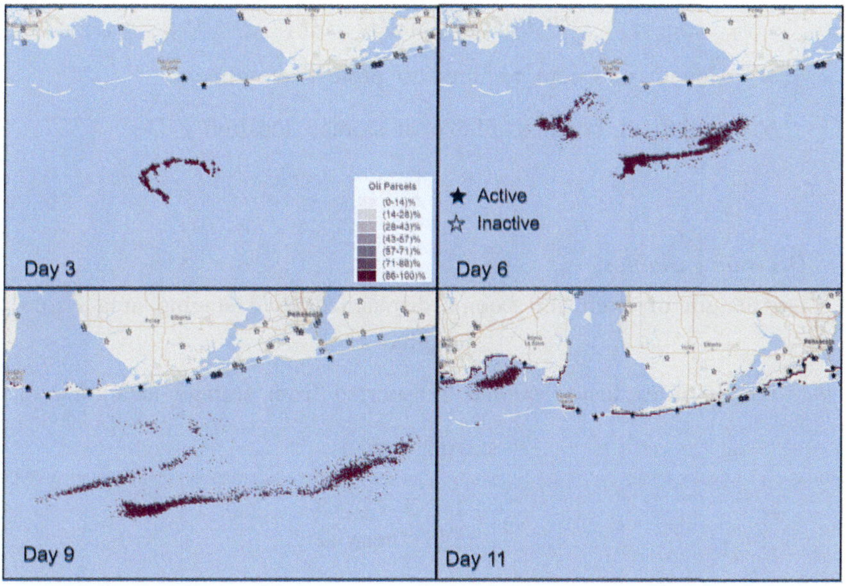

FIGURE 8.9 Temporal variation of the oil slick and the locations of the most optimal staging areas to dispatch response equipment from. Filled-in stars are the calculated optimal locations.

parameters below that the model contains many different factors which, besides the costs to deploy booms and the speed of response vessels, are based on the geographic data used by the model, including the oil spill model outputs.

Indices:

I = set of exclusive boom staging areas indexed by i

J = set of exclusive boom storage locations indexed by j

K = set of vulnerable shoreline segments indexed by k

Parameters:

S_k = potential impacts of oil spill on shoreline k

D^b_{ik} = time to transport booms from staging area i to shoreling k

D^t_{ji} = mileage from storage location j to staging area i

T_k = time for oil spill to hit shoreline k

f^b = vessel operating cost, per hour

c^b = cost for deploying one linear foot of boom

c^t = cost for transporting one linear foot of boom from storage location to staging area, per mile

Ψ_k = set of staging areas that can reach shoreline k before $T_k = \{i | D^b_{ik} \leq T_k\}$

M_k = length of boom required to protect shoreline k

N_j = length of boom available at storage location j

$$\beta = \sum_{j \in J} N_j$$

Decision variables:

x_{ik} = length of exclusion booms dispatched from staging area i to shoreline k

u_{ji} = length of exclusion booms transported from storage location j to staging area i

$$y_{ik} = \begin{cases} 1; \text{if } x_{ik} > 0 \\ 0; \text{otherwise} \end{cases}$$

$$z_k = \begin{cases} 1; \text{if shoreline } k \text{ is not protected by exclusion booms} \\ 0; \text{ otherwise} \end{cases}$$

The EBAM has two objective functions. The first [Eq. (8.17)] is to minimize the total spill impact on the vulnerable shorelines and the second [Eq. (8.18)] is to minimize the cost of dispatching the booms. In other words, the goal of this model is to protect sensitive coastal environments in the most economically efficient way. The model prioritizes the most sensitive coastal environments, while also deciding on how to make the best use of limited resources. The advantage of a bi-objective model is the ability to explore the trade-off between impacts and response. In some cases, objective [Eq. (8.17)] can be transformed and used as a final constraint Eq. (8.24) with the constraint method (Cohon, 2013). After transformation, we have the ability to vary Φ in constraint (8.24) to reflect different impact targets in terms of percent of total impact to evaluate the feasibility of different response options.

EBAM for coastal protection:

$$\min \sum_{k \in K} z_k S_k \tag{8.17}$$

$$\min \sum_{i \in I} \sum_{k \in K} \left(c^b x_{ik} + f^b D^b_{ik} y_{ik}\right) + \sum_{j} \sum_{i} c^t D^t_{ji} u_{ji} \tag{8.18}$$

Subject to:

$$\sum_{i \in I} u_{ji} \leq N_j, \forall j \in I \tag{8.19}$$

$$\sum_{k \in K} x_{ik} \leq \sum_{j \in J} u_{ji}, \forall i \in I \tag{8.20}$$

$$\sum_{i \in \Psi_k} x_{ik} - M_k + \beta z_k \geq 0, \forall k \in K \tag{8.21}$$

$$M_k y_{ik} \geq x_{ik}, \forall i, k \tag{8.22}$$

$$x_{ik} \geq 0, \forall i, k \tag{8.23}$$

$$u_{ji} \geq 0, \forall j, i$$

$$\sum_{k \in K} z_k S_k \leq \Phi \tag{8.24}$$

EBAM is subject to several constraints. Eq. (8.19) makes certain that the length of boom transported from a staging location to the shoreline is no more than the total length of boom available at that staging location. Eq. (8.20) ensures that the total length of boom transported from the storage facility to the staging location is less than or equal to the amount available at the storage facility. Eq. (8.21) makes sure that the length of boom dispatched to a specific shoreline segment is deployed before oil making landfall.

Eq. (8.22) makes sure the amount/length of boom dispatched to the shoreline is enough to cover the affected shoreline, and Eq. (8.23) simply imposes a positive values on some of the decision variables. To summarize, the model makes sure that sensitive areas are protected by booms before oil makes landfall, and that enough boom is brought to each sensitive area. If the nearest staging location to the impacted shoreline does not have enough boom, the model will choose the next nearest, and continue iterating through the staging areas until there are enough resources to ensure adequate shoreline protection before moving to the next sensitive shoreline location. Then, the model will go through the same iterative process again.

To illustrate the described procedure, the section below will provide a walkthrough of the example problem presented in Grubesic et al. (2018). The problem was formulated as follows. A nearshore surface spill was simulated just south of Mobile Bay, Alabama, in the GOM using the GNOME oil spill model (Beegle-Krause, 2001). The spill lasted 48 hours and over that time released 50,000 bbl of oil into the environment. It took 72 hours for all the oil from the spill to beach, with much of it making its way into the Bay. Staging area locations were obtained from the Florida Department of Fish and Wildlife (FWC, 2017) and each was given a prespecified length of boom which was randomly generated to be between 500 and 3000 ft. For this analysis, costs were monetized. Each vessel was assumed to cost $250 per hour to operate plus an additional $28 per foot of boom deployed (Etkin, 2004, 2009).

The oil spill parcels were tracked over the duration of the simulated period. The time of first beaching and final amount of oil was recorded for each grid cell in the impact grid. Total impacts were calculated using this information in the general equation for impacts detailed in Section 8.4.1 (Fig. 8.10). Another important attribute recorded for each grid cell was the total length of shoreline contained within a that grid cell. This value was used to determine the length of boom required to protect the sensitive shoreline within each of the affected grid cells. EBAM utilized this information, along with the staging location data, to determine boom deployment prioritization.

The results of this analysis illustrate several important points. First, given the known locations of equipment and how much equipment (vessels, length of boom) is located at each staging area, the model illustrates how much boom is necessary to completely protect sensitive coastal sites. Second, the results of this analysis identify where and in what order the response resources should be deployed in the most optimal way. The most sensitive and/or highly impacted areas will be prioritized, followed by less sensitive/lower impact areas. In other words, extra equipment will not be sent to some areas at the expense of not protecting more sensitive areas elsewhere. This can aid in avoiding costly mistakes. Third, response coordinators can determine where resources are coming from and then identify the locations where

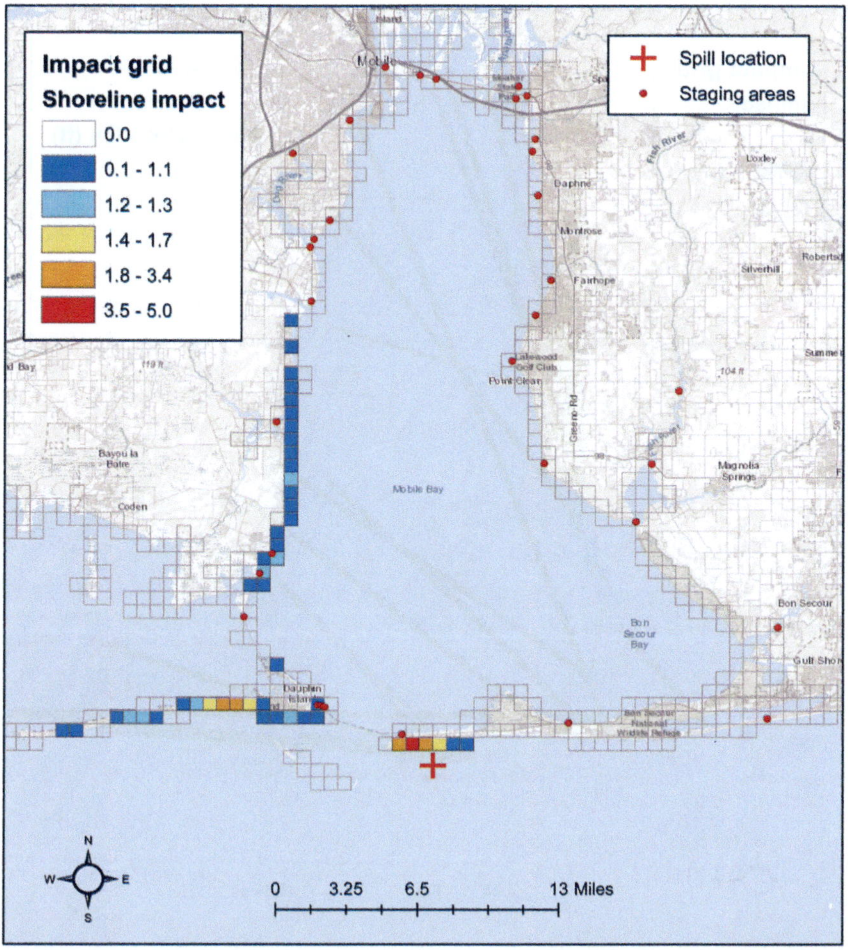

FIGURE 8.10 The staging locations and the impact grid used to inform the EBAM model. Equipment is sent from the staging areas to the areas that are predicted to have the highest impacts first, followed by the other locations in an optimal allocation with time. *EBAM*, Exclusion Boom Allocation Model.

there are resource inadequacies. That is, at some locations, in the event of a spill, the allocation of resources would not be sufficient to ensure that the sensitive receptors are protected in a timely manner. The locations where this might occur can be identified by EBAM.

Figs. 8.11 and 8.12 show the EBAM output in terms of the length of boom coming from each of the staging areas and which grid cell they are dispatched to. The thickness of the line indicates how much boom was required from each staging area. Thin lines indicate smaller amounts and thick lines indicate larger amounts. The staging area at the southernmost

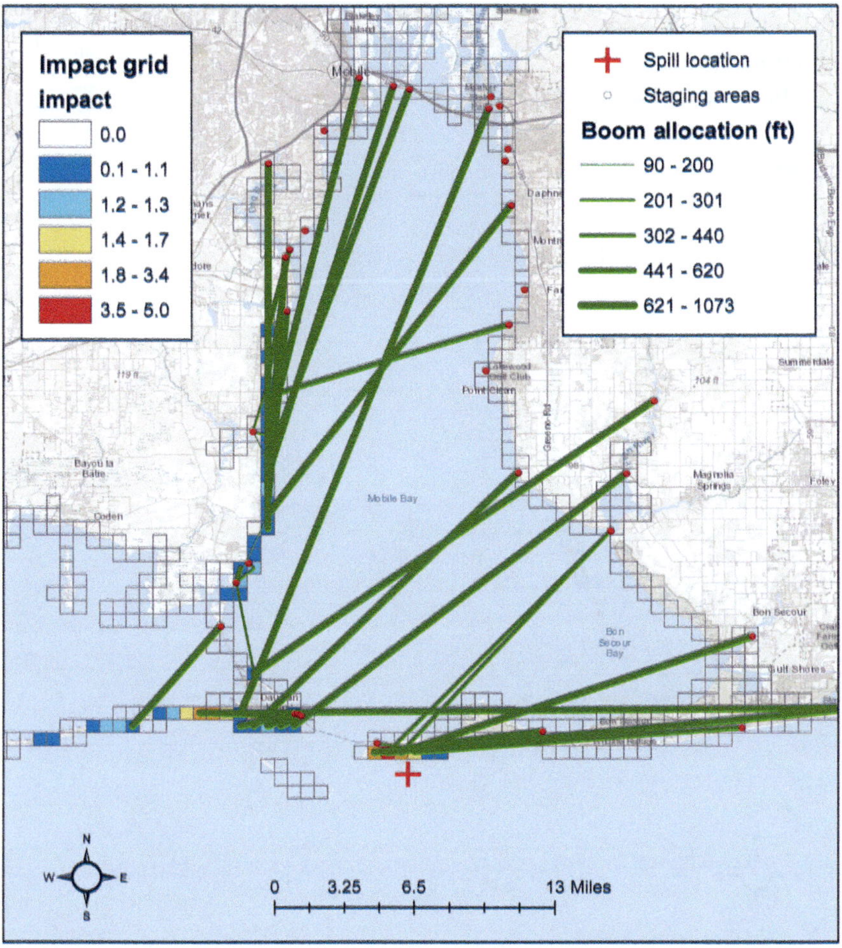

FIGURE 8.11 The results of the EBAM model indicating where booms need to come from to be most optimal while also ensuring that there are enough booms to protect the length of the shoreline within each grid cell. Line thickness indicates more boom being transported to the grid cell. This result is the most aggressive response where the goal is to achieve the highest reduction in impacts, no matter the cost. *EBAM*, Exclusion Boom Allocation Model.

portion of Mobile Bay, just to the east of the inlet is a good example of what having insufficient resources at a location in proximity to the spill means for response. The lines emanating from that staging area are thin, indicating that the length of boom available is small. As a result, it is not able to completely cover the impacted areas in the proximal grid cells. To completely cover those area, boom must be transported from staging areas further north (shown by the thick lines) which is costlier from both an economic and time standpoint.

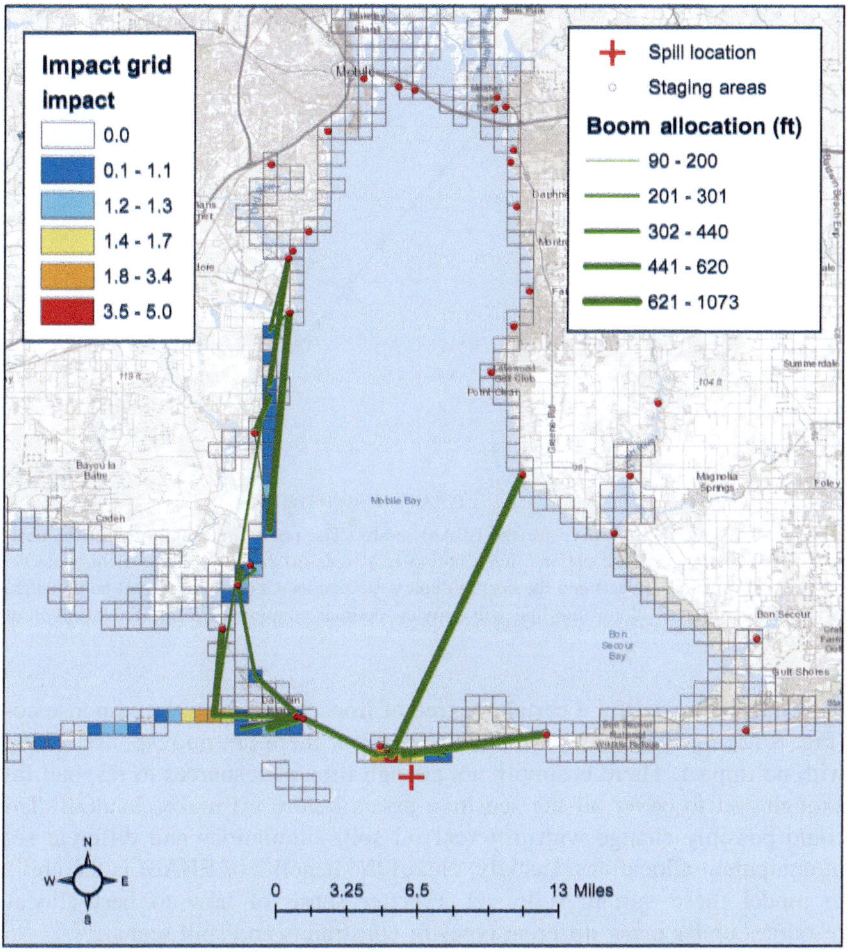

FIGURE 8.12 EBAM results for a balanced response between shoreline protection and costs to deploy the booms. Much fewer staging areas are needed to achieve the desired protection level. *EBAM*, Exclusion Boom Allocation Model.

Fig. 8.13 shows the trade-off curve from the EBAM set of solutions. Several response options corresponding to Figs. 8.11 and 8.12 are indicated. At the left-most point on the graph is the aggressive response option (Fig. 8.11). The optimal allocation of booms (determined by a random mix of equipment at staging areas) is deployed to the spill without any restrictions, resulting in the fewest impacts but the highest associated response cost. On the low end of the graph is benign neglect—no equipment is dispatched to the spill. The response cost is obviously the lowest (free!), but impacts are the highest. In the middle is a balanced approach where the

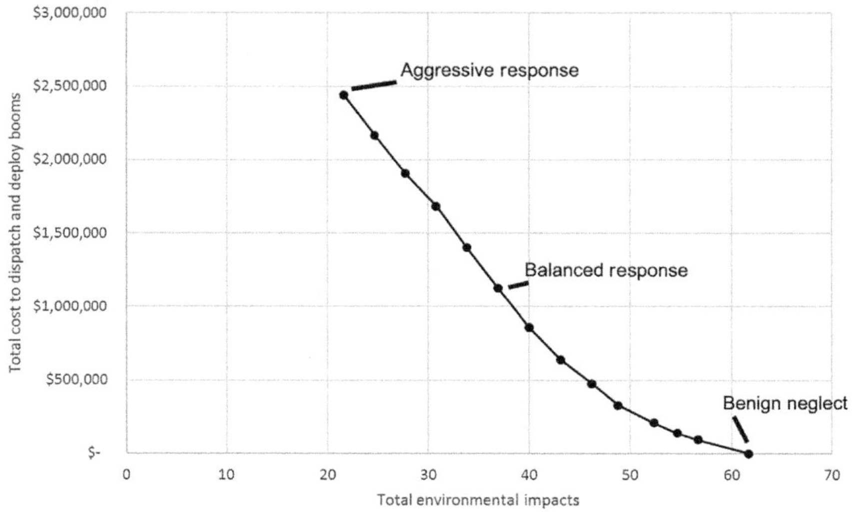

FIGURE 8.13 Trade-off curve for the EBAM model. The points indicated on the line correspond to different response options. Each option is a trade-off between the level of protection (total environmental impact) and the cost to deploy the booms. Choosing different combinations of impacts and costs along this line will provide optimal solutions. *EBAM*, Exclusion Boom Allocation Model.

responders allowed for a certain degree of impact to reduce the response cost (Fig. 8.12). As you may have already noticed, there are no response options with no impact. There is simply not enough time or resources to respond fast enough and to cover all the sensitive assets before oil makes landfall. This could possibly change with different oil spill simulations and different sets of equipment allocations. Luckily, one of the benefits of EBAM is the ability to model these variations to get a better sense of how to best allocate resources under many different types of constraints and spill scenarios.

With the geospatial intelligence provided by EBAM, responders are able to visualize and weigh the trade-offs between cost and impact. As mentioned previously, they are also able to identify the best-case and worst-case scenarios and determine what that means with respect to where and how resources are/should be allocated. Armed with this information, responders can take preemptive action by allocating their resources to the locations where modeled oil spills frequently impact, or they can model the oil spill in real time to help making operational response decisions.

8.5 Future work and knowledge gaps

The continued reliance on fossil energy means that exploration and production will also continue. This goes hand in hand with the increasing volumes of dangerous and toxic substances being routinely shipped around the globe.

To respond to these challenges, a great deal of work has been done within pollutant spill impact, risk, and response analysis (Nelson and Grubesic, 2020). This chapter has provided an overview of each, although it is not exhaustive. Great strides have been taken to improve algorithms and models, while data collection efforts have vastly enhanced our ability to represent vulnerable shorelines (Nelson and Grubesic, 2018a). As advances continue to be made on the methodological front, computational capabilities continue to increase. Faster computers that can handle huge data sets and run models at lightning speed are on the rise. This is highly advantageous due to the nature of particle-based spill models. With the ability to simulate hundreds of thousands of spills at once, the ability to develop probability curves for enhanced risk estimations increases. With that comes a better understanding of where the most likely impacts will occur across space and time. The behavior of the spill as well as the vulnerability and sensitivity of open ocean and coastal assets will likely change with the seasons. Understanding these variations will come from massive simulation efforts directed at modeling ambient conditions, seasonality, and, where applicable, the detailed geography and topography of the coastal environments.

In a related vein, response operations will be increasingly aware of how to respond to spills at different locations and at different times of year. Building a broad knowledge base can support changes in response resource stockpiles for different times of the year. Evaluating where a spill could occur and ensuring that there are enough response resources at the most vulnerable locations given the predicted movement and landfall of a spill for that location can be obtained through the models that were detailed above. Continued refinement of these models in varied locations helps to reduce the risk of deleterious effects and enhances our ability to respond effectively in the event of another spill.

8.6 Acknowledgment and disclaimer

This research was supported in part by appointments to the National Energy Technology Laboratory Research Participation Program, sponsored by the US Department of Energy and administered by the Oak Ridge Institute for Science and Education.

This report was prepared as an account of work sponsored by an agency of the US Government. Neither the US Government nor any agency thereof, nor any of their employees, makes any warranty, express or implied, or assumes any legal liability or responsibility for the accuracy, completeness, or usefulness of any information, apparatus, product, or process disclosed or represents that its use would not infringe privately owned rights. Reference therein to any specific commercial product, process, or service by trade name, trademark, manufacturer, or otherwise does not necessarily constitute or imply its endorsement, recommendation, or favoring by the US

Government or any agency thereof. The views and opinions of authors expressed therein do not necessarily state or reflect those of the US Government or any agency thereof.

References

Adger, W. N. (2006). Vulnerability. *Global Environmental Change, 16*(3), 268–281, Elsevier.

Al Shami, A., Harik, G., Alameddine, I., Bruschi, D., Garcia, D. A., & El-Fadel, M. (2017). Risk assessment of oil spills along the Mediterranean coast: A sensitivity analysis of the choice of hazard quantification. *Science of the Total Environment, 574*, 234–245. Available from http://doi.org/10.1016/j.scitotenv.2016.09.064.

Azevedo, A., Fortunato, A. B., Epifânio, B., den Boer, S., Oliveira, E. R., Alves, F. L., ... Oliveira, A. (2017). An oil risk management system based on high-resolution hazard and vulnerability calculations. *Ocean & Coastal Management, 136*, 1–18. Available from https://doi.org/10.1016/j.ocecoaman.2016.11.014.

Azevedo, A., Oliveira, A., Fortunato, A. B., Zhang, J., & Baptista, A. M. (2014). A cross-scale numerical modeling system for management support of oil spill accidents. *Marine Pollution Bulletin, 80*(1–2), 132–147, Elsevier.

Barker, C. H. (1999). The NOAA trajectory analysis planner: TAP 11. In *OCEANS'99 MTS/IEEE. Riding the crest into the 21st century*, 1999, pp. 1256–1261. IEEE.

Bauer, J. R., & Rose, K. (2015). Variable grid method: An intuitive approach for simultaneously quantifying and visualizing spatial data and uncertainty. *Transactions in GIS, 19*(3), 377–397, Wiley-Blackwell. Available from http://10.0.4.87/tgis.12158.

Beegle-Krause J. (2001). General NOAA Oil Modeling Environment (GNOME): A new spill trajectory model. In International *oil spill conference p*roceedings 2001(2). American Petroleum Institute: 865–871. https://doi.org/10.7901/2169-3358-2001-2-865.

Belardo, S., Harrald, J., Wallace, W. A., & Ward, J. (1984). A partial covering approach to siting response resources for major maritime oil spills. *Management Science, 30*(10), 1184–1196. Available from https://doi.org/10.1287/mnsc.30.10.1184, INFORMS.

Boer, S., den, Azevedo, A., Vaz, L., Costa, R., Fortunato, A. B., Oliveira, A., ... Rodrigues, M. (2014). Development of an oil spill hazard scenarios database for risk assessment. *Journal of Coastal Research, 70*(sp1), 539–544, Coastal Education and Research Foundation.

BSEE. (2018). Oil Spill Response Plans. Available at: https://www.bsee.gov/what-we-do/oil-spill-preparedness/preparedness-activities/oil-spill-response-plans.

Cai, L., Yan, L., Ni, J., & Wang, C. (2015). Assessment of ecological vulnerability under oil spill stress. *Sustainability*. Available from https://doi.org/10.3390/su71013073.

Canu, D. M., Solidoro, C., Bandelj, V., Quattrocchi, G., Sorgente, R., Olita, A., ... Cucco, A. (2015). Assessment of oil slick hazard and risk at vulnerable coastal sites. *Marine Pollution Bulletin, 94*(1), 84–95, Elsevier.

Castanedo, S., Juanes, J. A., Medina, R., Puente, A., Fernandez, F., Olabarrieta, M., & Pombo, C. (2009). Oil spill vulnerability assessment integrating physical, biological and socio-economical aspects: Application to the Cantabrian coast (Bay of Biscay, Spain). *Journal of Environmental Management, 91*(1), 149–159, Elsevier.

Cohon, J. L. (2013). Multiobjective programming and planning. Courier Corporation.

Copeland, C. (1999). Clean Water Act: A summary of the law. In 1999. Congressional Research Service, Library of Congress Washington, DC.

EIA (2018). Gulf of Mexico Fact Sheet. Available at: https://www.eia.gov/special/gulf_of_mexico/ Accessed 20.08.18.

Etkin D. S. (2004). Modeling oil spill response and damage costs. In *Proceedings of the fifth biennial freshwater spills symposium*, 2004.

Etkin D. S. (2009). Effectiveness of larger-area exclusion booming to protect sensitive sites in San Francisco Bay. Cortlandt Manor. Available at: http://tinyurl.com/y7p5hb59.

Fattal, P., Maanan, M., Tillier, I., Rollo, N., Robin, M., & Pottier, P. (2010). Coastal vulnerability to oil spill pollution: The case of Noirmoutier Island (France). *Journal of Coastal Research*, 879–887. Available from https://doi.org/10.2112/08-1159.1, Coastal Education and Research Foundation.

Fernández-Macho, J. (2016). Risk assessment for marine spills along European coastlines. *Marine Pollution Bulletin*, *113*(1–2), 200–210. Available from https://doi.org/10.1016/j.marpolbul.2016.09.015.

Fingas, M. (2016). *Oil spill science and technology*. Gulf Professional Publishing.

French-McCay, D. P. (2004). Oil spill impact modeling: Development and validation. *Environmental Toxicology and Chemistry*, *23*(10), 2441–2456, Wiley Online Library.

FWC (2017). Geographic Response Plan (GRP) staging areas. Available at: https://tinyurl.com/yctvlq6z Accessed 04.11.18.

Galt, J. A., & Payton, D. L. (1997). The development of a quantitative basis for optimal spill response planning. NOAA Technical Memorandum NOS ORCA 102. Seattle: Hazardous Materials Response and Assessment Division. *National Oceanic and Atmospheric Administration, 36*.

Galt, J. A., & Payton, D. L. (1999). Development of quantitative methods for spill response planning: A trajectory analysis planner. *Spill Science & Technology Bulletin*, *5*(1), 17–28, Elsevier.

Gasparotti, C. (2010). Risk assessment of marine oil spills. *Environmental Engineering & Management Journal*, *9*(4), 527–534. Available from http://search.ebscohost.com/login.aspx?direct = true&db = eih&AN = 52768930&site = ehost-live.

Grubesic, T., Wei, R., & Nelson, J. (2018). Protecting sensitive coastal areas with exclusion booms during oil spill events. *Environmental Modeling & Assessment*. Available from https://doi.org/10.1007/s10666-018-9634-2.

Grubesic, T. H., Wei, R., & Nelson, J. (2017). Optimizing oil spill cleanup efforts: A tactical approach and evaluation framework. *Marine Pollution Bulletin*, *125*(1–2), 318–329. Available from https://doi.org/10.1016/j.marpolbul.2017.09.012.

Guillen, G., Rainey, G., & Morin, M. (2004). A simple rapid approach using coupled multivariate statistical methods, GIS and trajectory models to delineate areas of common oil spill risk. *Journal of Marine Systems*, *45*(3–4), 221–235. Available from http://doi.org/10.1016/j.jmarsys.2003.11.006.

Gundlach, E. R., & Hayes, M. O. (1978). Vulnerability of coastal environments to oil spill impacts. *Marine Technology Society Journal*. Available from https://doi.org/10.1038/271164a0.

Gkonis, K. G., Kakalis, N. M., Ventikos, N. P., Ventikos, Y., & Psaraftis, H. N. (2007). A model-based approach for tactical decision making in oil spill response. International symposium on maritime safety, security & eenviromental protection. Athens, Greece, 2007.

IOGP (2015). Oil spill preparedness and response: An introduction. London. Available at: http://www.oilspillresponseproject.org/wp-content/uploads/2017/01/Oil_Spill_Preparedness_Response_Introduction_2016.pdf.

ITOPF (2019). International Tanker Owners Pollution Federation. Available at: http://www.itopf.org/. Accessed 14.01.19.

Iakovou, E., Ip, C. M., Douligeris, C., & Korde, A. (1996). Optimal location and capacity of emergency cleanup equipment for oil spill response. *European Journal of Operational Research, 96*(1), 72–80. Available from https://doi.org/10.1016/S0377-2217(96)00106-3.

Jensen, J. R., Ramsey, E. W., III, Holmes, J. M., Michel, J. E., Savitsky, B., & Davis, B. A. (1990). Environmental sensitivity index (ESI) mapping for oil spills using remote sensing and geographic information system technology. *International Journal of Geographical Information System, 4*(2), 181–201.

Kaplan, S., & Garrick, B. J. (1981). On the quantitative definition of risk. *Risk Analysis, 1*(1), 11–27, Wiley Online Library.

Lan, D., Liang, B., Bao, C., Ma, M., Xu, Y., & Yu, C. (2015). Marine oil spill risk mapping for accidental pollution and its application in a coastal city. *Marine Pollution Bulletin, 96*(1–2), 220–225. Available from http://doi.org/10.1016/j.marpolbul.2015.05.023.

Lee, M., & Jung, J.-Y. (2015). Pollution risk assessment of oil spill accidents in Garorim Bay of Korea. *Marine Pollution Bulletin, 100*(1), 297–303. Available from http://doi.org/10.1016/j.marpolbul.2015.08.037.

Li, P., Cai, Q., Lin, W., Chen, B., & Zhang, B. (2016). Offshore oil spill response practices and emerging challenges. *Marine Pollution Bulletin, 110*(1), 6–27. Available from http://doi.org/10.1016/j.marpolbul.2016.06.020.

Liu, X., Meng, R., Xing, Q., Lou, M., Chao, H., & Bing, L. (2015). Assessing oil spill risk in the Chinese Bohai Sea: A case study for both ship and platform related oil spills. *Ocean & Coastal Management, 108*, 140–146. Available from http://doi.org/10.1016/j.ocecoaman.2014.08.016.

Liu, Y., MacFadyen, A., Ji, Z. G., & Weisberg, R. H. (2013). *Monitoring and modeling the Deepwater Horizon oil spill: A record breaking enterprise*. John Wiley & Sons.

Mokhtari, S., Hosseini, S. M., Danehkar, A., Azad, M. T., Kadlec, J., Jolma, A., & Naimi, B. (2015). Inferring spatial distribution of oil spill risks from proxies: Case study in the north of the Persian Gulf. *Ocean & Coastal Management, 116*, 504–511. Available from http://doi.org/10.1016/j.ocecoaman.2015.08.017.

Nelson, J., & Grubesic, T. (2018). Oil spill modeling: Risk, spatial vulnerability, and impact assessment. *Progress in Physical Geography, 42*(1), 112–127. Available from https://doi.org/10.1177/0309133317744737.

Nelson, J., Grubesic, T., Sim, L., Rose, K., & Graham, J. (2015). Approach for assessing coastal vulnerability to oil spills for prevention and readiness using GIS and the Blowout and Spill Occurrence Model. *Ocean & Coastal Management, 112*, 1–11. Available from http://doi.org/10.1016/j.ocecoaman.2015.04.014.

Nelson, J. R., & Grubesic, T. H. (2017b). A repeated sampling method for oil spill impact uncertainty and interpolation. *International Journal of Disaster Risk Reduction, 22*, 420–430. Available from http://doi.org/10.1016/j.ijdrr.2017.01.014.

Nelson, J. R., Grubesic, T. H., Sim, L., & Rose, K. (2018). A geospatial evaluation of oil spill impact potential on coastal tourism in the Gulf of Mexico. *Computers, Environment and Urban Systems, 68*, 26–36. Available from https://doi.org/10.1016/j.compenvurbsys.2017.10.001.

Nelson, J. R., & Grubesic, T. H. (2018a). Oil spill modeling: Computational tools, analytical frameworks, and emerging technologies. *Progress in Physical Geography: Earth and Environment*. Available from http://doi.org/10.1177/0309133318804977.

Nelson, J. R., & Grubesic, T. H. (2018b). The implications of oil exploration off the Gulf Coast of Florida. *Journal of Marine Science and Engineering*, *6*(2). Available from https://doi.org/10.3390/jmse6020030.

Nelson, J. R., & Grubesic, T. H. (2020). Oil spill modeling: Mapping the knowledge domain. *Progress in Physical Geography: Earth and Environment*, 030913331989 7503. Available from http://doi.org/10.1177/0309133319897503.

Nelson, J. R., & Grubesic, T. H. (2021). *A spatiotemporal analysis of oil spill severity using a multi-criteria decision framework. Ocean & Coastal Management*, *199*, 105410.

NOAA (2013). Shoreline assessment manual. 4th Edition. Seattle, WA. Available at: https://response.restoration.noaa.gov/sites/default/files/manual_shore_assess_aug2013.pdf.

Olita, A., Cucco, A., Simeone, S., Ribotti, A., Fazioli, L., Sorgente, B., & Sorgente, R. (2012). Oil spill hazard and risk assessment for the shorelines of a Mediterranean coastal archipelago. *Ocean and Coastal Management*, *57*, 44–52. Available from http://doi.org/10.1016/j.ocecoaman.2011.11.006.

Psaraftis, H. N., & Ziogas, B. O. (1985). A tactical decision algorithm for the optimal dispatching of oil spill cleanup equipment. *Management Science*, *31*(12), 1475–1491. Available from https://doi.org/10.1287/mnsc.31.12.1475, INFORMS.

Sammarco, P. W., Kolian, S. R., Warby, R. A. F., Bouldin, J. L., Subra, W. A., & Porter, S. A. (2013). Distribution and concentrations of petroleum hydrocarbons associated with the BP/Deepwater Horizon Oil Spill, Gulf of Mexico. *Marine Pollution Bulletin*, *73*(1), 129–143. Available from http://doi.org/10.1016/j.marpolbul.2013.05.029.

Sebastiao, P., & Soares, C. G. (2006). Uncertainty in predictions of oil spill trajectories in a coastal zone. *Journal of Marine Systems*, *63*(3), 257–269, Elsevier.

Sebastião, P., & Soares, C. G. (2007). Uncertainty in predictions of oil spill trajectories in open sea. *Ocean Engineering*, *34*(3), 576–584, Elsevier.

Sepp Neves, A. A., Pinardi, N., & Martins, F. (2016). IT-OSRA: Applying ensemble simulations to estimate the oil spill risk associated to operational and accidental oil spills. *Ocean Dynamics*, *66*(8), 939–954. Available from https://doi.org/10.1007/s10236-016-0960-0.

Sepp Neves, A. A., Pinardi, N., Martins, F., Janeiro, J., Samaras, A., Zodiatis, G., & De Dominicis, M. (2015). Towards a common oil spill risk assessment framework—Adapting ISO 31000 and addressing uncertainties. *Journal of Environmental Management*, *159*, 158–168. Available from http://doi.org/10.1016/j.jenvman.2015.04.044.

Sepp Neves., Pinardi, N., Navarra, A., & Trotta, F. (2020). A general methodology for beached oil spill hazard mapping. *Frontiers in Marine Science*, *7*, 65.

Spaulding, M. L. (2017). State of the art review and future directions in oil spill modeling. *Marine Pollution Bulletin*, *115*(1–2), 7–19. Available from http://doi.org/10.1016/j.marpolbul.2017.01.001.

Shuohui, Z., Xuejing, S., Shuang, Z., & Xuan, C. (2006). Risk analysis methods in oil spill contingency plans. In Proceedings of the 7th Annual General Assembly and Conference AGA-7, 2006, 410–417.

SRA (2015). *Society for risk analysis glossary*. Available at: http://www.sra.org/sites/default/files/pdf/SRA_glossary_20150622.pdf.

Tkalin, A. V. (1986). Evaporation of petroleum hydrocarbons from films on a smooth sea surface. *Oceanology ONLGAE*, *26*(4).

Tong, D., & Murray, A. T. (2012). Spatial Optimization in Geography. *Annals of the Association of American Geographers*, *102*(6), 1290–1309. Available from https://doi.org/10.1080/00045608.2012.685044, Routledge.

USCG (2011). *On scene coordinator report: Deepwater Horizon oil spill.* Available at: http://docs.lib.noaa.gov/noaa_documents/NOAA_related_docs/oil_spills/on-scene_DWH_Report_Sep2011.pdf.

Verma, M., Gendreau, M., & Laporte, G. (2013). Optimal location and capability of oil-spill response facilities for the south coast of Newfoundland. *Omega, 41*(5), 856–867. Available from https://doi.org/10.1016/j/omega.2012.10.007.

Wilhelm, W. E., & Srinivasa, A. V. (1997). Prescribing tactical response for oil spill clean up operations. *Management Science, 43*(3), 386–402, INFORMS.

Wilhelm, W. E., Srinivasa, A. V., & Wilhelm, W. E. (1997). A procedure for optimizing tactical response in oil spill clean up operations. *Management Science, 102*(3), 554–574, Elsevier.

Zhang, D. F., Easton, A. K., & Steiner, J. M. (1997). Simulation of coastal oil spills using the random walk particle method with Gaussian kernel weighting. *Spill Science and Technology Bulletin, 4*(2), 71–88. Available from https://doi.org/10.1016/s1353-2561(98)00003-6.

Zhong, Z., & You, F. (2011). Oil spill response planning with consideration of physicochemical evolution of the oil slick: A multiobjective optimization approach. *Computers & Chemical Engineering, 35*(8), 1614–1630.

Chapter 9

Decision support tools for managing marine hydrocarbon spills in island environments

José Ramón Bergueiro López[1,2,3], José Manuel Calvilla Quintero[3], Kevin Soler Carracedo[4], Eloy Calvilla Quintero[5] and George Zodiatis[6]

[1]*Department of Chemical Sciences, University of Santiago de Compostela, Santiago, Spain,* [2]*Department of Chemical Engineering, University of the Balearic Islands, Palma, Spain,* [3]*Research group: CONSEMAR - Pollution and Marine Safety, University of La Laguna, San Cristóbal de La Laguna, Spain,* [4]*Science Faculty, Physics Department, University of La Laguna, San Cristóbal de La Laguna, Spain,* [5]*Marine Engineering, Navy and Naval Radioelectronics, Section of the Higher Polytechnic School of Engineering, University of La Laguna, San Cristóbal de La Laguna, Spain,* [6]*Coastal & Marine Research Lab, Institute of Applied and Computational Mathematics, Foundation for Research and Technology—Hellas, Heraklion, Greece*

9.1 Why marine hydrocarbon spills are a problem?

Marine pollution is a major problem for humanity as a whole because of the great influence that the sea has on life on Earth. Within marine pollution, hydrocarbon spills occupy a preeminent place due to its peculiar characteristics. Society all around the world is nowadays linked to a high consumption of hydrocarbons and to the remoteness of the places of production of hydrocarbons from those of greater consumption, so incidental spills may not be ruled out. These incidents, which may represent a small percentage of the total amount of oil that reaches the oceans, are nevertheless very serious because they impact local ecosystems and environment in a wider sense. Good information on oil tanker accidents that caused oil spills and their possible treatment can be obtained from the International Tanker Owners Pollution Federation Limited (ITOPF) (Bergueiro, Moreno, Martí, & Díaz, 2011a; Bergueiro, Moreno, Martí, & Díaz, 2011b; Bergueiro, Moreno, Martí, & Díaz, 2011c; ITOPF, 2020).

Marine Hydrocarbon Spill Assessments. DOI: https://doi.org/10.1016/B978-0-12-819354-9.00008-9

To minimize any spill impacts, it is necessary to follow a series of guidelines and practical steps. A pathway for such actions must first consider everything concerning the containment operations by means of barriers, fences, and interceptors, to prevent the continuous spread of hydrocarbons. Next, everything related to the process of hydrocarbon spreading must be considered using numerical simulation models. These models must take into account the effects of winds, currents, and waves; changes of the physical properties of the spilled hydrocarbon, including the evaporated, dispersed, and emulsified quantities; the quantities of hydrocarbons deposited on the coast; and hydrocarbons that remain in the water column and on the sea surface. Then, everything related to the recovery of the hydrocarbons (e.g., by means of skimmers and adsorbents) and the possible treatment with dispersants, as well as assessments of biodegradation of the hydrocarbons must be dealt with. Elimination of hydrocarbons by burning should also be addressed, although it is a technique that is not usually highly recommended. Finally, the recovery of the hydrocarbons that have impacted the coast must be considered, then their storage, and lastly their treatment/reuse/disposal.

All these actions must be coordinated by means of a spill response and/or contingency plan, identifying the resources destined to efficiently combat the effects of a spill. The plan must be developed to address an emergency situation and to identify required actions and possible solutions, because to be effective in the event of a spill, the decisions must be made quickly.

The objectives of the contingency plan are to establish a procedure that indicates the actions to be followed in the face of certain risks, so the impacts of a can be mitigated. The use of resources must be optimized, and adequate controls must be implemented to comply with the legislation, policies, and procedures.

The first hypothesis of a contingency plan is that no two oil spills are exactly the same. The second is that the treatment of the same oil spill at the beginning and at the end of the incident can be very different. The third hypothesis is that each oil spill should serve to validate the simulation of the oil spill models. In the absence of incidental spills, the emergency drills may be used to validate the simulation models and it is necessary to carry out at least 10 drills for achieving satisfactory training results. Another important principle is that a small error can be managed, a failure is a risk, and there should not be more failures than those that are unavoidable. It must be remembered that all improvisation is a risk, so all interventions must always be planned and nothing should ever be improvised. It is then essential to anticipate what may happen in an emergency, based on the most accurate information of potential incidents and their foreseeable developments. An analytical process may be followed:

1. Identification of marine hazards according to their typology, as well as the nature and quantity of polluting substances that could be involved in an incident

2. Modeling of oil spill trajectory taking into account the wind, sea currents, waves, bathymetry, shoreline of the area, and physical–chemical characteristics of the spilled substances
3. Estimation of the probability of incident occurrence
4. Evaluation of the environmental consequences and impacts of the spill
5. Risk characterization based on the assessment of incident probability and consequences.

Once the different risks have been identified and evaluated, the next step is to assess whether there are sufficient human and material resources available in the area to quickly, efficiently, and safely control such risks. To this end, the availability and capacity of the following means of combating marine pollution should be assessed:

1. Human factor in risk awareness and training for management, coordination, intervention, and assistance in an emergency situation
2. Material resources for the intervention and personal protective equipment
3. Alert and communication systems
4. Resources and centers to control and coordinate the emergency situation and engaged resources.

With respect to the organization facing an emergency, the Operational Plan to Combat Accidental Marine Pollution must establish an organizational structure that guarantees the provision of services and completion of recovery missions. This structure must have the following sections:

1. Direction and coordination of all actions and mobilized resources
2. Direct onsite intervention
3. Health care and evacuation of affected personnel
4. Onsite environmental management
5. Logistical support to guarantee the supply of materials and equipment and the operation of essential services during the emergency
6. Relations and coordination with stakeholders, such as authorities, media, and public
7. Management of maritime and terrestrial traffic.

The following information must be provided to safely and efficiently control each type of accident:

1. Risks associated with the emergency.
2. Personal protection equipment and necessary means of intervention.
3. Precise instructions for communication, risk control, and minimization of consequences for people and environment.
4. Main effects derived from the risk and recommendations for the application of first aid, tending to minimize the consequences in case of exposure of the personnel involved in the interventions.
5. Health and Safety Plan for the personnel involved in the actions.
6. Outside notification criteria and request for assistance.

Such management can be carried out by the means of SONIA (Operational System of Notification of Environmental Impacts) and SIROCO (Island Systems of Response and Operations to Ocean Pollutants) models, implemented in Spain and described in more detail below.

9.1.1 SONIA model

The SONIA model (El Modelo SONIA, 2009) is an integrated system of numerical simulation models formed by several modules that can work in an integrated or independent manner. When they are used in an integrated manner, they provide an exhaustive knowledge of the evolution of an oil spill at sea and of the oil-aging processes. The integrated system allows to choose the most suitable measures to minimize the impact of hydrocarbons on the coast and the environment in general, including the health and safety of all personnel involved in the contingency measures and in the cleaning and restoration of all coastal environments affected by the incident. The system connects to databases on hydrocarbon volatility, toxicity, and explosiveness, on types of booms for the containment, their form of deployment, and anchorage. It also has access to information on skimmers and other recovery systems, on dispersants, and on their application conditions and limitations to their use. The model addresses everything related to the elimination of hydrocarbons spilled on the sea surface through burning techniques. The model deals with all aspects related to the vulnerability, resilience, and induced recovery of coastal environments that have been affected by hydrocarbons. The model can access data concerning the coastal environments of the Balearic Islands and the fauna of these islands, which are susceptible to being affected by spills.

9.1.2 SIROCO model

The objective of the SIROCO model (Calvilla-Quintero J.M., 2008) is a set of management tools to effectively and efficiently manage any aspect of a hydrocarbon spill in the marine environment adapted to the coastal environment of the Canary Islands (Fig. 9.1).

The SIROCO model is based on oceanographic, meteorological, and GIS (Geographic Information System) databases and allows reproduction and interpretation of the behavior and trajectory of spilled oils. This integral system can be used operationally in emergencies, to plan the response to an oil spill at sea, and in preparatory emergency drills. It provides support in training the emergency response personnel who are involved in coastal cleanup and remediation/restoration tasks.

9.2 Response times

Whenever there is an oil spill at sea, the first part of a contingency plan must address the response time to collect onsite information, the time needed

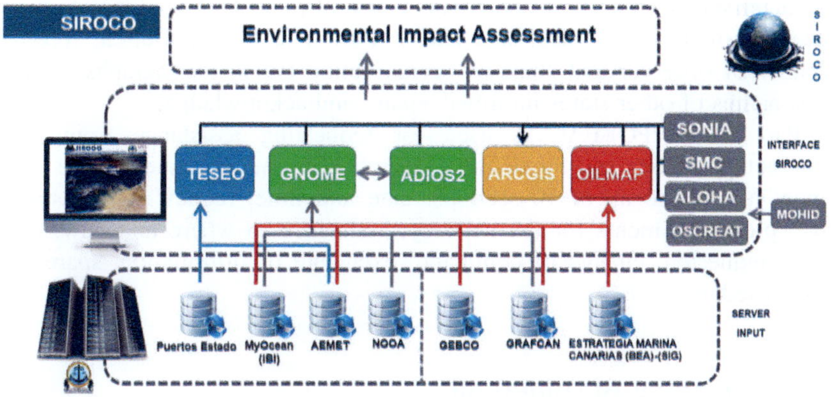

FIGURE 9.1 SIROCO model interface.

for the rescue of anybody affected, and the time to mount a response campaign. These tasks can be addressed using the TIMES (Model for calculating response times) model (Martí Moreno, 2008). As inputs, this model needs information on the spill location, time, distances from the spill to the response resources, and on the number of people to be rescued, in case there are any. After the spill coordinates are provided, the model indicates the site on the Balearic Islands where the closest response resources are located and the type of resources available there.

The output of the model gives the times needed to access the spill site by a plane or a helicopter. Then the model estimates the time needed to rescue any affected people using a helicopter or a tugboat and calculates the time needed for the transport of the response gear. Lastly, the model calculates the costs of the resources (plane, helicopter, tug) that can be used. The model also shows the notification procedure recommended by the International Maritime Organization (IMO) according to the resolution A.648 (ITOPF, 2011a, 2011b, 2011c). This resolution aims to harmonize the different notification systems and shows the general principles to which the notification systems and requirements for ships should conform, including guidelines for reporting events involving dangerous goods (IMO, 1999). An example of this type of notification procedure can be found in the figure of the POLREP (Pollution Reporting System) (REMPEC, 2020).

The POLREP consists of three reports, the POLWARN (Pollution Warning), the POLINF (POLlution INFormation), and the POLFAC (POLlution FACilities).

Part I or POLWARN gives the information or warning of the pollution or the threat. It is made up of five sections: date and time, position, incident, outflow, and acknowledge.

Part II or POLINF gives detailed supplementary report, as well as situation reports. It is made up of 15 sections: date and time, position, characteristics of pollution, source and cause of pollution, wind direction and speed, current or tide, sea state and visibility, drift of pollution, forecast, identity of observer and ships on scene, action taken, photographs or samples, names of other states informed, spare, and acknowledge.

Part III or POLFAC is used for requesting assistance from other Contracting Parties and for defining operational matters related to the assistance. It is made up of 10 sections: date and time, request for assistance, cost, prearrangements for the delivery, assistance to where and how, other states requested, change of command, exchange of information, spare, and acknowledge.

9.3 Spreading of spilled oil

Immediately after a petroleum substance is spilled on the water surface, the spreading begins. In the first stage, spreading is governed by the gradient of the difference in density between the water and the spilled hydrocarbons, reducing the importance of this effect as the hydrocarbon film becomes thinner. The forces that oppose the spreading at this time are those due to the inertia of the system, or inertia of the hydrocarbon layer, which decreases as it becomes thinner. This stage is known as the "Gravitational-inertial" stage. When the inertia of the hydrocarbon layer becomes very small, there are still other forces that retard the spreading; its friction on the water surface is related to the viscosity, which increases with time. These friction–viscosity forces oppose the gravitational force and correspond to the second stage, called "gravitational-viscous." In addition, at the edge of the hydrocarbon film, there is an imbalance between the air–hydrocarbon, air–water, and water–hydrocarbon interfacial tensions, expressed by the spreading coefficient. When the gravitational effect decreases, these uncompensated surface tensions become the driving force of spreading, which oppose the friction–viscosity forces. This is the third stage, called "viscosity-strains." Finally, depending only on the volume of hydrocarbons spilled, there is the fourth stage called "maximum area." For more detail, see Bergueiro and Domínguez (2001).

By means of the proposed AREAS (Model for calculating oil spill spreading areas) model (Martí Moreno, 2008), the values of the four spreading areas and their respective times can be assessed. The input parameters to the AREAS model are the following:

ρ_0 Oil density, kg/m^3

ρ_w Seawater density, kg/m^3

g Gravitational acceleration, m/s^2

V Volume of hydrocarbons spilled, m^3

υ Kinematic viscosity, m^2/s

σ Coefficient of oil spreading in seawater, mN/m

t time, s

The output parameters of the AREAS model are:

1. Areas of each of the four stages: gravitational-inertial, gravitational-viscous, viscous stresses, and maximum area
2. Time to achieve them

9.4 Evaporation of crude oil and derivatives

When a crude oil or a derivative product is spilled, the first process that takes place is evaporation, which begins at the same time as the spill and ends when all the volatile fractions have evaporated. The evaporation process of these hydrocarbon mixtures takes place through a complex process of matter and energy transfer. This transfer occurs first in the liquid phase, then through the interface, and finally in the gas phase. Complete and detailed information on the evaporation of crude oil and derivative mixtures can be obtained from the works of Bergueiro and Domínguez (1996, 2001). The management of the evaporation processes of crude oil and derived products can be carried out through the AVE (Assessment of Velocity of Evaporation) model (Martí Moreno, 2008). Through the application of this model, the equations of the evaporation curves of crude oil and derived products, spilled on the water surface and on beaches of different granulometries, for different air and water temperatures and wind conditions can be obtained. The equations of the evaporation curves of 464 crude oil and derived products can be obtained from the Environment Canada, Environmental Science and Technology Centre (Environment Canada, 2008) and from the work of Jokuty et al. (1996), among others.

The experimental data on evaporation rate as a function of time, for a specific temperature and wind speed, can be approximated by a number of expressions, although the most usual is of the type:

$$Fm = a\text{Ln}(1 + b{\cdot}t) \tag{9.1}$$

where Fm is mass evaporated fraction (%), t is elapsed time (min), and a and b are adjustment parameters from experimental data.

Likewise, for a given air speed, the evaporated fraction can be correlated with temperature and time by means of equation (Bergueiro et al., 2011a) from Environment Canada (Environment Canada, 2008).

$$Fv(\%) = (a + b \cdot T)\text{Ln}t \tag{9.2}$$

where Fv is evaporated fraction (%), T is temperature (°C), t is time (min), and a and b are adjustment parameters.

9.5 Containment by barriers, fences, and interceptors

Since an oil spill at sea occurs, the main actions that need to be taken must be aimed at preventing the mass of oil from spreading, followed by recovering most of the spilled oil. Containment systems are used to prevent the spreading of spilled oil. The term "spill containment" includes all those actions that avoid the extension and confine the spill to the smallest possible area. Sometimes the term containment is linked to the protection of a specific place or area, vulnerable sites and valuable infrastructure, such as water intake for desalination plant, fish farm, popular beach, etc.

9.5.1 Floating barriers/booms: elements, efficiency

A floating barrier/boom is a mechanical device designed to be deployed on the sea surface to surround a mass of spilled hydrocarbon. It must have high flexibility to be deployed, retrieved, and stored on reels. The basic elements of a containment barrier are the skirt, the freeboard, the ballast, the linkage or connection elements, and the tension or longitudinal traction elements (Fig. 9.2).

The functions of the boom components are as follows: (1) function of the freeboard is to prevent hydrocarbons from passing over the barrier, (2) function of the skirt (subsurface) is to prevent hydrocarbons from passing under the barrier, (3) function of the float (air, foam, or other floating material) is to prevent the barrier from sinking, (4) function of the longitudinal tension member is to resist the force of winds, waves, and currents, and (5) function of the ballast is to keep the barrier vertical.

The main characteristics of a hydrocarbon containment barrier should meet the following criteria:

1. Adapt to the movements of the water surface.
2. Block surface currents, but not waves.
3. Avoid leaks, above the freeboard and below the skirt with winds of 20 m/s, currents of 1 m/s, and waves of 3 m.
4. Resist drying out.
5. Resist torsional forces.
6. Resist the action of hydrocarbons.

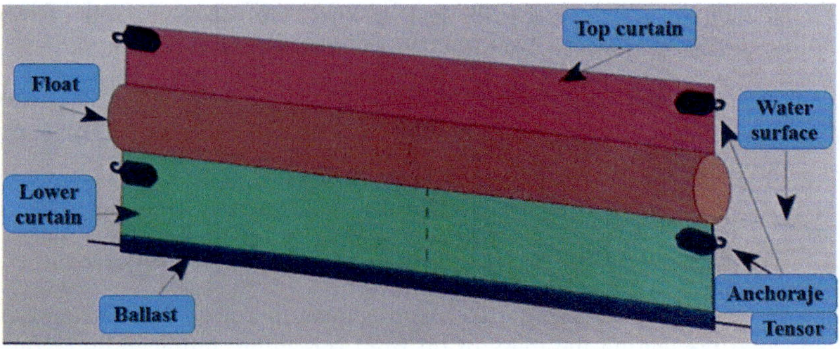

FIGURE 9.2 Elements of a hydrocarbon containment barrier/boom.

7. Resist the effect of high temperatures and ultraviolet rays.
8. Resist abrasion and the action of floating debris.
9. Be easy and safe to manipulate.
10. Be modular and expandable in length.
11. Need little logistical support.
12. Occupy a reduced storage volume.
13. Require little maintenance and long service life.

Three types of barriers/boom can be distinguished, depending on where they are to be used: coast, open sea, and rivers and canals.

The effectiveness of floating barriers/booms against some environmental variables is shown in Table 9.1. Where the term "Value" shows the ranges where the containment barriers have a normal behavior and the term "Maximum resistance" refers to the intervals that, if surpassed, the barriers/booms would be no longer effective.

Sometimes it is possible to resort to the containment of spills by means of improvised barriers/booms and fences, by means of adsorbent barriers, by means of chemical barriers, and by means of bubble barriers. When the spilled hydrocarbons are highly aged, and therefore of very high viscosity, their containment and recovery can be carried out by means of net barriers.

9.5.2 Bubble barriers

When the amount of hydrocarbons spilled is small and when it is necessary to protect ports or docks from the arrival of an oil slick without interfering with the maritime traffic, air bubble barriers, also known as pneumatic barriers, can be used. These barriers are a curtain of air bubbles produced by a perforated conductor submerged under the surface of the spill. This curtain of air bubbles, when it reaches the surface, prevents the hydrocarbons from continuing to spread (Fig. 9.3).

TABLE 9.1 Effectiveness of floating barriers/booms against some environmental variables.

Environmental variable	Value	Maximum resistance
Currents (m/s)	0.25–0.80	2.0–3.1
Winds (m/s)	<10	10–20
Waves (m/s)	0.3–1.8	2.5–4.5
Temperature (°C)	−40 to 60	

FIGURE 9.3 Air bubble barrier. *From Hydrotechnik Lübeck GMBH-Nauticexpo (2021).*

9.5.3 Adsorbent barriers

It is a barrier that is useful for both containment and recovery of hydrocarbons. Generally, these barriers are made of a rigid material covered with another adsorbent material, or of an adsorbent material strong enough to withstand the mechanical loads and stresses. It must have a flotation element to avoid sinking when the adsorbent material is saturated with oil. Such barriers are typically used when the hydrocarbon layer is very thin; otherwise the adsorbent material will quickly become saturated, which results in reduced recovery efficiency. Fig. 9.4 shows this type of barrier.

9.5.4 Net barrier

To contain and recover highly aged tarballs a net barrier can be used, as shown in Fig. 9.5.

9.5.5 Chemical barriers

These barriers do not exist physically as such but are formed "in situ" by the addition of a series of chemical compounds to the slick front. Their mission is to modify the surface tension and viscosity of the spilled oil. As a consequence, the spreading of spilled oil is inhibited. When these chemicals are spread on the surface of the water near the spilled oil, the oil is pushed back because the added chemical has a higher spreading coefficient than the floating hydrocarbons. In practice, these chemicals are applied by spraying from a boat or helicopter so that the spilled hydrocarbons are surrounded and concentrated in a thicker phase or layer, which is easier to recover. Its effect lasts only a few hours, so the recovery of the hydrocarbons must be done immediately after the application. This type of barrier is not very effective with very viscous crude oil, in very cold waters, and in areas where wind, currents, or wave action are intensive.

FIGURE 9.4 Adsorbent barriers used to contain a spill. *From (A) Haléco-Haladjian and (B) CONTEROL Seguridad y Medio Ambiente S.L.U.*

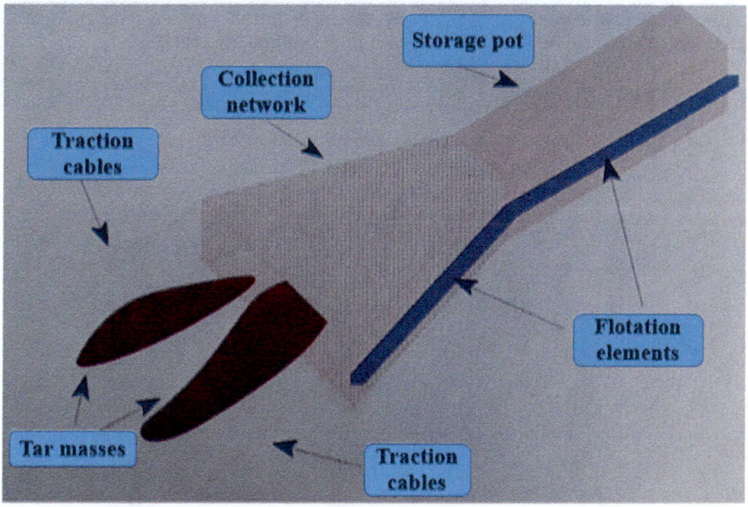

FIGURE 9.5 Net barriers to contain very viscous hydrocarbons.

9.5.6 Example configurations of containment barriers/booms

Fig. 9.6 shows the configuration of a barrier/booms, including its curvature.

The barriers/booms can be deployed in a V, U, or J shape, as shown in Fig. 9.7.

The main tasks for which a barrier/boom can be used are:

1. Concentration and containment of hydrocarbons.
2. Deviation of hydrocarbons.
3. Coast protection.

Fig. 9.8 shows various configurations of barriers/booms, fences, and interceptors, deployed at sea and in a channel.

FIGURE 9.6 Configuration of a barrier/booms to contain hydrocarbons.

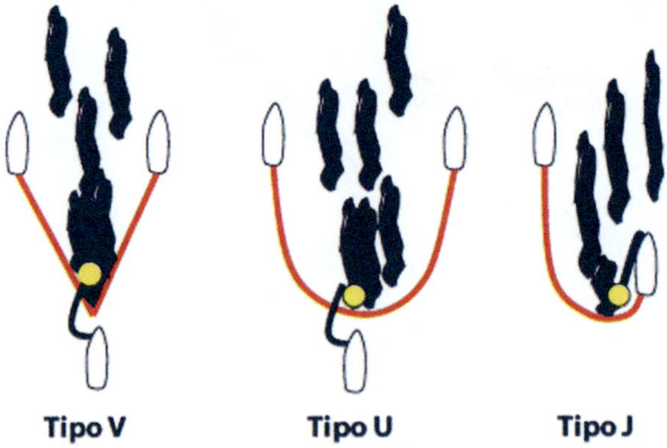

FIGURE 9.7 Deployment of a hydrocarbon containment barrier/booms in a V, U, or J shape. *From http://www.cethus.org/mar_limpio/conservacion_s1.html.*

Before deploying a containment fence, it is necessary to do the following:

1. Select the trained or skilled personnel to act in the operations of containment and recovery of the spill.
2. Select the priority areas where the available barriers can be used with maximum efficiency.

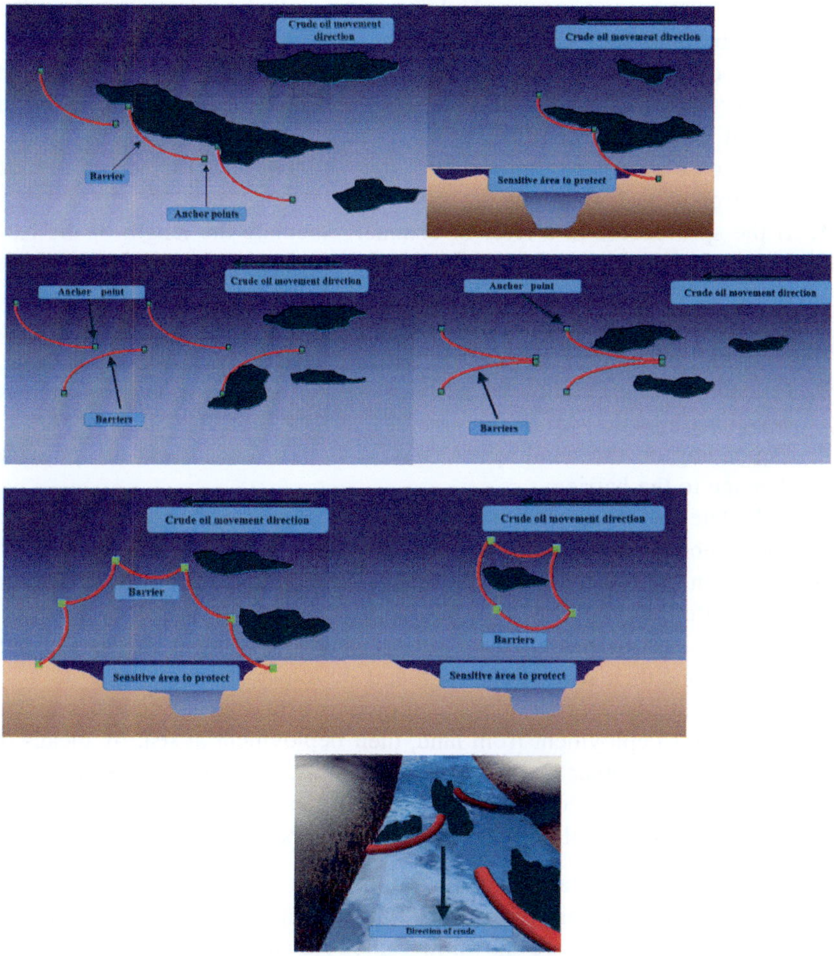

FIGURE 9.8 Different configurations of barriers/booms, fences, and interceptors, deployed at sea and in a channel.

3. Decide if it is possible to protect these areas with barriers and whether the barriers should be towed or moored.
4. Obtain information on winds, waves, currents, and tides at the time of deployment of the barriers and during the processes of containment and recovery of hydrocarbons, as well as recovery of the barriers itself.
5. Calculate the length of the necessary barriers/booms, including their curvature, once the length of the areas to be protected is known.
6. Calculate the cost of the barriers and the reels needed to store them.
7. Calculate the lengths of the anchor and buoy lines.

8. If it is necessary to use a double barrier, calculate the separation distance between the two.
9. In the case of a ship that has caused the spill, calculate the length of the barrier necessary to surround the ship and prevent the displacement of the oil.
10. In the case of using skimmers, calculate the length of the barrier needed to surround each skimmer.
11. If the spill occurs in a river or channel, calculate the length of the barrier to carry out the detour of hydrocarbons to areas where recovery is possible.
12. Calculate the barrier length needed to contain a spill in a port, knowing the width of the port and the speed of the current.
13. Calculate the efficiency of each of the barriers as a function of the wavelength of the waves.
14. Depending on the wind speed, waves, and current, assess the possible damage to the barriers.
15. Calculate the forces that winds, waves, and currents exert on the skirt and freeboard of the barriers.
16. Based on the above data and once the possible damage that the barriers may suffer has been assessed, select the most appropriate ones.
17. Analyze the reliability, ease, and speed of deployment of the selected barriers and their recovery, repair, and subsequent storage.
18. Analyze all the logistics for the transfer of the barriers to the selected areas, their deployment from land, their deployment at sea, by means of boats and any other operation that is necessary while the operations of containment and recovery of the hydrocarbons last.
19. Know in advance all the limitations that the barriers may have.

Most of these calculations can be done with the ALFONSO (Oil spill boom model) model (Martí Moreno, 2008). The input and output parameters of the ALFONSO model are shown in Tables 9.2 and 9.3.

All aspects concerning the use of barriers in response to oil pollution can be obtained from the ITOPF (2011a, 2011b, 2011c).

The selection of a barrier can be done using a matrix shown in Table 9.4.

The effectiveness of a barrier depends on the nature of the spilled petroleum, the way they reach the coast, and the types of places they reach as shown in Table 9.5.

9.6 Recovery by skimmers

Once a spill is contained, the spilled substance may be recovered by means of skimmers and adsorbent materials (ITOPF, 2012a, 2012b). A skimmer is a mechanical device through which hydrocarbons can be recovered from the water surface. To have a good performance in the recovery of hydrocarbons,

TABLE 9.2 Input and output parameters of ALFONSO model.

Input parameters to the model				
Data on the area affected by hydrocarbons	Environmental data		Barrier data	
Width of the area to be protected (m)	Current velocity (m/s)	Wind speed (m/s)	Skirt height (m)	Freeboard height (m)
45	0.1	8	0.5	0.5
Output parameters to the model				
Required barrier length (m)	Barrier angle of inclination (degrees)	Force exerted by the current and wind on the barrier skirt and freeboard (N)		
56	30	1.451		
Possible wind damage to the barrier		**Possible damage caused by the current on the barrier**		
No damage		No damage		
Barrier and reel costs				
Cost of the barrier (€)		Cost of the reel (€)		
Maximum	Minimum	30.000		
28.000	16.800			
Separation distance between two containment barriers				
1.1 m				

it is necessary to account for the viscosity of the spilled mixture, its power of adhesion, and variation with time due to the aging processes. The possibility of using several skimmers throughout the recovery process should be considered.

Depending on the procedure and the material used to recover the hydrocarbons from the sea surface, the skimmers are of two types: with oleophilic and nonoleophilic components. Among the former, there are the disk, oleophilic rope, drum, brush, and belt skimmers. Among the latter, there are the suction, the dump, the belt, and the drum skimmers.

To select a good skimmer several aspects must be considered:

1. Quantity and type of hydrocarbons spilled.
2. Physical properties of the hydrocarbons to be recovered, at the time of the spill and its variation as they age.
3. Recovery rate of the different types of skimmers.

TABLE 9.3 Other input and output parameters of the ALFONSO model.

Input parameters		Output parameters
Calculation of the length of a barrier surrounding a leaking oil tanker		
Length of boat (m): 80		Required barrier length (m): 262.50
Beam of the boat (m): 25		
Calculation of the barrier length required to contain a spill in a port		
Width of the port (m): 200		Required barrier length (m): 201.50
Current velocity (m): 1		
Calculation of the barrier length required to surround a skimmer (m)		
Number of skimmers: 1		Required barrier length (m): 600
Calculation of the barrier length required to contain or divert a spill in a river or channel		
Width of the river or channel (m): 50		Required barrier length (m): 200
Efficiency of a barrier as a function of the wavelength of the waves		
Wavelength (m): 1		The effectiveness of the barrier is good
Possible damage caused by the sea current in a barrier		
Current speed (km/h): 1		No damage to the barrier
Possible wind damage to a barrier		
Wind speed (km/h)	From 0 to 19	No damage to the barrier
	From 20 to 36	Small damage to the barrier
	From 37 onward	Great damage to the barrier

4. Sea state during the recovery.
5. Presence of floating debris (number of objects and dimensions) and need for accessories for the proper functioning of each of the skimmers.

When using skimmers, it is also necessary to take into account the containers for the temporary storage of the recovered hydrocarbon and water mixtures. These containers must allow the separation of the hydrocarbons from the water, by means of a simple decantation process. When the spill is not in the vicinity of the coast, there may be problems with the storage of the recovered mixtures; inflatable containers or barges can be used for the temporary storage. In all cases, it is necessary to have the systems for discharging the mixtures into the land containers. The hydrocarbons recovered and separated from water do not usually present additional problems, whereas water can present legal problems when it is returned to the sea, as it is never totally free of hydrocarbons.

TABLE 9.4 Matrix for the selection of a barrier.

Key		Type of barrier				
1. Well2. Regular3. Poor		With internal foam float	Self-inflating	With external extension element	Inflatable	Fence
Environmental conditions	Offshore with waste $A0 > 1$ m $V < 1$ knot	2	2	1	1	3
	Port $A0 < 1$ m $V < 1$ knot	1	1	1	2	2
	Calm waters $A0 < 30$ cm $V < 1$ knot	1	1	1	2	1
	Troubled waters $V > 1$ knot	2	3	2	1	3
	Surface water $P < 30$ cm	1	2	2	3	4
Functional characteristics	Waste behavior	1	3	2	3	2
	Excessive buoyancy	2	1	1	2	3
	Response to waves	2	2	1	1	3
	Solidity	2	3	1	1	1
Operational characteristics	Ease of use	2	1	2	3	2
	Ease of cleaning	1	1	1	3	1
	Ease of recovery	3	1	1	2	3
	Cost/m: I, Low; II, Medium; III, High	I	III	II	III	II

A denotes wave height (m); *V*, current speed at the surface (knots), and *P* depth of the water layer (cm).

TABLE 9.5 Effectiveness of a barrier depending on the nature of the hydrocarbons, the way in which they arrive, and the types of places they reach.

		Type of barrier according to its place of use	
Nature of hydrocarbons	Medium or large viscosity	Simple	Oceanic
	Very high viscosity	Yes	Yes
	Reverse emulsions	Yes	Yes
	Hydrocarbons with residues	Yes	Yes
	High dirt waste	Yes	Yes
	Dispersions	Yes	Yes
How the hydrocarbons arrive	Massive	Yes	Yes
	Diffuse	Yes	Yes
	In very thin films	Yes	Yes
Types of places where hydrocarbons arrive	Sand (> 30 m)	Yes	Yes
	Water (height < 50 cm)	Yes	Yes
	Rocky areas	No	Yes
	Wave zones	No	No

9.6.1 SIRA (Skimmer effectiveness model) model

By means of the SIRA model (Martí Moreno, 2008), it is possible to assess skimmer effectiveness for recovery of spilled hydrocarbons. The SIRA model classifies skimmers into two groups and only covers the first one. This first group is related to the skimmers used to recover hydrocarbons spilled on or near the coast. The second one corresponds to skimmers that are used to recover large quantities of oil that are spilled in the open areas not limited by coasts. With respect to the skimmers of the first group, the model allows access to a database of 208 skimmers. For each skimmer, this database shows its commercial name, its photography, the hydrocarbon recovery flow, the type of device used for the recovery of the same ones and the supplying house. The skimmers have been classified according to the device they use for the recovery of hydrocarbons. Table 9.6 shows the input and output parameters of the SIRA model.

TABLE 9.6 Input and output parameters of the SIRA model.

Input and output parameters of the SIRA model	
Input parameters	Volume of hydrocarbon spilled (m^3)
	Type of skimmers available
	Number of available skimmers
	Hydrocarbon–water separator available
Output parameters	Estimated time for the recovery of the sea surface from the hydrocarbon–water mixture
	Total volume of hydrocarbons and water recovered
	Volume of hydrocarbons separated in the decanter
	Separate water volume in the decanter
	Volume of hydrocarbons not recovered from the sea surface
	Best skimmer in the database
	Best hydrocarbon–water separator in the database
	Total time savings in the performance of the best skimmers and decanters in the database with respect to those that have been used.
	Estimated costs

9.7 Treatment of hydrocarbons with adsorbent materials

9.7.1 Adsorption and absorption

Absorption is a physical phenomenon in which one or more components of a gaseous mixture diffuse into a liquid. It does not involve chemical changes, so the process is reversible. Adsorption is a physical phenomenon, where one or more compounds (adsorbates) come into contact with a solid (adsorbent) adhering to the surface of the same through a physical force, which is known as the London dispersion force (ITOPF, 2014a, 2014b, 2014c, 2014d). This process does not involve an exchange of electrons, which makes it reversible. In some cases, there may be chemisorptions, which implies modifications in the chemical structure of the adsorbent and adsorbate, so the process is irreversible.

9.7.2 Characteristics of the adsorbents

The main characteristics that a good hydrocarbon adsorbent must have are the following:

1. High adsorbent power.
2. Minimum toxicity to the environment.

3. Safe to use.
4. Minimal cost.
5. Easy to apply.
6. Easy to recover.
7. Unalterable during long periods of storage.
8. Easy desorption of the retained hydrocarbons.
9. Tested by an accredited company.
10. Environmentally friendly.

9.7.3 Adsorbent materials

Some adsorbent materials of natural origin are the peat, the sawdust, the cork, and the straw. Some inorganic adsorbents are vermiculite, pumice, polypropylene, and other polymers.

There are various ways in which the different adsorbents are commercialized. Some of them are in the form of loose material, particles, enclosed in a net in the shape of a pillow, cushions, socks, sausages or barriers, leaves, mats, cloths, rolls or mats, and as loose fibers.

Some hydrophobic adsorbent materials have been tested by CEDRE (Centre de documentation, de recherche et d'expérimentations sur les pollutions accidentelles des eaux) according to the standard norm to determine the adsorbent power of a product (NFT 90–360), using Light Arabian crude oil (heated to 110°C and viscosity 42–45 cP at 20°C) to recover hydrocarbons spilled both at sea and in inland waters (CEDRE, 2020a, 2020b).

Adsorbent Type A (bulk and in "spaghetti"). The adsorbents were tested taking into account the adsorption capacity, which allows a comparison of the performances of the different tested adsorbents, and the nature of the adsorbent because it is an essential element to define the conditions of storage and disposal of the product, such as incineration. Only the products that meet the following criteria are listed:

1. Adsorbent capacity. Adsorbent capacity in weight greater than 5 or, in volume, greater than 0.5 (calculated according to the apparent density of the product).
2. Hydrophobia. Water retention capacity/hydrocarbon retention capacity equal to or less than 0.25.
3. Stability. The product must remain stable to maintain its properties.

The criteria for determining the adsorbent power were based on economic data, the theoretical price per treated liter, and combining the retention capacity in weight (adsorbent capacity) with the price of the adsorbent.

Adsorbent Type B and C (sheets, rolls, and blankets). The criteria followed for their study were the same as those used for type A adsorbents.

Adsorbent Type D and E (pillows, socks, and feathers).

Type G (special products).

1. Adsorbent capacity. Adsorbent capacity in weight greater than 10.
2. Hydrophobia. Water retention capacity/hydrocarbon retention capacity equal to or less than 0.25
3. Stability. The product must remain stable to maintain its properties.

9.7.4 Spraying of adsorbents on a spill

By means of the ROSA (Oil spill adsorbents model) model (Martí Moreno, 2008), a series of calculations can be made on the spraying of adsorbents on spilled crude oil. The ROSA model is a management tool made up with a series of folders in which the following information is stored:

1. Theoretical concepts concerning adsorbents.
2. Considerations when using adsorbents.
3. Types of adsorbents and their main characteristics.
4. Specifications of a good adsorbent.
5. Adaptation of adsorbents for different types of environments contaminated by hydrocarbons.
6. Distribution of adsorbents on hydrocarbons:
   a. Manual distribution.
   b. Distribution with spray guns. Cannon spraying rate: $200 < Q < 500$, L/min.
7. Experimental procedures to calculate the adsorbent power of a product
8. Criteria to be followed for the evaluation of adsorbents:
   a. Adsorption capacity in weight and volume.
   b. Hydrophobic power of the adsorbent.
   c. Ease to recover the adsorbent from the sea surface.
   d. Ease of desorption of the hydrocarbons obtained.
   e. Possibility of reusing the adsorbent.
   f. Means of disposal of the adsorbent contaminated by hydrocarbons:
      i. Treatment in incinerators.
      ii. Treatment in cement plants.
      iii. Stabilization and storage2u.
9. Equations to calculate the adsorbent power.
10. Adsorption capacity:
    a. Hydrocarbon initial.
    b. Hydrocarbon end.
    c. Of water.
11. Application techniques of the adsorbents:
    a. Shape of the adsorbent.
    b. Application technique.
12. Leading global institutions that have tested adsorbents.
13. Main industries that supply adsorbent materials:
    a. Address and contact person.

   b. Types of adsorbents they supply:
      i. Aspect of the adsorbents.
      ii. Main characteristics.
      iii. Cost of adsorbents.
14. Matrix for the selection of the adsorbents according to the type of adsorbent (synthetic, organic, or inorganic), the type of material that makes it up, and the considerations for its use.
15. Application logic of the adsorbents depending on:
   a. Time
   b. Quantity required/day.
   c. Treatment requirements.
   d. Manufacturer suggestions.

The input parameters for the ROSA model, once a specific adsorbent material has been selected, are the following: mass of spilled hydrocarbon (kg), area occupied by the spill (m^2), spraying rate of the system used (L/min), and price of the adsorbent (€). By default, once a certain adsorbent has been selected, the model indicates its price, which can be modified at any time to keep it updated.

The output parameters of the ROSA Model are:

1. Amount of adsorbent to be used (kg)
   1.1 For a fast treatment.
   1.2 For a comprehensive treatment.
2. Application time (min).
3. Scope of the downwind spraying system (m).
4. Spraying system coverage speed (m^2/min).
5. Cost of the adsorbent material (€).
6. Percentage of water retained by the adsorbent (%).

The model selects the most appropriate adsorbent from its database, as well as the savings obtained with the selected adsorbent.

9.8 Treatment of crude oil by dispersants

Before dealing with dispersants, it is appropriate to clarify two particular misconceptions about them. The first erroneous concept is that the dispersants task is to sink hydrocarbons, forming a lethal layer on the bottom of the sea. The second misconception refers to the fact that dispersants have a higher toxicity than oil itself. The today third-generation dispersants are surface-active concentrates, which are miscible with water. They can be applied in their pure or prediluted state and are very useful for the treatment of hydrocarbon spills, for oil from low to intermediate viscosity.

A crude oil dispersant is a compound containing polar groups, compatible with hydrocarbons, and a polar group, compatible with water. The task of dispersants is to decrease the surface tension, so that, once the hydrocarbon droplets have formed, they do not tend to congregate. Since hydrocarbon-degrading microorganisms are found in seawater and not in hydrocarbons, the biodegradation process occurs on the surface of the formed droplets. Therefore, by increasing the surface area of the droplets the biodegradation process is increased. For example, a droplet of 1 mm in diameter, when supplied with high mixing energy, can disperse and form 10,000 new drops, with diameters ranging from 1 to 70 μm (ITOPF, 2014b). Laboratory studies show that the most stable dispersions are those with a diameter of less than 45 μm. If the average size of the newly formed particles was about 45 μm, the area would now be 20 times larger (if compared to the initial droplet size), which would considerably increase the biodegradation process.

9.8.1 Types of dispersants

The first generation of dispersants dates from the 1960s and they were mere degreasers. Due to their high toxicity, their use was promptly discontinued. The second generation of dispersants, known as Type I dispersants, contains a solvent with a low content of aromatic hydrocarbons and the active surface agent in a concentration between 15% and 25%. They are usually sprayed undiluted in a dispersant to hydrocarbon ratio of 1:1 to 1:3. Although they have a lower toxicity than those of the first generation, their use has been discontinued in some countries. The dispersants of the third generation are formed by two or three surface-active agents, with glycol and solvents from the distillate of light type crude oil. The most commonly used dispersants are the anionic type, such as sodium alkyl sulfosuccinate, and the nonionic type, such as fatty acid esters and ethoxylated fatty acid esters. The concentration of the active surface agents in them ranges from 25% to 65%.

Two types of the third-generation dispersants can be distinguished today (ITOPF, 2011a, 2011b, 2011c; CEDRE, 2014): those of Type II and Type III (the Type I were discarded due to their high toxicity). Type II dispersants are usually diluted in seawater before being sprayed on the oil spill to a concentration of the order of 10%. To achieve high efficiency, the recommended dispersant/water-to-hydrocarbon ratio ranges from 1:5 to 2:1. Type III dispersants are commonly used undiluted and can be sprayed from aircraft or ships. Recommended doses of pure dispersant to hydrocarbon range from 1:5 to 1:50. Ideally, the dispersant's effectiveness should be tested with the hydrocarbon mixture just before application.

9.8.2 Efficiency of dispersants

Information on dispersants tested by CEDRE can be obtained from the CEDRE website (Grote et al., 2016). There is information on the dispersant's efficiency, toxicity, and biodegradability according to the French NF T

90–345, NF T 90–349, and NF T 90–346. The acceptance criteria have been defined by CEDRE and used since January 1, 1988. This approval procedure is carried out without prejudice to the procedures prescribed by French law No. 77–771 of July 12, 1977, as amended by French law No. 82–905 of October 21, 1982 on the control of chemicals and its implementing provisions (CEDRE, 2014).

9.8.3 Dispersant application methods

When the spilled products have a very high viscosity, such as fuel oils, very old crudes or emulsions, the effectiveness of dispersants is minimized, so their use is not recommended. If the spilled products are gasoline or small quantities of gas oils, which remain on the surface of the sea as thin films, it is also not recommended to apply dispersants.

The three stages of the chemical dispersion process, of a mixture of hydrocarbons, are shown in Fig. 9.9 (Bergueiro & Domínguez, 1992). In the first stage, the dispersant is sprayed on the hydrocarbon mixture and it is supplied with mixing energy. In the second stage, the molecules of the dispersant, due to the mixing energy and the decrease of interfacial tension, migrate toward the hydrocarbon–water interface, causing the crude to break into small drops that do not tend to mix together. In the third stage, due to the effect of microorganisms, the hydrocarbons are degraded to carbon dioxide, water, nutrients, and cell masses. The first two stages are relatively short in time while the third stage usually lasts for days or even years.

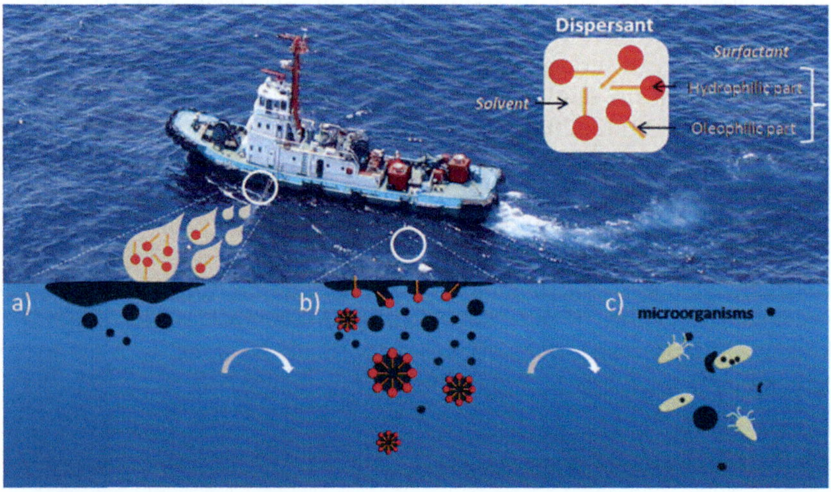

FIGURE 9.9 Stages in the process of chemical dispersion of a hydrocarbon mixture. (a) Oil spill without dispersion (b) Dispersant action on oil spill (c) Microorganisms action on dispersed oil spill.

Fig. 9.10A shows some dispersion of the crude oil, due to the natural dispersants that crude oil has. Fig. 9.10B shows the dispersion of the hydrocarbons, moments after a dispersant has been sprayed and the mixing energy has been supplied. Finally, Fig. 9.10C shows the aspect of the dispersion formed after some time.

The application method is selected depending on the volume of hydrocarbons to be dispersed and their location. When it comes to dispersing small quantities of hydrocarbons, the dispersants can be sprayed manually by means of hoses equipped with spray heads. Fig. 9.11 shows an operator spraying dispersant on a small quantity of hydrocarbon.

Dispersant can also be sprayed from a boat, using spray arms, as shown in Fig. 9.12.

When the dimensions of the spill are large, it is recommended that the dispersant be sprayed from a light aircraft, as shown in Fig. 9.13.

For large spills at sea, the dispersant is usually applied from large aircrafts, as shown in Fig. 9.14.

For oil spills in areas that present difficult access for boats, dispersants can be sprayed from helicopters, as shown in Fig. 9.15. This technique is usually very effective since, in addition to spraying the dispersant at very specific points, the helicopter propellers provide sufficient mixing energy for effective dispersion of the hydrocarbons.

FIGURE 9.10 Aspect of the partial dispersion of the hydrocarbons (Above), intermediate dispersion (Bottom-left), and final dispersion (Bottom-right) *From ITOPF. (2011a).* Effects of oil pollution on fisheries and mariculture. Technical information paper (TIPS). *London: International Tanker Owners Pollution Federation Limited (ITOPF); ITOPF. (2011b).* Uso de dispersantes en el tratamiento de derrames de hidrocarburos. Documento de Información Técnica. *London: International Tanker Owners Pollution Federation Limited (ITOPF); ITOPF. (2011c).* Use of booms in oil pollution response. Technical information paper (TIPS). *London: International Tanker Owners Pollution Federation Limited (ITOPF).*

FIGURE 9.11 Operator spraying dispersant by means of a hose. *From National Oceanic and Atmospheric Administration (NOAA).*

The most modern techniques to use dispersant involve channeling the hydrocarbons spilled at sea through barriers/booms, which causes the hydrocarbons to concentrate at the end of the barrier. Fig. 9.16 shows the application of dispersants over an oil spill, previously channeled to the end of the barrier.

9.8.4 Selection of a good dispersant

There is a large amount of bibliographic data on the dispersing power of most dispersants on the market. Before the application of a particular dispersant, it is recommended to perform laboratory tests to determine its efficiency against crude oil to be treated at sea, in different degrees of aging. Extreme caution is recommended when extrapolating laboratory results to the sea, given the great difficulty of accurately replicating sea conditions in a laboratory.

When dispersing crude oil with Type III dispersants, the most widely used at the present time, a 1:20 dispersant–crude ratio is usually used. When the crude oil has a low viscosity and is not aged, this dose can be reduced. Similarly, when medium viscosity crudes are involved or are aged, this ratio should be increased. In some cases, it is also recommended to repeat the process several times.

To estimate the amount of dispersant needed, the average thickness of the spill and the area of the spill must be known. If some time has elapsed since the spill event took place, the average thickness of a slick can be estimated

FIGURE 9.12 Spraying of dispersant from a spray arm placed on a boat. *From ITOPF. (2011a).* Effects of oil pollution on fisheries and mariculture. Technical information paper (TIPS). *London: International Tanker Owners Pollution Federation Limited (ITOPF); ITOPF. (2011b).* Uso de dispersantes en el tratamiento de derrames de hidrocarburos. Documento de Información Técnica. *London: International Tanker Owners Pollution Federation Limited (ITOPF); ITOPF. (2011c).* Use of booms in oil pollution response. Technical information paper (TIPS). *London: International Tanker Owners Pollution Federation Limited (ITOPF).*

at about 0.1 mm. Therefore, in a hectare, there will be 1000 L. If the ratio is 1:20, the amount of dispersant to be sprayed is 50 L (ITOPF, 2014b).

The dispersant spray rate (discharge rate) (L/s) can be calculated by multiplying the application rate (L/m^2) by the aircraft speed (m/s) and by the spray arm width (m). The same calculations apply when spraying from a boat or helicopter.

The main problem that arises when calculating the amount of dispersant needed resides in the estimation of oil layer thickness, since this varies from the center of the spill to its edges. The practical and most efficient solution is to make the calculations with the thickness of the thickest part of the spill, since if the hydrocarbons in that area are dispersed, those in the thinnest areas of the spill will also be dispersed.

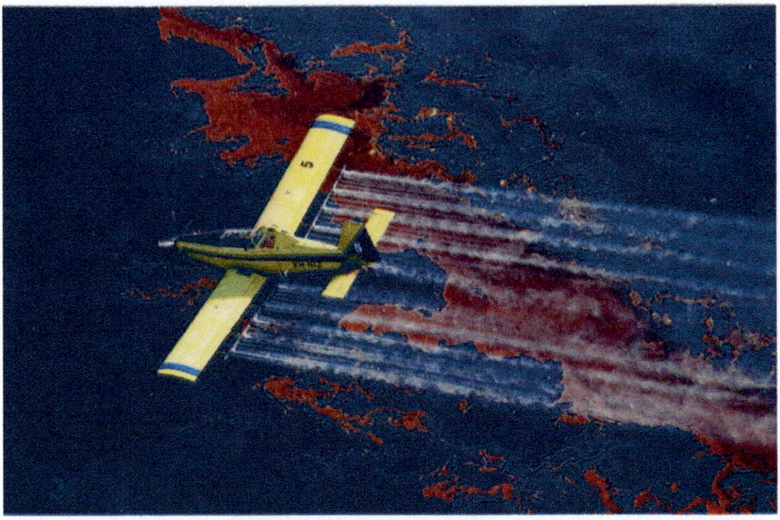

FIGURE 9.13 Spraying of dispersant from a light aircraft. *From Australian Maritime Safety Authority (AMSA) (2020).*

FIGURE 9.14 Spraying of dispersant from an aircraft (United States Coast Guard, 2018). *From United States Coast Guard. (2018).* Coast guard's joint maritime test facility reopens after completion of final tests. *United States Coast Guard United States Department of Homeland Security. <https://www.dcms.uscg.mil/Our-Organization/Assistant-Commandant-for-Acquisitions-CG-9/Newsroom/Latest-Acquisition-News/Article/1682615/coast-guards-joint-maritime-test-facility-reopens-after-completion-of-final-tes/>.*

FIGURE 9.15 Spraying of dispersant from a helicopter. *https://www.sciencedirect.com/science/article/pii/B9781856179430100164. M. Fingas (2011). Oil Spill Science and Technology. A Practical Guide to Chemical Dispersion for Oil Spills. Chapter 16, pp. 583—610. https://doi.org/10.1016/B978-1-85617-943-0.10016-4.*

9.8.5 Limitations of dispersant application

Before making the decision to use dispersants for the treatment of a specific oil slick, it is necessary to know the amount of oil spilled, the water depth, the distance of the oil to the coast, the presence of sensitive areas, as well as the presence of birds, fish, crustaceans, and marine mollusks.

Generally, dispersants are allowed to be used when the spill is far from sensitive areas, is more than two miles from the coast and the water depth is greater than 50 m. Fig. 9.17 shows the areas where spraying of dispersants is not recommended, for example, in the vicinity of the Cabrera Archipelago, for oil spills of 10, 100, and 1000 tons (Bergueiro et al., 2010).

The DISPERSANT (Oil spill dispersant model) model (Martí Moreno, 2008) is a simulation model through which various aspects related to the treatment of crude oil by means of dispersants can be addressed, depending on the type of crude oil spilled, its viscosity, the degree of aging, and the efficiency ratio of the dispersant. As output parameters, the model calculates the spraying time, the spraying flow, and the cost of spraying, from large and small aircraft, from helicopters and from boats. Likewise, the model allows for the calculation of the variation in the concentration of the dispersant and the dispersed crude oil in the water column. Table 9.7 shows the data input and output screens of the DISPERSANT model.

(a)

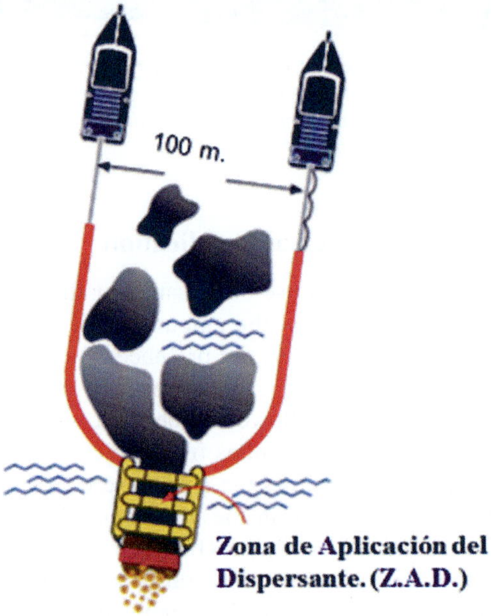

(b)

FIGURE 9.16 Photo (above) and scheme (below) of the modern technique used to apply a dispersant, channeling the hydrocarbons through containment barriers. *From (A) Suzhou Jiahe Non-Woven Products Co., Ltd. 2020. (B) Authors.*

Table 9.8 shows the data input and output screen of the DISPERSANT model for the calculation of time and spray rate and the cost of operating a large aircraft.

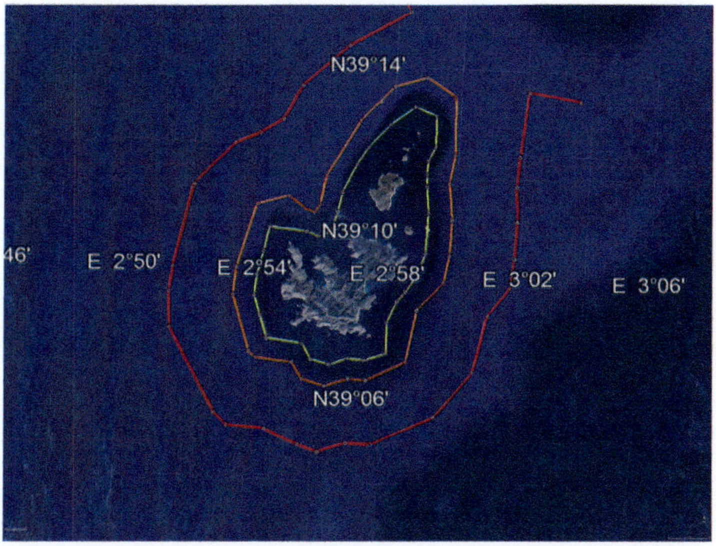

FIGURE 9.17 Application limits of dispersants in the Cabrera Archipelago, for of 10, 100, and 1000 ton pills.

Table 9.9 shows the input and output data of the DISPERSANT model for the calculation of the distribution, in the water column, of the spilled hydrocarbon mass.

Another section of the DISPERSANT model allows to estimate the suitability of using dispersants in different coastal environments. The input parameters to the model are the location of the application point, in the Mediterranean Sea or the Atlantic Ocean, the distance to the coast, and the depth of the water layer. Fig. 9.18 shows the input and output data of a simulation.

9.9 Elimination of crude oil by bacterial degradation

One of the most common procedures used to remove oil spilled on the surface of the sea is through bacterial degradation. For such degradation to be effective, it is necessary that the hydrocarbons are dispersed in the seawater surrounding the spill. By applying dispersants and the necessary mixing energy, the mass of hydrocarbons is "broken" into small droplets, which do not tend to mix together due to the effect of surface tension. The surface of droplets formed can be a million times larger than that of the original crude oil slick. Given that the bacterial degradation processes are surface processes, the increase in the area of the hydrocarbons brings as a consequence a notable increase in the process of elimination of the hydrocarbons.

Once the crude oil has been dispersed in the sea, bacterial degradation processes can begin, provided that a sufficient number and type of microorganisms, capable of metabolizing the dispersed hydrocarbons, are present in

TABLE 9.7 DISPERSANT Model data input and output screen.

Previous considerations					
Spilled product					**Comments**
95 and 98 Octane automotive gasoline. Aviation gasoline 100 Ll Gas oils. JET A1. JP8. Fuel oilsCrude oil aged with $\mu > 1.000$ mPa.s					Do not use dispersants, given their low efficiency
Input parameters to the DISPERSANT model					
Indicate the type of oil spilled					
Enter your viscosity, mPa.s					
Enter the amount of spilled oil, *t*					
Indicate the degree of hydrocarbon aging (%)	5	10	15	20	25
Proposed efficiency ratio, hydrocarbons/dispersant	25:1	20:1	15:1	10:1	5:1
	Select another efficiency ratio				
Indicate the dispersant to be used					
Indicate your cost (€/kg)	Use the default cost indicated in the database				
Model output parameters					
Quantity of dispersant to be sprayed (kg)	Cost of the used dispersant (€)				
Amount of the best dispersant in the database (kg)	Cost of the used dispersant (€)				
	Savings achieved (%)				

the marine environment, using them as a source of matter and energy. Among the marine microorganisms that are capable of metabolizing crude oil, we can highlight bacteria, yeasts, molds, unicellular algae, and protozoa. These microorganisms are usually found throughout the world seas, increasing in those marine environments where there has been a previous process of contamination or in coastal areas that receive untreated urban wastewater and/or industrial waste.

The most influential factors for the speed and degree of degradation of an organic substrate are the type of substrate, the presence of nutrients (nitrogen, phosphorus, and potassium), the concentration of oxygen and the temperature. Through the process of degradation, the organic substrate is transformed into carbon dioxide, water, nutrients, and bacterial cell mass.

Based on the existing available knowledge, no single microorganism is capable of degrading all the compounds present in a crude oil so, sometimes, it is necessary to resort to "bacterial consortiums" so that the degradation

TABLE 9.8 Second data output screen of the DISPERSANT model.

Input parameters to the DISPERSANT model	
Select one of the following means for spraying the dispersant	Large aircraft
	Small aircraft
	Helicopters
	Boats
Enter the following data	Volume of dispersant to be sprayed (L)
	Width of spill (m)
	Spill length (m)
	Spray arm width (m)
	Speed of the spraying medium (km/h)
	Cost of renting the medium (€/h)
Output parameters to the DISPERSANT model	
Spraying time (min)	
Spraying rate (L/min)	
Cost of the spraying (€) (Minimum 1 h)	

TABLE 9.9 Distribution of a crude oil and a dispersant, in the water column.

Input parameters to the DISPERSANT model		
Crude oil to be dispersed	Dispersant used	
Mass of hydrocarbons to be used (kg)	Mass of dispersant to be sprayed (kg)	
	Spill area (m^2)	
	Model output parameters	
	Depth range (m)	Concentration (ppm)
Crude oil	Between 0 and 1	Concentration between 0 and 1
	Between 1 and 2	Concentration between 1 and 2
	Between 2 and 4	Concentration between 2 and 4
	Between 4 and 8	Concentration between 4 and 8

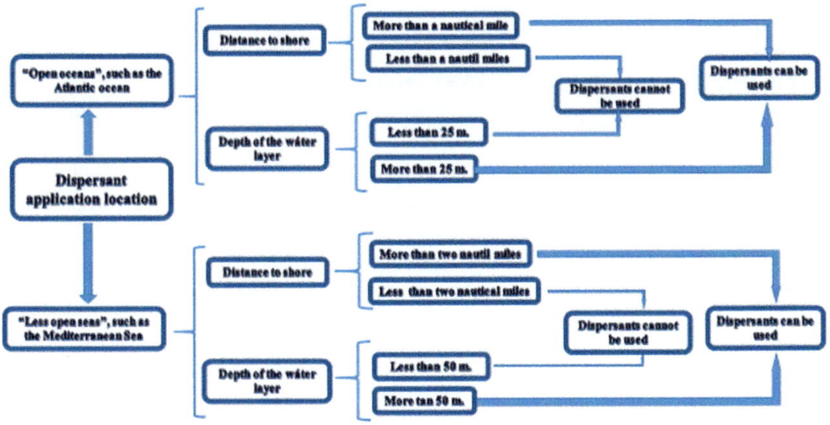

FIGURE 9.18 Considerations for the use or prohibition of dispersants in seas such as the Mediterranean or the Atlantic Ocean.

process can occur with high speed and intensity. Likewise, "biodegradation accelerators" are often used to supplement the deficit of nitrogen, phosphorus, and potassium that exists in some marine environments.

Bergueiro and Domínguez (2001) have studied the degradation of crude oil and dispersants used in the treatment of oil slicks, both in the presence and in the absence of biodegradation accelerators. They have also studied the degradation of crude oil dispersed in seawater in the presence of "bacterial consortia" called "biodegradation activators." The bacterial consortium used was formed by bacteria of the family Pseudomonadaceae, of the genus Pseudomonas, of the species *Pseudomonas pútida*. These are nonpathogenic bacteria that normally exist in natural environments, such as freshwater basins and in the seawater. They can grow in temperatures between 22°C and 40°C and hibernate at below 4°C temperatures. The biodegradation activator is marketed in the form of freeze-dried powder formed by the bacterial consortium itself, which is accompanied by an inert material such as kaolin, a source of carbon and salts of nitrogen, phosphorus, and potassium. Before using the biodegradation activator, it is necessary to activate it by mixing the product with water at a temperature between 10°C and 30°C, adding air to make it brew for 3–6 hours.

Once the hydrocarbons are dispersed in the water, the biodegradation accelerator is supplied, consisting of a carbon source, easily biodegradable, and of an oleophilic nature, to form a single phase with the hydrocarbons. The carbon source is usually oleic acid. The second component that goes in the accelerator is formed by a source of phosphorus. One of the first phosphorus sources used was the trilauryl phosphate. The third component is urea, which provides the nitrogen source. Finally, the fourth component of the biodegradation accelerators is in charge of decreasing the viscosity of the mixture and reducing the interfacial tension, thus avoiding the coalescence of the hydrocarbon droplets formed by the effect of the dispersants and the mixing energy supplied. One of the first compounds of this type was butoxyethanol. Due to the toxicity that some of these compounds may have, it is necessary to continue looking for new compounds, more compatible with the marine environment and allowing a greater decrease in the surface tension and viscosity of the mixture. In a single oleophilic phase, there are together the crude oil, the easily biodegradable carbon source, the sources of nitrogen, phosphorus, and potassium, and the agent in charge of reducing the viscosity and surface tension. It has been proven that, even in cold climates, the formed oleophilic microemulsion allows the rapid growth of microorganisms by consuming the carbon source present in the accelerator. Once that all the carbon in the accelerator has been almost consumed, the amount of hydrocarbon-degrading bacteria has increased considerably and biodegradation of the hydrocarbons can begin. It must be taken into account when crude oil affects the coast, they can end up deposited in high tides areas, where there is usually a deficit of both moisture and nutrients.

It is known from experimental studies (Rajendran et al., 2021) that some large and complex molecules, as well as highly branched hydrocarbons present in crude oil are very difficult to degrade, so they usually appear after an oil spill in the form of high-density tarry masses. These masses end up mixing with algae, sand, and water forming so-called tarballs, which in some cases end up sinking to the seabed.

Information concerning the degradation of crude oil at sea can be obtained from ITOPF Technical Information Document No. 2 (ITOPF, 2014b).

Similarly, Bergueiro and Domínguez (2001) have studied the degradation of a Kirkuk type crude oil dispersed in seawater, due to the effect of the dispersants present in the crude oil. The dispersion has been generated by agitating 2.5 g of crude oil with 3 L of seawater. The mixture was agitated at 4000 rpm for 5 minutes by means of a three-blade axial flow agitator with a 1:1 propeller pitch with a location not centered in the reactor.

When Bergueiro and Domínguez (2001) analyzed the results obtained from the representation of the variation of hydrocarbon concentration as a function of time, they observed that the sequence of points was difficult to adjust to a single equation. However, if the process is divided into two phases or stages, a good adjustment is achieved. For the Phase I, the results fit well into an experimental equation, while the Phase II fit into a linear equation. Table 9.10 shows the results obtained in the degradation of a Kirkuk type crude oil. Subscripts 1 and 2 indicate a validity for the constants of Phase I and II, respectively. The values correspond to the end of the first stage and the beginning of the second stage and are obtained by an iterative method. Similar results to those obtained by these authors were obtained in the degradation of crude oil from the "Amoco Cadiz" tanker accident.

TABLE 9.10 Constants of the equations representing the different stages of the degradation process of a Kirkuk type crude oil, dispersed in seawater.

Stage	Equation	Coefficient *a*	Coefficient *b*	t_0 (days)	Recovered crude oil (%)	Correlation coefficient (*r*2)
I	$Cs = a_1 . \exp(-b_1 \cdot t)$	29.23 ppm	3.75×10^{-2} day^{-1}	0.00	85.7	0.972
II	$Cs = a_2 - b_2(t - t_0)$	4.16 ppm	1.58 ppm day^{-1}	52.07	95.7	0.999

Table 9.11 shows the results of the degradation process of a Kirkuk type crude oil dispersed in seawater by means of the Renex 706 dispersant.

Likewise, Bergueiro and Domínguez (2001) have studied the degradation of the active ingredients of the Renex 706 dispersant. The variation of the concentration with time fits into Eq. (9.3). The final results are shown in Table 9.12.

$$\mathrm{Cs} = C_{\mathrm{so}} \cdot e^{K_2(t-t0)2} \tag{9.3}$$

Further information on the degradation process of crude oil spilled at sea can be obtained in chapter 9 of Bergueiro and Domínguez (2001).

9.10 Filmogens

Filmogens (Hyland and Schneider, 1976) are a series of products used to reduce the adhesion capacity of hydrocarbons on solid surfaces, such as rocks, concrete walls, stony beaches, etc. Those on the market are mainly made up of algae extracts (alginates) which, once they have been sprayed onto solid surfaces, reduce the ability of hydrocarbons to adhere to these substrates.

The first important characteristic that these products must have is that they do not damage the flora and the fauna that lives on the treated surfaces. The main procedure and characteristics of a standard product are the following: Prepare a solution with a concentration of about 5 g/L, stirring for about 20 minutes. It is recommended to spray the solution about 24 hours before the hydrocarbons impact on the solid surfaces, applying about 0.3 L/m^2. Then the product must be left to dry for about 2–3 hours. It is recommended to spray the product preferably on the areas affected by high tide and in the intertidal areas. The time that a filmogen should remain active should be at least 3–4 days. The protection of the filmogen usually lasts a week.

Once the hydrocarbons have impacted the areas protected by the filmogens, the hydrocarbons can be separated from the affected surfaces by means of low-pressure water jets, avoiding the impact of pressurized water on the flora and fauna living on the treated surface. To separate the hydrocarbons, it is recommended to wash them as quickly as possible or in less than 5 days.

To avoid contamination with the separated hydrocarbons, the area will be protected by barriers and fences. Subsequently, the hydrocarbons will be recovered by means of skimmers and adsorbent materials. Although filmogens do not usually have side effects, it is recommended that gloves and goggles be worn during preparation and spraying to avoid skin irritation.

By means of simulation models of hydrocarbon drift in the sea, we can determine the areas that will be affected, as well as the time of the event. Of course, the filmogen should not be applied in areas exposed to waves, since it will not remain on the substrate. Likewise, all necessary means must be in

TABLE 9.11 Constants of the representative equations of the different stages of the degradation process of a Kirkuk crude, dispersed in seawater by means of the Renex 706 dispersant.

Stage	Equation	Coefficient *a*	Coefficient *b*	t_0 (days)	Recovered crude oil (%)	Correlation coefficient (*r*2)
I	$Cs = a_1 \cdot \exp(-b_1 t)$	256.11 ppm	0.0177 day^{-1}	0.00	40.0	0.850
II	$Cs = a_2 - b_2(t - t_0)$	166.04 ppm	0.7776 ppm day^{-1}	24.40	89.1	0.980

TABLE 9.12 Kinetic coefficients in the degradation process of the nonionic dispersants of Renex 706.

C_{so} (ppm)	$K_2.10^{-5}$ (days^{-2})	t_0 (days)	t (days)	Coefficient regression (r^2)
49.57	8.13	0	78.5	0.988

place to recover the hydrocarbons that have not adhered to the substrate and therefore return to the sea.

In the event of a massive arrival of hydrocarbons in a very large coastal area, the most sensitive areas should be selected for preferential protection.

The deposition of all the deposited product must be done with jets of cold water at low pressure, although in the case that a deep cleaning is needed, it can be necessary to use jets of water at higher pressure. In the case that the filmogenis not successfully deposited in the first application, it is recommended to repeat the process a second time.

When the spilled hydrocarbon mixture has one of the two automotive gasolines (95 and 98 Octane), no treatment with the filmogen should be carried out, since both gasolines evaporate completely in less than 24 hours. The same conclusion is reached when the hydrocarbon mixtures spilled has JET A1, JP8, or any of the gas/oils used for automotive, fishing, and agricultural purposes.

As a mere guide, the amount of Light Arabian crude oil that can be deposited on a rocky surface is estimated at about 340 g/m^2. When this surface has been treated with a good filmogen, the amount deposited can be reduced to 68 g/m^2.

Environment Canada provides information on adhesion values (g/m^2), as a function of the evaporated fraction, for crude oil and derivatives. Further information on filmogens can be obtained from the following website (Environment Canada, 2008). The points addressed in this paper are the following: Principles, Material and personnel for distribution, How to proceed, and Cautions.

9.11 Incineration of spills

As a general rule, the elimination of hydrocarbons spilled on the sea surface by burning is not usually recommended, since the advantages of this method usually are less significant compared to the disadvantages. It should be noted that crude oil usually has sulfur derivatives, which produce sulfur dioxide during the burning process. Carbon monoxide, sulfur oxides, smoke, suspended particles, and soot are also often produced. For these reasons, the technique of burning is not usually implemented when the spill is near the coast and there is a high probability that the gases originated during the burning process affect the population living in that coastal areas. In areas far from the coast, it is a technique that was applied under strict control and safety conditions during the spill of the Deepwater Horizon platform in the Gulf of Mexico. Fig. 9.19A and B shows a controlled fire from a spill from the Deepwater Horizon platform on May 6, 2010, in the Gulf of Mexico, under the supervision of the US Coast Guard.

Through the FIRE model (Martí Moreno, 2008), issues related to the burning can be assessed. Initially the spill is surrounded by containment barriers deployed in an open or closed U shape. The longitudinal part of these barriers is of a conventional type, while the curved barrier is a fireproof type, to resist high temperatures.

The safety regulations require that the vessels where the supervisor and the head of operations are located must be at a distance of more than 50 m, while the vessel where the skimmers are located for the recovery of unburned fractions must be at a minimum distance of 200 m. The involved personnel must be provided with safety suits and masks, which they will wear during the entire time of the burning operation.

Once all the safety measures have been checked, the spill is sprayed with gelled gasoline and incendiary darts are thrown to start the combustion process. In some cases, the fire may extinguish, and it may be necessary to repeat the previous steps of ignition again. In other causes, the beginning of the burning process is initiated manually by means of a flare, although it is not usually recommended because of the high risk to the personnel who perform it.

The input parameters to the FIRE model are the diameter of the spill and the type of U deployment of the barrier: open U or closed U. The length of the spill should be 1/3 or 1/2 of the half length of the fire barrier. This condition is imposed so that the thickness of the hydrocarbon layer is always greater than 2 mm (Fig. 9.20).

The data shown below are the result of the simulation of the burning process of a crude oil spilled in the sea with a diameter of 200 m. This spill is contained by an open U-shaped barrier. The length of the slick contained between the barriers is *d*/3, where *d* is the length of the fireproof semibarrier. The output parameters of the model are:

Safety distance between the tugs: 210 m
Total length of the barrier: 666.67 m
Length of the fire-resistant barrier section: 133.33 m
Length of the conventional section (two sections): 266.67 m
Area occupied by the spill: 25−473.23 m^2
Volume occupied by the spill: 50.95 m^3
Minimum burning area: 9819.11 m^2
Burning speed: 4.83 m^3/day
Removal speed: 0.49048 L/(m^{2}*day)

(a)

(b)

FIGURE 9.19 Process of burning an oil spill from the Deepwater Horizon platform in the Gulf of Mexico (A) and (B). *Photo from Associated Press/Gerald Herbert (2010)*

Burning time: 4.08 days
Safety distance: 6110 m
Heat flow: 713384.51 kW
Temperature: 950°C

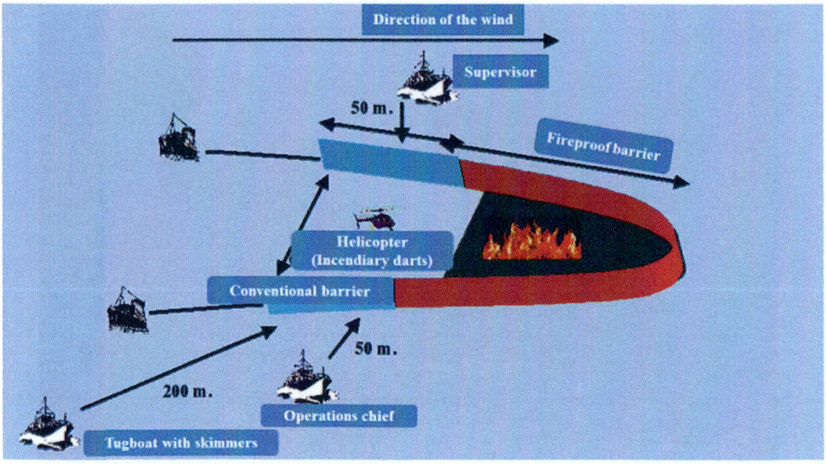

FIGURE 9.20 Operating conditions of the process of burning oil spilled at sea.

Among the advantages of "in situ" burning are:

1. Approximately 90% of the spilled oil at sea surface can be quickly eliminated.
2. Any mixture of hydrocarbons can be burned, although very old crudes and fuel oils pose ignition problems.
3. The process requires a minimum of staff and logistical input.
4. Hydrocarbons can be removed safely and effectively.
5. The final unburned waste can be recovered by traditional mechanical procedures.
6. The column of smoke usually dissipates quickly and does not cause problems for the environment.

Among the disadvantages of "in situ" burning are:

1. It generates a large amount of smoke with great visual impact.
2. Smoke can reduce visibility.
3. The water below the slick can be heated due to the combustion energy.

Among the limitations of "in situ" burning are:

1. The thickness of the hydrocarbon layer must be at least 2 mm, although higher thicknesses are recommended for effective combustion.
2. If the hydrocarbons are emulsified, the ignition is very difficult.
3. If the diameter of the slick is greater than 225 m, it is recommended to separate it into smaller slicks.

In the event of a boat catching fire, the management of the extinguishing can be carried out by the FOAMS model (Martí Moreno, 2008). The input parameters for the FOAMS model are the length of the vessel, its beam, the width of the largest tank, and the price of the foam.

The output parameter of the model is the foam flow to be sprayed. This flow rate is estimated as a function of the total area of the vessel, the area of the largest tank, and the area swept by the largest tank. From all these parameters, the one with the higher value is usually selected. It is recommended that the spraying rate should never be less than 1250 L/h. The FOAMS model allows access to a series of considerations about treatment with foams. Among all the theoretical considerations, we can highlight the following:

1. Considerations regarding the extinguishing of fuel fires, by means of the application of foams.
2. Structure of a flame.
3. Flame extinction, as a function of:
   The blocking oxygen supplied to the fuel.
   The isolation between the vapors and the fuel.
   The cooling of the fuel with the water from the foam.
4. Classification of concentrated foams: types and characteristics.
5. Main characteristics that the foams must have.
6. Foam expansion ratios.
7. Expansion process of a foam.
8. Matrix for the selection of foams that minimizes or suppresses the release of toxic products or flammable vapors.
9. Theoretical considerations regarding the calculation of the speed needed to apply the foams.
10. Optimal speed for the application of a foam.

9.12 Tarred balls

Since the treatment processes for oil spills at sea are usually not one hundred percent effective, it is common that a series of tarry residues known as "tarball" are left behind. These residues usually remain in the water for a long time, because their bacterial degradation is usually minimal or nonexistent, due to their high viscosity.

By means of the BOLAS (Tarball model) model (Martí Moreno, 2008), the evolution of these tarballs in the seawater can be assessed. The input parameters to the BOLAS model are shown in Table 9.13.

The model allows access to general information, such as the equations used to calculate the density of tarballs, their origin, and the causes of their formation and to the logistic requirements for the recovery of tarballs, necessary equipment and materials, speed of cleanup, fuel requirements for the vehicles used, and the requirements for access to contaminated areas.

TABLE 9.13 Input and output parameters of the BOLAS model.

Input parameters of "BOLAS" model	
Parameter	**Usual values**
Percentage of sand in the tarball	0%–45%
Percentage of organic matter in the tarball	0%–45%
Density of the hydrocarbon mixture	700–950 kg/m^3
Density of the sand that makes up the tarball	2.300 kg/m^3
Density of organic matter	900 kg/m^3
Degree of aging of the mixture	0%–45%
Output parameter of the "BOLAS" model	
Calculation of the density of the tarball	
Considerations to take into account	
If the density of the tarball is greater than 1.033 g/cm^3	The ball sinks into the sea
If the density of the tarball is less than 1.033 g/cm^3	The ball floats in the sea
If the density of the tarball is equal to 1.033 g/cm^3	The ball remains "gabled"

9.13 Toxicity of crude oil, dispersants, and of the mixture

One of the parameters to be considered before deciding on a specific spill treatment is toxicity. To measure toxicity, the "Lethal Dose" and "Lethal Concentration" expressed as LD50 and LC50, respectively, are often used (CCOHS, 2018).

The LD50 is defined as the amount of a given product that can cause the death of 50% of a group of test animals. It is a way of measuring the potential short-term poisoning (acute toxicity) of a product. It is usually expressed in mg/kg. The LC50 is the concentration of a substance that can cause the death of 50% of a group of test animals. It is usually expressed in ppm. The key difference between LD50 and LC50 is that LD50 represents the lethal dose, while LC50 represents the lethal concentration.

There are different types of toxicity tests, which can be performed on various species. The different species differ in their susceptibility to chemicals, most likely due to differences in accessibility, metabolic rate, excretion rate, genetic factors, dietary factors, age, sex, health, and stress level of the organism.

As standard species for the determination of the toxicity of crude oil, derivatives, and dispersants, the following animals are used:

1. *Daphnia magna*. Known as "water flea." It is a species of planktonic crustacean of the suborder of cladocerae that, as an adult, measures up to 5 or 6 mm.

2. *Patella vulgata*. Known as "common limpet." It is a gastropod mollusk from the Patellidae family.
3. *Crangon crangon*. Known as "brown shrimp" or "common shrimp." It is a species of shrimp from the family Crangonidae, order Decapoda

As defined by the American Society for Testing and Materials (ASTM International) (ASTM, 2021), these species are typically selected based on its availability, commercial, recreational and ecological importance, successful past use and regulatory use.

Normally, for fish the toxicity is expressed as LC_{50}^{96}, for the crustaceans as $LC_{50}^{48}/EC_{50}^{96}$ and for the inhibition of microalgae growth as EC_{50}^{972}.

For fish:

$$\text{Log } LC_{50} = -0.94 \text{ Log } P_{OW} + 0.94(0.000068\, P_{OW} + 1) - 1.25 \qquad (9.4)$$

where Log P_{OW} is the logarithm of the octanol/water distribution coefficient (KOW) and Bioconcentration factor (BCF) the potential of a substance to bioaccumulate.

The relationship between the logarithm of the distribution coefficient (Log P_{OW}) and the BCF is shown in Table 9.14.

9.13.1 Danger zones

The Lethal Concentration values serve to define the different danger zones. The Type A danger zone is defined as one where the LC50 is equal to or less than 300 ppm. In the Type B danger zone, the LC50 is higher than 200 ppm, but lower than 1000 ppm. In the Type C danger zone, the LC50 is higher than 1000 ppm, but lower than 3000 ppm. Finally, in type D danger zone, the LC50 is higher than 3000 ppm, but lower than 5000 ppm (CCOHS, 2018).

TABLE 9.14 Relationship between the logarithm of the distribution coefficient and the bioconcentration factor (BCF).

Value	Effect	Value of BCF
Log $P_{OW} < 1$	Not bioaccumulate	Not measurable BCF
Log $P_{OW} < 2$	Very low potential to bioaccumulate	$1 < BCF < 10$
Log $P_{OW} < 3$	Low potential to bioaccumulate	$10 < BCF < 100$
Log $P_{OW} < 4$	Moderate potential to bioaccumulate	$100 < BCF < 500$
Log $P_{OW} < 5$	High potential to bioaccumulate	$500 < BCF < 4000$
Log $P_{OW} > 5$	Very high potential to bioaccumulate	$BCF > 4000$

9.13.2 Standardized analysis methods

A variety of acceptable standardized analytical methods have been published. Some of the most widely accepted agencies for publishing these methods in the United States are the American Public Health Association, the US Environmental Protection Agency (EPA), ASTM International, the International Organization for Standardization (SCC, 2021), the Canadian Ministry of Environment and Climate Change (APHA, 2021), and the Organisation for Economic Cooperation and Development. Standardized methods (OECD, 2021) offer the ability to compare the results obtained with these methods between different laboratories.

There are many types of toxicity tests widely accepted by the scientific world and regulatory agencies. The type of method used depends on many factors: the specific regulatory agency performing the test, the resources available, the physical and chemical characteristics of the environment, the type of toxic product, the species available in the environment, the laboratory and field analyses, the selection of the endpoint and the time, and resources available to perform the analyses are some of the most common influencing factors in the design of the tests.

Lethality is the most common effect used in toxicology and used as an endpoint for acute toxicity testing. While chronic toxicity tests are performed, sublethal effects are the final indicators observed. These endpoints include behavioral, psychological, biochemical, and histological changes. There are several effects that occur when an organism is simultaneously exposed to two or more toxic products. These effects include additive effects, synergisms, potentiating effects, and antagonistic effects.

The additive effect happens when the combination of the effects is equal to the sum of the individual effects.

The synergistic effect occurs when the combination of the effects is much greater than the sum of the individual effects.

Potentiating effect occurs when an individual chemical that has no effect is added to a toxic product, and its combination has a greater effect than the toxic product alone.

The antagonistic effect occurs when a combination of chemicals has less effect than the sum of their individual effects.

Table 9.15 shows the classification given by GESAMP (2013) to catalog toxicity by grades.

9.13.3 Assessing of toxicity using the EVA (Hydrocarbon Evaporation Rate model) model

The flowchart of the EVA model (Martí Moreno, 2008) is as follows. First, the model allows access to folders, containing all the basic information about the mean Lethal Concentration (LC50), the mean Effective Concentration (EC50), and the mean Tolerance Limit (TLm).

TABLE 9.15 Lethal dose and its meaning according to the GESAMP.

GESAMP number	Meaning	LC50 96 h (mg/L)
5	Extremely toxic	<0.01
4	Very toxic	0.01–1
3	Moderately toxic	1–10
2	Slightly toxic	10–100
1	Virtually nontoxic	100–1000
0	Nothing dangerous	>1000

TABLE 9.16 Input and output values of the EVA model for the toxicity of the Decane.

EVA model input data				
Compound selected	CH3 groups		CH2 groups	
	Number of groups	Toxicity contribution factor value	Number of groups	Toxicity contribution factor value
Decane	2	0.702	8	0.527
Output parameters of EVA model				
Bioaccumulation factor (Log *P*)				5.62
DL 50 (mg/L)				1.00
Toxicity according to Veith and Brooke's criteria				Low toxic and low bioaccumulative product
Qualification according to GESAMP				2, slightly toxic

The EVA model allows determining the lethal dose of a compound from its molecular structure through the procedure of group contribution and determining the lethal dose of crude oil and derived products of dispersants and of crude oil mixtures and dispersants. Table 9.16 shows the input and output values of the EVA model, for the calculation of the toxicity of the Decane hydrocarbon (Veith, Call, & Brooke, 1983).

EVA model also allows obtaining the toxicity of crude oil and its derivatives, of dispersants and of the mixture of both. Table 9.17 shows the input and output data of the toxicity of a Light Arabian crude oil, of the dispersant

TABLE 9.17 Toxicity of Light Arabian crude oil, BP 100X dispersants, and a mixture of Fuel oil and Corexit 7664.

Input parameters of EVA model		Output parameters of EVA model	
Type of compound	Tested species	Toxicity value (48 h LC50)	Toxicity value (96 h LC50)
Middle Arabia	*Daphnia magna*	7.4	
Dispersant BP 1100X	Patella *vulgata*	3700	3700
Heavy fuel oil	*Pecten opercularis*		120
Corexit 7664			

BP 1100X and of a mixture of Heavy Fuel Oil and the dispersant Corexit 7664.

9.13.4 Innovative ecofriendly biosolvents for combating oil pollution

The need to use natural dispersants that could be used instead of the chemical one without harming the marine environment, led the development of an innovative plant extract ecologically friendly, nontoxic solution, known as MSL aqua control for sea surface oil spill cleanup operations for semi closed sea areas, as well as for sedimented oil (Theodorou 2017, 2016). The MSL solution effectiveness based on the high and fast reduction of the oil tension at sea surface and the resulting drastic evaporation, following the interaction of the MSL solution with the spilled oil. In the case of sedimented oil, the MSL solution converts it to liquid in the form of droplets, and then smooths the progress of their dispersion in the water column and their rapid evaporation when they rise at the sea surface.

A detailed laboratory examination of the MSL solution regarding its efficiency, degradation, toxicity, capability to remove the total petroleum hydrocarbon, and several other contaminant parameters (biological oxygen demand, chemical oxygen demand, fats oils grease, total phosphorous, and total Kjeldahl nitrogen), were carried out using seawater samples after the in situ deployments of the MSL solution in several polluted coastal areas and semi closed resources in Cyprus (Kaniklides and Costa, 2017), as well as in Greece and Egypt. Based on these extended laboratory examinations, the efficiency of the MSL solution was found to be more than 80%, that is, higher by more than 20% than the chemical dispersant efficiency recommended by EPA, while the biodegradation rate of the Arabian crude oil for example was as high as 90% after 4 days from the MSL solution deployment. The MSL solution toxicity was tested for various marine organisms, such as shrimps and fishes and was varied between 200 and 600 ppm, depending on the used species.

The MSL solution, as demonstrated in Fig. 9.21 from its initial deployment at the oily polluted sea surface until the oil biodegradation, acts during several phases, which are visible first from the oil slick color change. The oil spill initially from black turns to brown and then to light brown with the formation of thin sheen due to the reduction of the oil tension and the fast evaporation, and at the final phase when the polluted sea area becomes clear from the oil slick. The MSL solution was used successfully for the cleaning and the maintenance of the good environmental condition of marinas, fisheries shelters, ports, tourist beaches, etc.

9.14 Slick trajectory models and their operational applications

Among the questions proposed by managers of oil spills at sea, once a spill occurs, the following are highlighted: Which trajectory the oil spill will follow at sea under the effect of the winds, sea currents, and waves? Which coastal areas may be affected by the oil? At what time, the oil will impact the coasts nearest to the spill? How the volume of oil spilled will vary over time? What amount of oil will impact the coasts? What amount will be retained on those coasts? All these questions can be answered through the application of oil spill models. Oil spill models consist of a number of modules that, by means of equations or algorithms with different complexities, depending on the oil spilled characteristics and the prevailed metocean conditions, allow to simulate at different time intervals the spatial evolution of

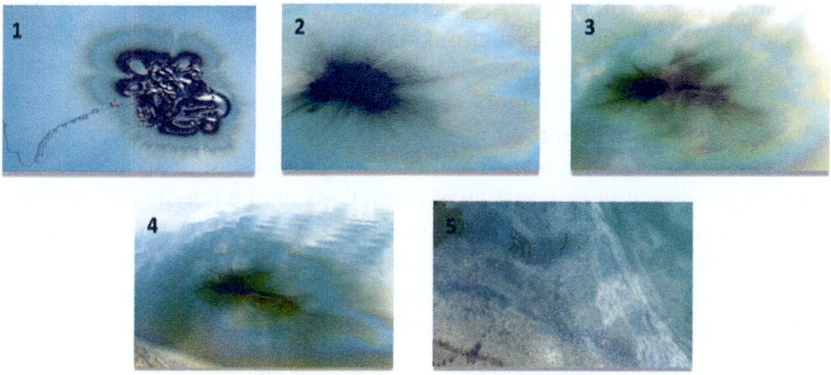

FIGURE 9.21 Example of the visible phases of an oil slick after the MSL solution deployment, the reduction of oil tension, and the formation of thin sheen to the clearness of the polluted area. *MEYDAN solutions.*

the spill and its properties. Their aim is to predict in advance the intensity of the phenomena affecting the spilled oil and, therefore, the existing risks in areas of interest, in assisting the response agencies. Most of these models allow for modifications over time, the changes of the metocean conditions. The data resulting from the oil spill modeling can be obtained in the form of data files, tables, or graphs, in which two or more variables are related to each other. In general, the oil spill models include all the phenomena that affect the oil spillage in the marine environment (spreading and displacement on water, evaporation, dispersion in air and water, emulsification, sedimentation, etc.).

Oil spill models are widely used for contingency planning, which are particularly useful for the personnel in charge of the decisions. By modeling the most likely oil spill scenarios, the decisions about the appropriate response measures become more reliable. Similarly, the most vulnerable marine and coastal areas can be identified and decisions can be made on the appropriate responses to be taken.

Applied to a fictitious problem, the models are used to carry out oil spill simulations and thus, train all the personnel that will later act in real spills in different circumstances. In this way, it will be possible to assist the response agencies to estimate the necessary resources to be used to minimize the effects of hydrocarbons on the environment in general.

The effective use of oil spill models, during a real emergency response, can have its complications since, in a very short time, it is necessary to know all the properties and characteristics of the spilled hydrocarbons as well as their evolution over time. The more realistic the data input into the models, the more accurate the simulation results will be. Most of the input parameters to the simulations vary spatially and temporally. The models can predict with some accuracy the changes that a crude oil spill at sea may undergo and indicate whether such hydrocarbon mixtures are likely to dissipate naturally or whether they are likely to reach the surrounding coast. Spill managers typically use this information to decide on the effectiveness response techniques to be employed at each point in the spill. It should be noted that the models are simply simulations and it will necessary to check the result of their predictions against real-time observations, if available.

In general terms, the main contribution of the oil spill models is their ability to provide at operational level predictions on the possible evolution of a spill and its aging. Therefore, before an actual spill, the oil spill models can be used to identify those sensitive areas susceptible to be affected by the spill and that consequently need greater and faster response measures. The software programs currently available cover the following fields:

1. Spill modeling: displacement and evolution over time of the properties of spilled hydrocarbon mixtures.
2. Databases of crude oil and derivative properties.

3. Storage and transmission of results.
4. Cartographic data of all the areas susceptible to be affected by hydrocarbons.
5. Learning and training of the personnel in charge of the fight against oil spills and the cleaning and restoration of the affected coastal environments.

The following section describes a few selected numerical spill assessment models.

9.14.1 The OILMAP model

The OILMAP model (RPS, 2021), developed by Applied Science Associates, incorporate a series of computer models that allows to manage emergency plans through real-time monitoring of oceanographic and meteorological conditions, by providing the prediction of the trajectory and aging processes of any oil spill, from simulations in different points of potential risk. Its modular system allows to characterize the oil-aging processes along the spill trajectory, both on the surface and in the water column; predicting the probability of interaction of the oil spill with the coast and, through "backtracking," it is possible to backtrack the spill to find out the source of the spill. The processes of spreading, evaporation, dispersion, and emulsification are implemented through the formulation proposed by MacKay (Lonardo, and Douglas, 1973). Oil interaction with the coast is formulated according to the retention capacity of each type of coast.

OILMAP also allows to represent georeferenced objects (GIS) and to determine the environmental impact on natural resources. Using its stochastic model, it is possible to take into consideration the possible variations of environmental conditions. In this module, the simulations are carried out by means of the random variation of the initial conditions of the spill, within a period for which meteorological and oceanographic data are available. By means of a study of multiple trajectories, contour curves are generated, which allow to evaluate the probability of finding a particle within the area of study. The different contours can be correlated, through a GIS database, to help in the assessment of the affected natural resources.

The basic models available at OILMAP are: deterministic model of drift and aging processes, the stochastic model (probabilistic-impact risk assessment in contingency plans), and the subsurface transport model. It also incorporates a backtrack model that allows predicting the origin of an oil spill by knowing information of its current position and experienced drift.

9.14.2 Trajectory and weathering

The main tasks of this model are the following:

1. Predict the trajectory of an oil spill, both instantaneous and continuous.

2. By means of a series of algorithms, deal with the phases of hydrocarbon aging and the interactions with the different types of coast (hydrocarbon-coast, hydrocarbon-reef, and hydrocarbon-ice).
3. Carry out the animation of the hydrocarbon movement, identifying the impact on the coast.
4. Represent the time evolution of the hydrocarbon-aging process and visualize the resources, represented in a GIS, impacted by the oil spill.

9.14.3 Stochastic model

It serves to perform the following functions:

1. Determine the probability of the different trajectories that a hydrocarbon spill can follow.
2. Obtain the percentages of probability of finding hydrocarbons in the surface of the sea and near the coast, as well as to give the contours of route of the hydrocarbons.
3. Determine the vulnerability of a particular area that has been affected by a hydrocarbon spill.
4. Determine the possible origin of an oil spill that has been detected previously.

9.14.4 Subsurface transport model

The objective of this model is to predict the transport of hydrocarbons in the water column, as well as their incorporation and dissolution in it.

9.14.5 TESEO model

TESEO (Castanedo et al., 2014) is a 3D numerical oil spill model to simulate the transport, aging, and spatial distribution of oil spills in the marine environment, both in the open sea and in bays, estuaries, and ports. It was developed by the Institute of Environmental Hydraulics, of the University of Cantabria, in the framework of different projects, such as the ESEOO (Spanish Operational Oceanography System), financed by the Ministry of Science and Technology from the United States. The management of the model is carried out through a friendly graphic interface that provides all the information required by the managers to deal with an emergency situation at sea. To achieve this objective, the model consists of a transport and weathering module to represent the evolution and natural processes that crude oil undergoes once it has been spilled in the marine environment. The transport module simulates the process of drifting of the spilled oil by tracking independent numerical particles representing the oil slick. The evolution of the particles is calculated by superimposing the transports induced by wind, waves, currents, and turbulent dispersion. The aging module incorporates dispersion, evaporation, emulsion, viscosity and density changes, dispersion in the water column, sedimentation, and adhesion to the coast. The numerical model consists of a transport model and a hydrocarbon-aging model.

For operational use, the system is connected via FTP (File Transfer Protocol) to ocean–meteorological data provider systems (wind, currents, and waves) both nationally and internationally. The model also provides hydrocarbon transport prediction, as well as *backtracking* to identify the origin of the pollution.

9.15 GNOME model

The GNOME (General NOAA Operational Modeling Environment) is a model developed by the National Oceanic and Atmospheric Administration (NOAA) and used by the Emergency Response Division in case of an oil spill in the marine environment, allowing an accurate estimation of the oil trajectory plume. The model estimates this trajectory, including an uncertainty analysis, generated by accidental oil spills in coastal areas, taking into consideration the influence of wind, tides, and currents.

The model output also provides specific data for postprocessing in a geographic information system or through the GNOME Analyst model. The application can be used in the development of contingency plans for accidental marine pollution and in the evaluation of ecological risks, and can help in the decision making process in the application of dispersants to combat a spill.

The application has been developed in C++ language, using a two-dimensional Eulerian/Lagrangian path model. Its development is based on elements of object-oriented programming and classes, allowing the model to be easily updated.

GNOME model uses a stochastic (Monte Carlo) motion engine, where data from the operational observing system (force data) is assimilated along with a set of parameters such as wind drag coefficient, turbulent diffusivity and uncertainty factors for each component.

The result is a time series of spot charts, representing a certain amount of probability. If a real emergency situation occurs, the trajectory is estimated by selecting "nonweathering" as the type of pollutant.

For detailed information on the aging processes, the model is supported by the ADIOS2 (Automated Data Inquiry for Oil Spills) tool.

9.15.1 Simulation of drift

The input parameters for the simulations performed with the OILMAP, GNOME, and TESEO models are shown in Table 9.18.

Fig. 9.22 shows the path and position of the spill after 228 h. The simulation has been carried in the vicinity of the Salinas of Ibiza (Spain), which

TABLE 9.18 Input parameters for the oil spill models.

Input parameter	Characteristic/intensity
Spilled oil type	Light Arabia crude oil
Oil spill models used	OILMAP, GNOME, TESEO
Location of the spill source	39°05′ 0.0″N; 0°37′ 50″E
Type of spill release	Discontinued
Spilled volume (m^3)	100
Duration of discharge (h)	96
Duration of the simulation (days)	10
Wind direction (°)	West–north–west (292.50)
Wind intensity (m/s)	8
Direction and intensity of sea currents	The predominant ones during all the time that the simulations last
Air temperature (°C)	20
Water temperature (°C)	15
Wind factor	3.5

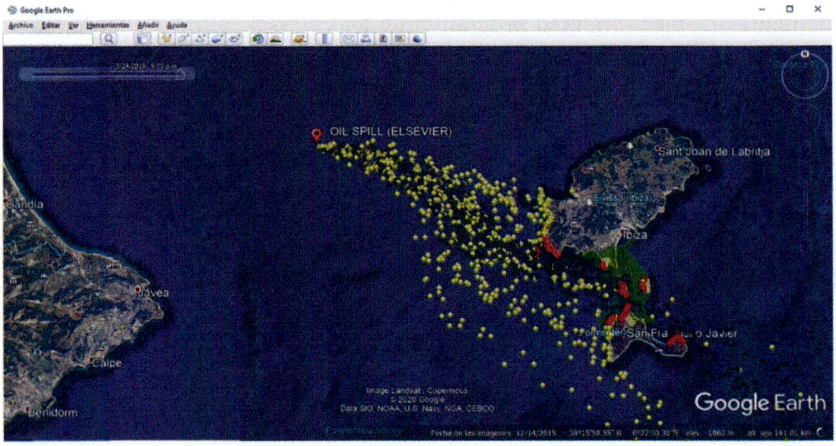

FIGURE 9.22 Path and position of the spill after 228 h from its origin. Yellow spots represent the marine areas likely to be affected by the spill, black spots are marine areas with a higher probability of being affected, and red spots are coastal areas impacted by hydrocarbons.

simultaneously are a National Park, a ZEPA zone (Special Protection Area for Birds), and an ANEY zone (Natural Area of Special Interest).

Fig. 9.23 shows the path followed by the spill and the points on the coast that have been impacted by the hydrocarbons. The red spots indicate the coastal areas where the hydrocarbons have impacted, whereas the yellow spots indicate the coastal and maritime areas with a high probability of being affected by the oil spill. Finally, the black spots represent the marine areas where there is still hydrocarbon on the sea surface.

In addition to the visualization of the trajectory followed by the spill and the coastal areas susceptible to be affected by the hydrocarbons, the model allows to estimate the thickness of the different layers of hydrocarbons, indicated by colors. When the area of the spill appears in blue, its thickness is greater than 0.0001 mm, but less than 0.001 mm. When the area of the spill appears in green, its thickness is greater than 0.001 mm and less than 0.01 mm. If it appears in yellow, its thickness is greater than 0.01 mm and less than 0.1 mm. Finally, if the representation appears in red, its thickness is greater than 0.1 mm, but less than 1 mm (Fig. 9.24).

The simulation model predicts the variation of the volume of hydrocarbons (m^3) that remain on the sea surface, the volume that has been dispersed into the water column, due to the effect wind/wave action, the volume that has been deposited on the coast, the volume that has been evaporated, and the volume that has not been able to be detected (Fig. 9.25).

The model also simulates the viscosity variation with time (Fig. 9.26), the thickness of the hydrocarbon layer, the total area of the spill, and the total volume of the emulsified oil remaining on the sea surface.

Fig. 9.27 shows the variation with time (h) of the thickness of the remaining hydrocarbon mixture (m).

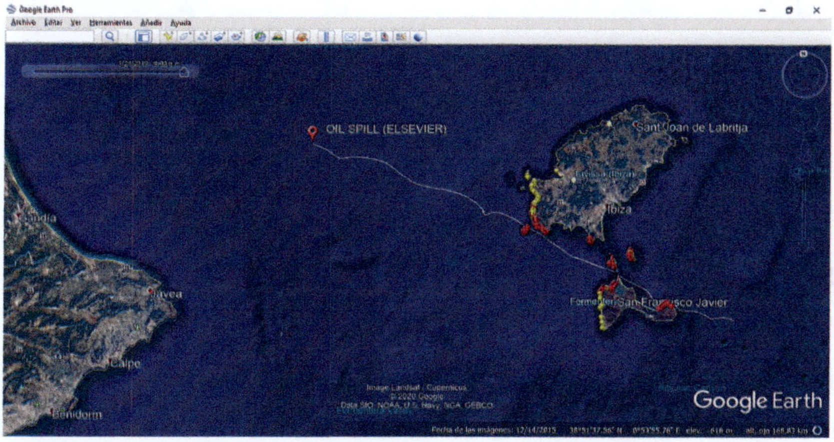

FIGURE 9.23 Trajectory through the spill and points of impact on the coast.

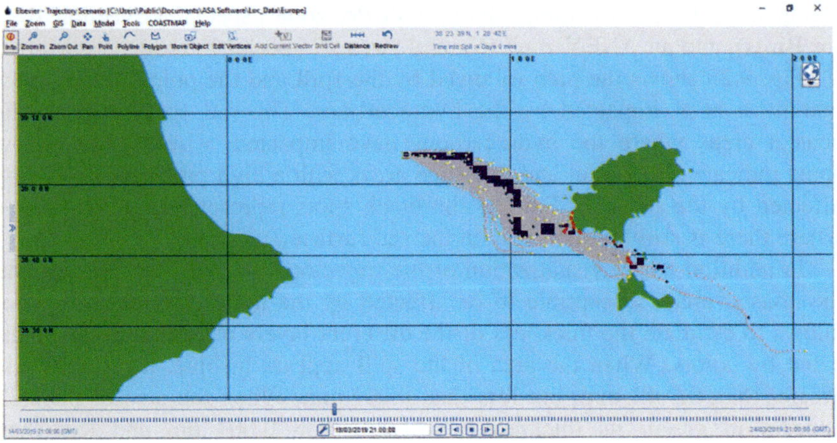

FIGURE 9.24 Indication of the thickness of the different hydrocarbon stains by means of colors.

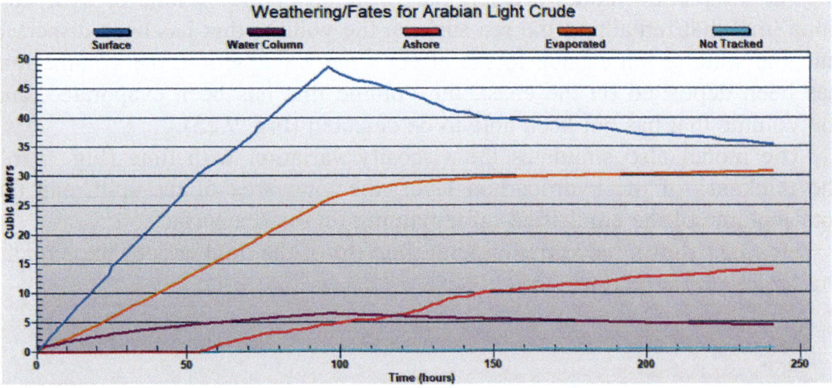

FIGURE 9.25 Variation with time of the volume of hydrocarbons that have been remained on the sea surface, dispersed into the water column, evaporated, and retained on the coast.

The variation with time (h) of the total area of the spill (m^2) that remains on the sea surface is shown in Fig. 9.28.

Fig. 9.29 shows the variation over time (h) of the total volume (m^3) of the oil slick including the emulsified.

9.15.2 Simulation results—TESEO

The input parameters for the simulation with TESEO are shown in Table 9.18.

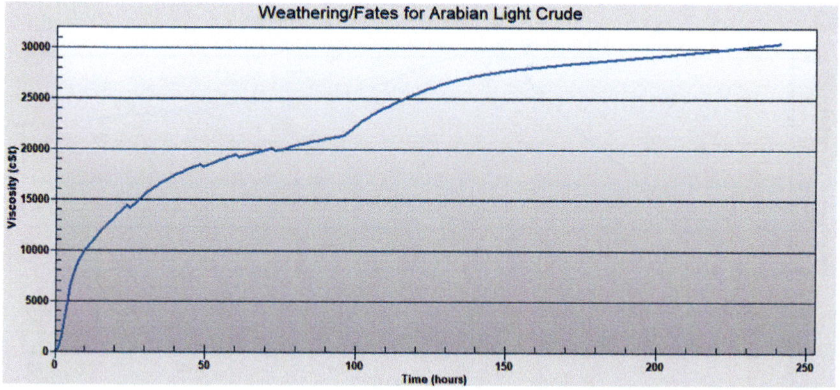

FIGURE 9.26 Variation of the viscosity of the mixture of hydrocarbons that remain on the sea surface.

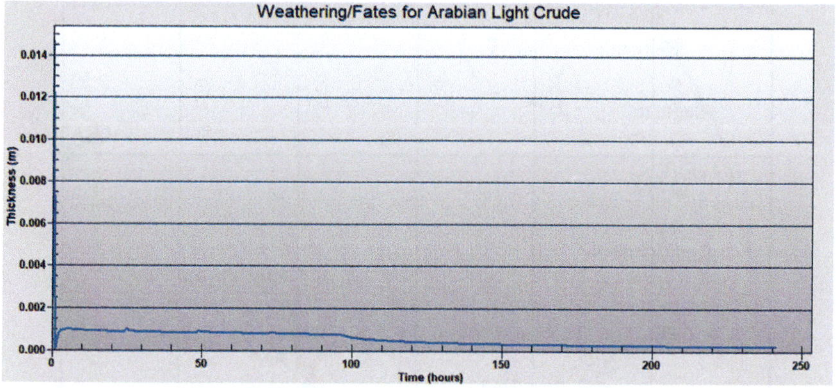

FIGURE 9.27 Variation with time of the thickness of the remaining hydrocarbon mixture.

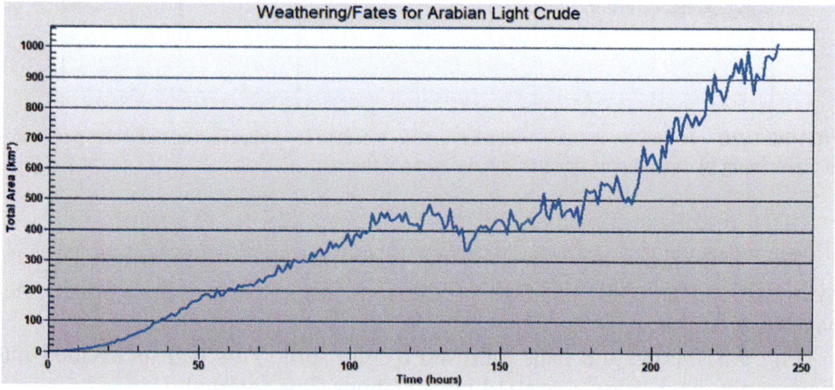

FIGURE 9.28 Variation over time of the total area of the spill.

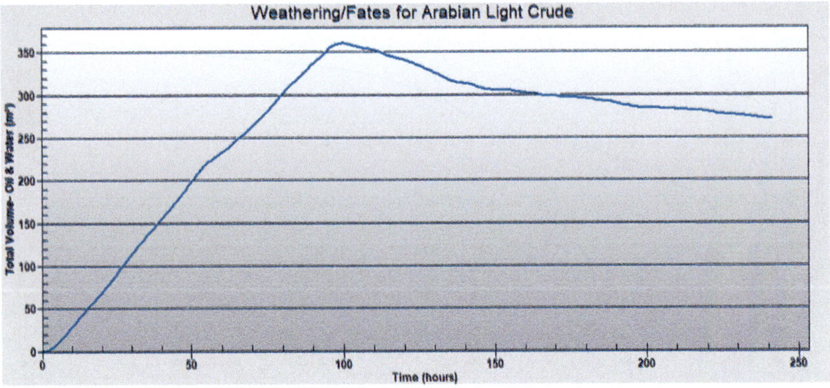

FIGURE 9.29 Variation over time of the total volume of oil including the emulsified.

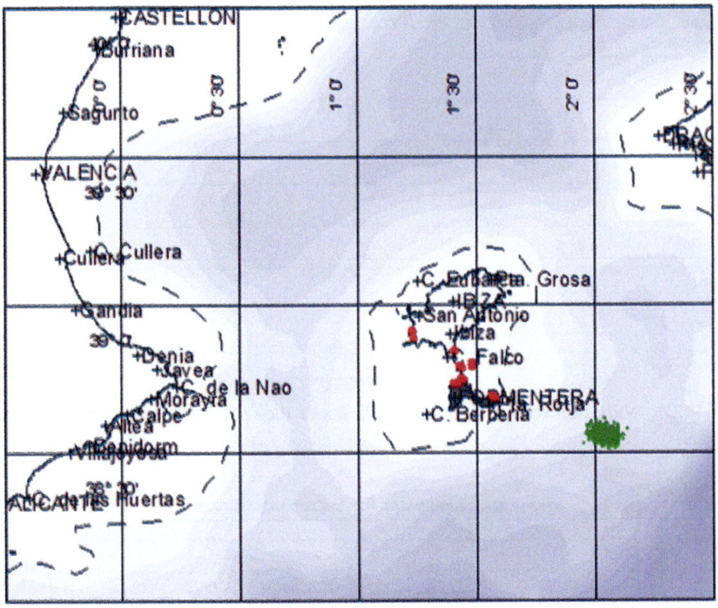

FIGURE 9.30 Points on the coast that have been affected by hydrocarbons and the position of the remaining hydrocarbons, once they have entered the sea.

Fig. 9.30 shows the points on the coast that have been affected by the hydrocarbons and the position of the remaining hydrocarbons, once they have entered the sea.

Fig. 9.31 shows the path followed by the spill. This path is divided into eight zones, each zone separated by a 12-hour time interval.

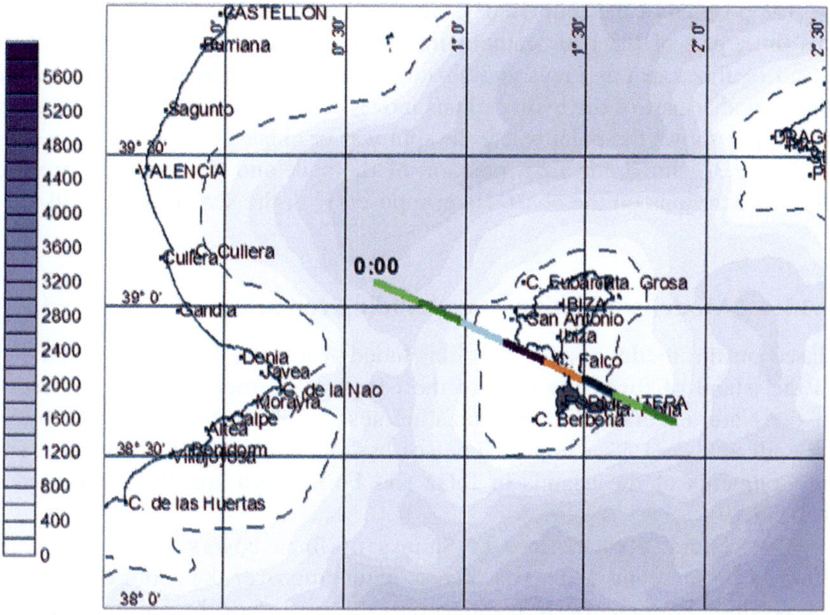

FIGURE 9.31 Oil spill trajectory modeling. Path followed by the spill, divided into eight zones, separated by 12-h intervals.

As results of the simulations, the model provides the following results:

1. Variation of density (kg/m^3) of the remaining hydrocarbons with time (h).
2. Variation of the evaporated fraction (%) with time (h).
3. Variation of the mass of evaporated hydrocarbons (t) with time (h).
4. Variation of the emulsified content (%) of the remaining hydrocarbon mixture with time (h).
5. Variation of the mass of remaining hydrocarbons (t), including the emulsified, over time (h).
6. Variation with time (h) of the percentage (%) of hydrocarbons that have been deposited on the coast.
7. Variation with time (h) of the mass (t) of hydrocarbons that has been deposited on the coast.
8. Viscosity variation (cSt.10^3) with time (h) of the remaining hydrocarbon mixture.

9.15.3 Simulation results—GNOME

The input parameters for GNOME are the same as for the previous simulations, shown in Table 9.18. GNOME provided the slick path in a general format or superimposed with the current flow field.

The position of the spill, 15 days after its origin, is shown in Fig. 9.32. After this time, part of the hydrocarbons impact to the west and south coast of the island of Ibiza, in Las Freus and also to the west of the island of Formentera, while another part of the hydrocarbons moves southward into the open sea. This figure also shows the point where the spill was originated (+).

Fig. 9.33 shows the final position of the spill and the impacted areas of the hydrocarbons on the coast, superimposed with the sea currents field.

9.15.4 Model simulations—concluding remarks

Based on the model results, it is concluded that most of the south and west of the island of Ibiza, the west of the island of Formentera, and the reserve of the Natural Park Salinas of Ibiza are susceptible of being affected by the oil spill analyzed. The impact times of hydrocarbons in the different coastal environments of the islands of Ibiza and Formentera are those indicated in Table 9.19.

The Natural Park reserve of Salinas of Ibiza covers an area of about 3000 ha on land and 13,000 ha on sea, extending over the south of Ibiza and the north of Formentera. It also occupies the section of the sea that separates

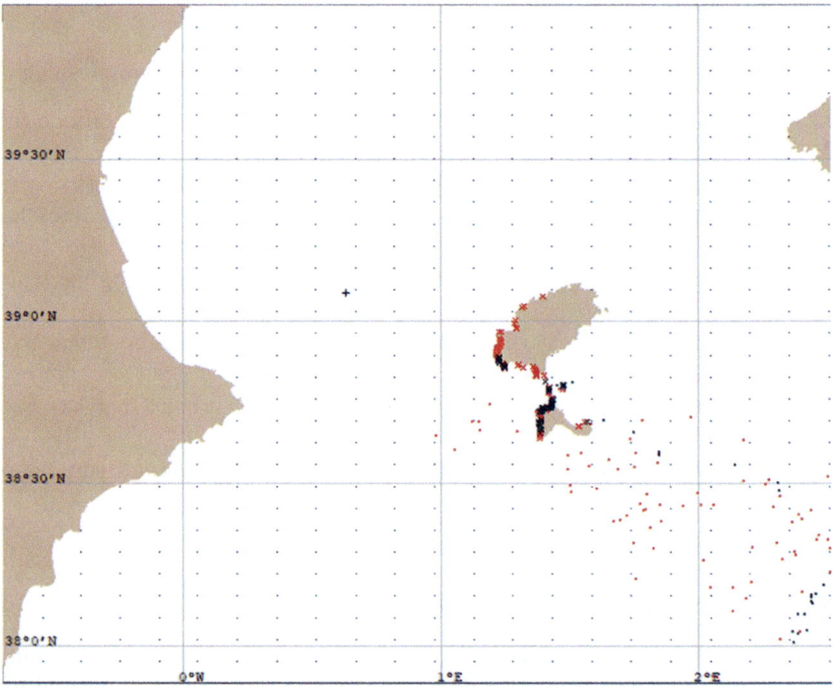

FIGURE 9.32 Position of the spill after 15 days of its origin.

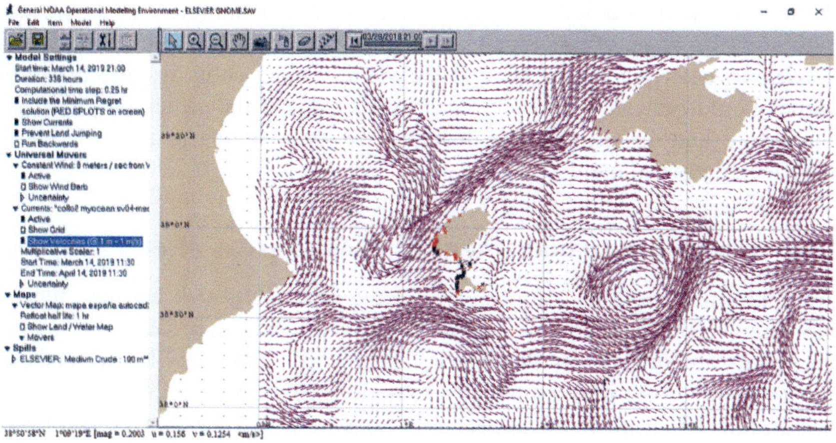

FIGURE 9.33 Final position of the discharge, superimposed with the sea currents field.

TABLE 9.19 Impact times of hydrocarbons in different coastal environments of the islands of Ibiza and Formentera.

Impacted area	Time of impact (h)
First impact on the island of Ibiza	55
First impact on the "Freus"	63
First impact on the island of Formentera	121
Hydrocarbons that go into the sea	240

them, home of a large part of the oceanic posidonia prairies that surround the islands, that was included in the declaration of Ibiza as a World Heritage Site in 1999.

The Salinas were declared a Natural Park in 2001. Furthermore, all these areas form the ANEY (Natural Area of Special Interest) and the ZEPA (Special Protection Areas for Birds), of ornithological interest for the European Union.

The reserve also includes a saltwater lagoon known as "Estany Pudent" and "Estany Pudent," which is connected to the sea by the "Punta de Sa Boca."

Knowing the trajectories of the slicks, contingency plans can be designed for their containment by barriers, fences, and interceptors. The recovery of the oil slicks can be carried out by means of skimmers and possible treatment with dispersants. Once the coastal areas susceptible to be affected by oil slicks have been identified, the response measures regarding to their protection can be studied, by means of containment barriers, cleaning, and restoration of these areas. Finally, with the results of the simulations, it is possible to determine the marine areas where fishing activities and maritime traffic should be restricted, both during the time that the oil slicks are on the sea surface and during the time dedicated to the cleaning and restoration of the affected environments.

9.16 Integrated management of coastal areas after a spill

Although the current techniques of oil containment, through barriers, fences, and interceptors, and the recovery of the same, through skimmers and/or adsorbents are sufficiently developed, usually part of the spilled oil ends up impacting the coast near the spill. The actions that need be taken in case of an oil spill, which ends up impacting a coast, is usually managed through a procedure called Integrated Management of Coastal Areas (IMCA). IMCA is defined as a dynamic process that, bringing together governments, societies in general, scientists, administrations and public and private interests, achieves the preparation and execution of a contingency plan, whose main objective is to minimize the impact of an oil spill on the marine environment (European Commission, 2016). The plan must also respond to the expectations of all the users of the coastal areas that may affected by the oil slick. Since the purpose of the plan is to minimize the impact of oil spills, it is necessary to manage the available environmental information in operational way. This is achieved through an interactive process between decision makers and experts who will give the final form to the contingency plan.

The process seeks to optimize all the possible actions related to the prevention of accidents with hydrocarbon spills, as well as those actions aimed at the containment and recovery of hydrocarbons spilled at the sea and the cleaning and restoration of coastal environments contaminated by hydrocarbons.

The architecture of an Integrated Management Plan of a coastal area is mainly based on what is called management stages, which shows the different steps to follow until reaching the definition of the plan. This management stages are fed by two ways of data entry. One way is the standard type route, which serves as methodological support, so it is called "Reference Elements." The second way, which is specific to the area under analysis, is called "Local Elements," allowing the method to be adapted to very different cases.

A good management system should consist of the following six stages:

Stage 1. Analysis of the problem.

Stage 2. Definition of the coherent management units.

Stage 3. Qualification of the coastal space.

Stage 4. Indicators and indexes.

Stage 5. Information systems.

Stage 6. Orientations, proposed improvements, and objectives to be achieved.

Therefore IMCA should begin with an exhaustive study of the problems presented. As the first stage the coastal area and its features (e.g., wetlands, lagoons, beaches of fine and coarse sand, pebble beaches, rocky areas, cliffs, and artificial structures), which may be impacted, should be assessed.

In the second stage, the extent of the coastal area affected by oil slicks must be assessed depending on the quantity and type of spilled oil that reach the coast, the metocean conditions (temperature, direction and intensity of winds, and currents), as well as the characteristics of the coast itself.

In the third stage, a classification of the coastal area must be carried out with a series of starting hypotheses. It should be noted that the analysis of a coastal area must be done based on the scientific knowledge describing the relationships between the different elements.

The fourth stage should address all aspects of indicators and indexes. Its primary mission is to transform the data operating in three phases. In the first phase, the parameters and criteria are transferred to the indicators to establish a linkage criterion. In the second phase, the indicators are transformed into indexes to prioritize the criteria. In the third phase, the indexes are compared to classify typologically the coherent management units. The main missions of the indicators should be the following:

1. They must be measured and quantified or elaborated from available and reliable data and they also must provide sufficient information on the current situation, which will be used as a reference situation.
2. They must be able to be used by all plan managers.
3. They must be of a general nature and allow control of the affected environment.

The fifth stage concerns the operational information systems providing the plan managers with the necessary data for a range of situations. A good information system should allow the acquisition of spatial and thematic data and the storage of reports in the form of accessible thematic maps. The following issues should be considered when preparing a GIS. Areas representing ecological interest, tourist and recreational facilities, fisheries and aquaculture, desalination plants, water intakes for cooling of thermal power plants, all of which may be potentially affected by oil slicks and/or from dispersants spray must be identified, to consider them as of high priority.

The sixth and final stage considers improvements and objectives to be achieved. Environmental management information should not be dissociated from the overall management, leading to the development of integrated coastal zone management models. This management should be shaped in the following stages:

Stage 1. Identification of the problems to be solved.

Stage 2. Analysis of the causes of these problems.

Stage 3. Identification of the area to which the Management/Contingency Plan will be applied.

Stage 4. Identify the most appropriate ways to solve all the problems presented.

Stage 5. Identification and analysis of all administrative permits that are necessary for the implementation of the Management/Contingency Plan.

Step 6. Use of all the experience acquired in real cases, carried out previously.

Stage 7. Proposal of improvements for the new plans.

References

APHA. (2021). *Highbeam research (Gale).* American Public Health Association. <Encyclopedia.com>; <https://www.encyclopedia.com/medicine/diseases-and-conditions/pathology/american-public-health-association>.

ASTM. (2021). Biological effects and environmental fate. In: *Industrial biotechnology* (ASTM Volume 11.06 Environmental). ASTM International. <https://www.astm.org/BOOKSTORE/BOS/1106.htm>.

Bergueiro, J. R., & Domínguez, F. (1992). Dispersantes Químicos y Mareas Negras (pp. 1–196). ISBN: 84–86682-06–01. DL: PM 960–1992. Universidad de las Islas Baleares.

Bergueiro, J. R., & Domínguez, F. (1996). *Evaporación de Mezclas de Hidrocarburos* (pp. 1–348). Editorial Bílbilis, ISBN: 84–86682-07X.

Bergueiro, J. R., & Domínguez, F. (2001). ISBN: 84–699–60571 *La Gestión de los Derrames de Hidrocarburos en el Mar* (pp. 1–822). Universidad Islas Baleares.

Bergueiro, J., Lindo, D., Moreno, S., Calvilla, J., Gómez, J., & González, J. (2010). Study of the geographical boundaries for the free use of dispersants. *Journal of Maritime Research, VII* (I), 41–54, ISSN: 1697–4840.

Bergueiro, J. R., Moreno, S., Martí, A., & Díaz, M. (2011a). *El Siniestro de la Plataforma "Deepwater Horizon" en el Golfo de México* (pp. 1–852). ISBN: 84–694–2255–0. DL: PM 399–2011.

Bergueiro, J. R., Moreno, S., Martí, A., & Díaz, M. (2011b). *Modelos de simulación y Gestión utilizados en el Vertido de la Plataforma "Deepwater Horizon" en el Golfo de México* (pp. 1–852). ISBN: 84–694–3381–2. DL: PM 675–2011.

Bergueiro, J. R., Moreno, S., Martí, A., & Díaz, M. (2011c). *Modelos Complementarios Utilizados en la Gestión del Vertido de la Plataforma "Deepwater Horizon"* (pp. 1–421). ISBN: 978–84–694–9689-3. DL: PM 1628–2011.

Castanedo, S., et al. (2014). A high resolution operational oil spill model at Santander Bay (Spain): Implementation and validation. *International Oil Spill Conference Proceedings, 2014*(1), 516–530. Available from https://doi.org/10.7901/2169-3358-2014.1.516.

CCOHS. (2018). *What is a LD50 and LC50?* CCOHS: Canada's National Centre for Occupational Health and Safety Information Canada Available at. Available from https://www.ccohs.ca/oshanswers/chemicals/ld50.html.

CEDRE. (2014). *Spill response products.* CEDRE. Centre de documentation, de recherche et d'exprimentations sur les pollutions acidentelles des eaux. <https://wwz.cedre.fr/en/Resources/Fact-files/Spill-response-20products/Dispersants>.

CEDRE. (2020a). *List of hydrophobic floating sorbent to be used at sea and in inland waters.* Tested by CEDRE. Centre de documentation, de recherche et d'exprimentations sur les pollutions acidentelles des eaux.

CEDRE. (2020b). *List of universal sorbentsfor use on land.* Tested by CEDRE. Centre de documentation, de recherche et d'exprimentations sur les pollutions acidentelles des eaux.

El Modelo SONIA. (2009). Su aplicación a la gestión de derrames de Hidrocarburos en el mar. In: *Jornadas de Seguridad y Contaminación Marina. Escuela técnica Superioer de Nautica, Máquinas y Radioelectrónica Naval.*

Environment Canada. (2008). *ESTC spills technology databases, oil properties database.* Environmental Science and Technology Centre, Environment Canada Available at. Available from http://www.etc-cte.ec.gc.ca/databases/OilProperties/oil_prop_e.html.

European Commission. (2016). *Additional tools integrated coastal management.* Belgium: European Commission. <https://ec.europa.eu/environment/iczm/index_en.htm#: ~ :text = Integrated%20coastal%20management%20aims%20for,of%20infrastructure%20and%20mitigation%20and>.

GESAMP. (2013). Revised GESAMP hazard evaluation procedure for chemical substances carried by ships. In: *GESAMP reports and studies no. 64* (2nd ed.). London: International Maritime Organization (IMO). ISSN: 1020–4873.

Grote, M., et al. (2016). *The use of dispersants to combat oil spills in Germany at sea.* Berlin: Federal Institute for Risk Assessment. BfR.

Hyland, J. L., & Schneider, E. D. (1976). Petroleum hydrocarbons and their effects on marine organisms, populations, communities, and ecosystems. In: *Sources, effects and sinks of hydrocarbons in the aquatic environment, symposium proceedings* (pp. 464–495). Arlington, VA: American Institute of Biological Sciences.

IMO. (1999). *Manual sobre la contaminación quimica parte 1—Evaluacion del problema y medios de respuesta* (p. 12). IMO (International Maritime Organization).

ITOPF. (2011a). *Effects of oil pollution on fisheries and mariculture. Technical information paper (TIPS).* London: International Tanker Owners Pollution Federation Limited (ITOPF).

ITOPF. (2011b). *Uso de dispersantes en el tratamiento de derrames de hidrocarburos. Documento de Información Técnica.* London: International Tanker Owners Pollution Federation Limited (ITOPF).

ITOPF. (2011c). *Use of booms in oil pollution response. Technical information paper (TIPS).* London: International Tanker Owners Pollution Federation Limited (ITOPF).

ITOPF. (2012a). *Use of skimmers in oil pollution response. Technical information paper (TIPS).* London: International Tanker Owners Pollution Federation Limited (ITOPF).

ITOPF. (2012b). *Uso de materiales adsorbentes en la respuestaa derrames de hidrocarburos. Documento de Información Técnica.* London: International Tanker Owners Pollution Federation Limited (ITOPF).

ITOPF. (2014a). *Use of sorbent materials in oil pollution response. Technical information paper (TIPS).* London: International Tanker Owners Pollution Federation Limited (ITOPF).

ITOPF. (2014b). *Use of dispersants to treat oil spills. Technical information paper (TIPS).* London: International Tanker Owners Pollution Federation Limited (ITOPF).

ITOPF. (2014c). *Fate of marine oil spills. Technical information paper (TIPS).* London: International Tanker Owners Pollution Federation Limited (ITOPF).

ITOPF. (2014d). *Use of dispersants to treat oil spills. Technical information paper (TIPS).* London: International Tanker Owners Pollution Federation Limited (ITOPF).

ITOPF. (2020). *Oil tanker spill statistics.* London: International Tanker Owners Pollution Federation Limited (ITOPF). <https://www.itopf.org/knowledge-resources/data-statistics/statistics/>.

Jokuty, P., Whiticar, S., Wang, Z., Fingas, M., Lambert, P., Fieldhouse, B., & Mullin, J. (1996). A catalogue of crude oil and oil product properties (Vols. 1–2). In: *Report series no. EE-157.* Environmental Technology Centre.

Kaniklides, S., & Costa, C. N. (2017). Plant extracts as an effective solution to mitigate marine pollution. *Global Journal of Science Frontier Research: E Marine Science*, *17* (1), 26–43.

Lonardo, G. D., & Douglas, A. E. (1973). The electronic spectrum of HF. I. The $B1\Sigma + -X1\Sigma +$ band system. *Canadian Journal of Physics*, *51*(4), 434–445. Available from https://doi.org/10.1139/p73-057.

Martí Moreno, A. (2008). *Sistemas informáticos para la minimización de vertidos de hidrocarburos en el mar: su aplicación al puerto de Alcudia* (Tesis doctoral, doctoral dissertation). Universidad de las Islas Baleares.

OECD. (2021). *Test guidelines for chemicals*. Paris: The Organisation for Economic Co-operation and Development (OECD). <https://www.oecd.org/env/ehs/testing/oecdguidelinesforthetestingofchemicals.htm>.

Rajendran, S., et al. (2021). WorldView-3 mapping of Tarmat deposits of the Ras Rakan Island, Northern Coast of Qatar: Environmental perspective. *Marine Pollution Bulletin*, *163*, 111988. Available from https://doi.org/10.1016/j.marpolbul.2021.111988.

REMPEC. (2020). *Pollution preparedness and response*. Regional Marine Pollution Emergency Response Centre for the Mediterranean Sea (REMPEC). <https://www.rempec.org/en/our-work/pollution-preparedness-and-response/emergency-response/emergency-response/polrep>.

RPS. (2021). *Web oil spill model and response system. OilmapWeb*. South Kingstown, RI: RPS ASA (Applied Science Associates). http://staging.asascience.com/software/oilmap/oilmap-web.shtml.

SCC. (2021). The Standards Council of Canada (SCC), Canada. https://www.iso.org/member/1619.html.

Theodorou, P. (2016). An innovative dispersant with very low toxicity and bio-accumulation, the experiment at the Ayia Napa fishing shelter in Cyprus. *Geophysical Research Abstracts*, *18*, EGU2016–18543.

Theodorou, P. (2017). Innovative eco-friendly bio-solvent for combating sea surface and sedimented oil pollution. *Geophysical Research Abstracts*, *19*, EGU2017–18582.

United States Coast Guard. (2018). *Coast guard's joint maritime test facility reopens after completion of final tests*. United States Coast Guard, United States Department of Homeland Security. <https://www.dcms.uscg.mil/Our-Organization/Assistant-Commandant-for-Acquisitions-CG-9/Newsroom/Latest-Acquisition-News/Article/1682615/coast-guards-joint-maritime-test-facility-reopens-after-completion-of-final-tes/>.

Veith, G. D., Call, D. J., & Brooke, L. T. (1983). Structure-toxicity relationships for the fathead minnow, *Pimephales promelas*: Narcotic industrial chemicals. *Canadian Journal of Fisheries and Aquatic Sciences*, *40*, 743–748, 1.983.

Further reading

Aamo, O. M., Reed, M., & Downing, K. (1997, April). Oil spill contingency and response (OSCAR) model system: Sensitivity studies. In: *International oil spill conference* (Vol. 1997, No. 1, pp. 429–438). American Petroleum Institute.

Alves, M. T., Eleni, K., George, Z., & Robin Lardner. (2016). Hindcast, GIS and susceptibility modelling to assist oil spill clean-up and mitigation on the southern coast of Cyprus (Eastern Mediterranean). *Deep Sea Research II*, *133*, 159–175. Available from https://doi.org/10.1016/j.dsr2.2015.07.017.

Bergueiro, J. R., & Moreno, S. (2002). *Limpieza y Restauración de Costas Contaminadas por Hidrocarburos* (pp. 1–422). ISBN: 699–8769-0. DL: PM 1203–2002.

Bergueiro, J. R., et al. (2004). La Gestión Integrada de Zonas Costeras Contaminadas por Hidrocarburos en las Islas Baleares (pp. 1–356). Govern de les Illes Balears. Conselleria d'Economía, Hisenda i Innovació. Direcció general de Reserca, Desembolupamente tecnològic i Innovació. ISBN: 84-688-8919–9. DL: PM 2349–2004.

Bergueiro, J. R., Serra, F., Santos, A., & Moreno, S. (2004). *La Gestión Integrada de las Zonas Costeras Contaminadas por Vertidos de Hidrocarburos en las Islas baleares* (pp. 1–355). ISBN: 84–688-8918–9. DL: PM 849–2349.

Bergueiro, J. R. (2012). *El Siniestro de la Plataforma Deepwater Horizon en el Golfo de México: Bibliografía Más Relevante* (pp. 1–692). ISBN: 84–95847-96–4. DL: PM 849–2012.

Bergueiro, J. R., Moreno, S., Martí, A., & Díaz, M. (2012). *Acontecimientos acaecidos durante Siniestro de la Plataforma "Deepwater Horizon" en el Golfo de México* (pp. 1–288). ISBN: 978–84-694–9690-9. DL: PM 109–2012.

Buist, I., Coe, T., Jensen, D., Potter, S., Anderson, E., Bitting, K., & Kurt Hansen, K. (2003). Oil spill response offshore, in-situ burn operations manual. In: *Report no. CG-D-06–03* (pp. 1–155). Washington, DC: United States Department of Homeland Security United States Coast Guard Marine Safety and Environmental Protection (G-M).

Calvilla Quintero, J. M., González Almeida, J. A., Gómez Gómez, J. I., Bergueiro López, J. R., et.al. (2008). Sistemas Insulares de Respuesta y Operaciones Ante Contaminantes Oceánicos (SIROCO). Congreso Nacional de Medio Ambiente. Cumbre del Desarrollo Sostenible. Madrid: Palacio Municipal de Congresos del Campo de las Naciones.

CEDRE. (2018). *List of dispersants convenient for use on casual oil spills products tested by Cedre (in open sea).* CEDRE. Centre de documentation, de recherche et d'exprimentations sur les pollutions acidentelles des eaux.

Delgado, L., Kumzerova, E., & Martynov, M. (2006). Simulation of oil spill behaviour and response operations in PISCES. *WIT Transactions on Ecology and the Environment, 88*, 279–292. Available from https://doi.org/10.2495/CENV060271.

ITOPF. (2018). *In-situ burning. Documents & guides response techniques.* London: International Tanker Owners Pollution Federation Limited (ITOPF). <https://www.itopf.org/knowledge-resources/documents-guides/response-techniques/in-situ-burning/>.

Keramea, P., Spanoudaki, K., Zodiatis, G., Gikas, G., & Sylaios, G. (2021). Oil spill modeling: A critical review on current trends, perspectives and challenges. *Journal of Marine Science and Engineering, 9*, 181. Available from https://doi.org/10.3390/jmse9020181.

Lardner, R. W., Zodiatis, G., Loizides, L., & Demetropoulos, A. (1998, October 5–9). An operational oil spill model in the Levantine Basin (Eastern Mediterranean Sea). In: *International symposium on marine pollution*, Monaco.

Prince, R. C. (2015). Oil spill dispersants: Boon or bane? *Environmental Science & Technology, 49*(11), 6376–6384. Available from https://doi.org/10.1021/acs.est.5b00961.

Reed, M., Daling, P., Brakstad, O., Singsaas, I., Faksness, L., Hetland, B., Ekrol, N. (2000). Oscar: A multi-component 3-dimensional oil spill contingency and response model. In: *Proceedings of the Arctic and Marine Oil Spill Program (AMOP) technical seminar* (pp. 663–680). Vancouver, CA.

SASEMAR-CEDRE. (2002). *Utilización de productos filmógenos. Sociedad de Salvamento y Seguridad Marítima (SASEMAR).* CEDRE. Centre de documentation, de recherche et d'exprimentations sur les pollutions acidentelles des eaux.

Zodiatis, G., Coppini, G., Perivoliotis, L., Lardner, R., Alves, T., Pinardi, N., ... Neves, A. A. S. (2017). *Numerical modeling of oil pollution in the Eastern Mediterranean Sea. Oil pollution in the Mediterranean Sea: Part I* (pp. 215–254). Springer. Available from http://dx.doi.org/10.1007/698_2017_131.

Zodiatis, G., Lardner, R., Spanoudaki, K., Sofianos, K., Radhakrishnan, H., Coppini, G., ... Drago, A. (2021). Oil spill modelling assessment. In O. Makarynskyy (Ed.), *Marine hydrocarbon spill assessments*. Elsevier, In press.

Index

Note: Page numbers followed by "*f*" and "*t*" refer to figures and tables, respectively.

C

D

E

O

CPI Antony Rowe
Eastbourne, UK
May 18, 2024